Sustainable Materials and Technology

Sustainable Materials and Technology (SMT) book series publishes research monographs (both edited and authored volumes) showcasing the latest developments in the field and comprehensively covering topics such as:

- Recycling of waste into useful material and their energy applications
- Catalytic action of Nano oxides for efficient carbon reforming process
- Sustainable technologies for plastic transformation
- Bifunctional nanoparticles for sustainable water splitting applications
- Sustainable dying and printing
- New materials from waste
- Sustainable Manure Management and Technology: Potentials, Uses and limitations
- Sustainable Mechanical Engineering Approach
- Sustainable biochemistry for the improvement of health
- Sustainable development of Mechanical recycling of automotive components
- Sustainable-waste recycling and conversion in useful materials for different applications
- Sustainable development of inexpensive Nano-photo catalysts
- Sustainable development of recycling of discarded lithium ion batteries
- Modern sustainable cement and concrete
- Sustainable adsorbent for hazardous removal
- Sustainable superior electromagnetic shielding materials
- Excellent sustainable nanostructured materials for energy storage device
- Sustainable development of heavy metal detoxification from water
- Carbon dioxide utilization for sustainable energy
- Sustainable development in green syntheses of materials
- Environment friendly and sustainable cloth for garments application
- Sustainable design and application of eco-materials
- Nanoparticles for sustainable environment applications
- Sustainable remediation of industrial contaminated water towards potential industrial applications
- Biomaterials for sustainable bioremediations

Asim Ali Yaqoob · Akil Ahmad
Editors

Microbial Electrolysis Cell Technology

Fundamental to Sustainable Applications

Springer

Editors
Asim Ali Yaqoob
City of Oxford College
Oxford, UK

Akil Ahmad
Department of Chemistry
College of Sciences and Humanities
Prince Sattam Bin Abdulaziz University
Al-Kharj, Saudi Arabia

ISSN 2731-0426 ISSN 2731-0434 (electronic)
Sustainable Materials and Technology
ISBN 978-981-97-3355-2 ISBN 978-981-97-3356-9 (eBook)
https://doi.org/10.1007/978-981-97-3356-9

This Springer imprint is published by the registered company Springer Nature Singapore Pte Ltd.
The registered company address is: 152 Beach Road, #21-01/04 Gateway East, Singapore 189721, Singapore

Preface

MEC is a new and significant way to producing hydrogen and methane, as well as being highly important in nutrient recovery. It has a broad range of applications, including energy generating and wastewater treatment. MEC is one of the bioelectrochemical systems that has recently received a lot of interest. The majority of researchers are working on MEC. Anodic bio-catalytic oxidation and cathodic reduction processes enable the technique. A MEC operation necessitates an external energy source, which makes the system less environmentally friendly.

A technique similar to microbial fuel cells (MFC) is a microbial electrolysis cell (MEC). While MFC use the microbial decomposition of organic molecules to generate an electric current, MEC partly reverses the process by using an electric current to generate hydrogen or methane from organic material. A sustainable energy source should preferably be used to generate the electric current. The generated hydrogen or methane can be used in an internal combustion engine or PEM fuel cell to generate energy. MEC systems depend on some kind of components, including: There are microorganisms attached to the anode. The kind of microorganisms used in the MEC impacts its output and effectiveness. Materials—carbon fabric, carbon paper, graphite felt, graphite granules, or graphite brushes—can all be used as the anode material in a MEC. In order to lower the overpotential necessary for hydrogen synthesis, platinum may be used as a catalyst. Research on biocathodes as an alternative is being sparked by the high price of platinum. The stainless-steel plates were employed as cathode and anode materials or as an alternate catalyst. Other materials include tubing, gas collecting systems, and membranes (although some MEC are membraneless). Its growing popularity can be seen by the volume of research articles published in the reputed journals.

This book covers the basic principle of microbial electrolysis cells and cathode and anode along with their synthesis, characterization, and potential applications in energy generation along with environmental pollutants removal. The book provides substantial attention of the commercial status, trends, and presentation of MECs.

We are significantly grateful to all competent researcher, scholar, and leading experts to contribute their valuable research work. The chapters provided cutting-edge up-to-date research findings on the MECs. We gathered all the material given by renowned researches and academicians on MECs and related research from Malaysia, Mexico, India, China, Pakistan, and Iran and finally compiled this project in a fruitful way.

Al-Kharj, Saudi Arabia Akil Ahmad
Oxford, UK Asim Ali Yaqoob

Contents

About the Editors

Dr. Asim Ali Yaqoob obtained his B.Sc. (2010), M.Sc. (2014), and M.Phil. (2018) from Mirpur University of Science and Technology Mirpur, Mirpur, 10250, AJK-Pakistan. Recently he completed his Ph.D. (2021) from School of Chemical Sciences, Universiti Sains Malaysia (USM), under the supervision of Dr. Mohamad Nasir Mohamad Ibrahim. His area of interest is energy generation coupled with wastewater treatment, and currently, his project is on microbial fuel cells. He has published 50 articles in high reputed journals and several book chapters. He is also Co-author for an academic book (*Graphene: A Versatile Advanced Material*) published by USM Press, Malaysia. He has completed M.Phil. in Material Chemistry (2018) with the topic of "Designing and characterizing of Ag@Polycarbazole nanocables" from Mirpur University of Science and Technology (MUST), AJK-Pakistan. His Google Scholar H-index and citation is 28 and 3598, respectively, and has filed one patent for an industrial process. He also invited reviewer of many reputed international journals such as *Nanotechnology Reviews*, *International Journal of Energy Research*, *Journal of Chemistry*, *PLOS ONE*, and many others. As a part of academic work, he has different collaborations with well-known research labs in the world. He enjoys sharing his industrial experience, where he spent one years worked as Research Assistant at Ekahala Resources Sdn Bhd, Malaysia.

Dr. Akil Ahmad is currently working at Prince Sattam bin Abdulaziz University, Saudi Arabia, as Assistant Professor in Chemistry and having the experience of seven years as Research Fellow, Teaching Fellow, Postdoc, and Visiting Researcher from Universiti Teknologi Malaysia, Universiti Sains Malaysia, University of KwaZulu-Natal, South Africa, and Universiti Kebangsaan Malaysia, Malaysia. He has completed Ph.D. in Analytical Chemistry (2011) with the topic "Modification of resin for their use in the separation, preconcentration and determination of metal ions" from Aligarh Muslim University (AMU), India. His research interest includes the areas of environmental pollutants and their safe removal, synthesis of nanoparticles and nano-sorbents (GO, CNT), photo-degradation and antimicrobial effects, water and wastewater treatment, and adsorption and ion exchange. He has published more than 150 research articles and chapters in the journals and publishers of international repute such as *Scientific Reports*, *Talanta*, *Chemical Engineering Journal*, *Journal of Industrial and Engineering Chemistry* and *Journal of Molecular Liquids*. He has also edited ten books of Springer and Elsevier. H-index and citation in Scopus are 35 and 5005, and in Google Scholar, H-index and citation are 40 and 6450. He is Guest Editor of many reputed journals, namely *Adsorption Science and Technology*-Hindawi, *Polymers* MDPI, and *Frontiers in Environmental Chemistry* and *Journal of Chemistry*, Hindawi.

Basic Electrochemical Concept of Microbial Electrolysis Cell

A. A. A. Mutalib, N. F. Jaafar, and T. A. M. Torlaema

Abstract Global concerns over the reliance on unsustainable fossil fuels have resulted in the demand for alternative resources. At the same time, the problem regarding wastewater management parallel to emerging industrialization is also becoming more serious. These two emerging scenarios led to the establishment of a multipurpose technique such as microbial electrolysis cell (MEC) with the ability to generate energy and applicable for wastewater treatment. The MEC is an environmentally friendly biochemical system based on waste-product conversion, designed to generate hydrogen or non-polluting by-products from biodegradable waste using electrochemically active bacteria. In this chapter, further basic introductory information of the MEC is presented. This chapter consists of two main sections including the first section which is dedicated to the definition, background, and importance or advantages of MEC. Meanwhile, the second section then outlines the brief concept in the MEC system particularly in terms of general components and basic electrochemical reactions that occurred during the process.

Keywords Electrochemical technology · Biohydrogen generation · Environmental sustainability · Wastewater treatments

1 Overview of Bio-Electrochemical System

Bio-electrochemical systems (BES) are a novel category of technology platforms that integrate the interdisciplinary areas of biology, chemistry, and electrochemistry. These systems utilize microbes, enzymes, and organic materials to facilitate electrochemical reactions, presenting a wide range of applications that have substantial environmental and technological ramifications [3]. BES technologies allow the implementation of organic waste as a viable fuel source, at the same time providing the opportunity for efficient wastewater treatment, and the extraction of energy [12].

A. A. A. Mutalib · N. F. Jaafar (✉) · T. A. M. Torlaema
School of Chemical Sciences, Universiti Sains Malaysia, 11800 USM Penang, Penang, Malaysia
e-mail: nurfarhana@usm.my

A. A. Yaqoob and A. Ahmad (eds.), *Microbial Electrolysis Cell Technology*, Sustainable Materials and Technology, https://doi.org/10.1007/978-981-97-3356-9_1

In the BES, microorganisms and enzymes function as biocatalysts, effectively facilitating redox processes and enabling the generation of electrical power. There are several variations on how these biological tools being employed for the different purpose leading to the development of different BES devices, namely microbial electrolysis cell (MEC), microbial fuel cell (MFC), microbial desalination cell (MDC), and enzymatic biofuel cell (EBFC). Microbial metabolism plays a crucial role in the functioning of microbial fuel cells (MFCs) and microbial electrolysis cells (MECs), facilitating the efficient conversion of organic materials into power or valuable chemical compounds. Enzymatic biofuel cells (EBFCs) utilize enzymes as catalysts for the oxidation of fuels and reduction of oxidants, hence facilitating the production of power. Moreover, microbial desalination cells (MDCs) utilize microorganisms to aid the transfer of ions and the process of desalination, simultaneously generating electrical power.

Unfortunately, there are ongoing issues in the aspects of efficiency, scalability, and material durability. These technologies encounter several obstacles, including comparatively lower energy conversion efficiencies, the requirement for optimization and scaling up, and the potential deterioration of electrodes and ion-exchange membranes over some time (De Vrieze et al. 2018) Therefore, the continuous investigation and practical implementations are propelling progress in BES, rendering it a prospective domain for tackling present-day environmental and energy-related issues. The present body of research BES is primarily directed toward the improvement of efficiency, scalability, and cost-effectiveness. Pilot initiatives and realistic implementations are showcasing the viability of these technologies in the domains of wastewater treatment and the generation of power on a smaller scale. As the development and advancement of these systems progress, they exhibit significant potential in effectively tackling critical environmental and energy issues faced by the global community.

1.1 Examples of Bio-Electrochemical Devices

1.1.1 Enzymatic Biofuel Cell

An enzymatic biofuel cell (EBFC) is a kind of bio-electrochemical device that produces electricity using enzymes as biocatalysts to catalyze the reduction of an oxidant, such as oxygen, at the cathode and the oxidation of a fuel, such as glucose, at the anode. In this process, chemical energy is transformed into electrical power by the transfer of electrons during reactions that are catalyzed by enzymes [30]. The potential of enzymatic biofuel cells in several applications has attracted interest. EBFCs work by using the catalytic activity of enzymes to promote redox reactions between an oxidant such as oxygen and fuel like glucose at two different electrodes [35].

One noteworthy use is in the biosensors industry, where EBFCs can be included to supply power to sensor devices [46]. These compact, self-sufficient power sources are especially helpful for environmental and medical monitoring since they can supply steady, dependable energy for sensor operation. EBFCs have the advantage of being efficient and specialized.

Compared to certain other bio-electrochemical systems, enzymes can produce a higher energy conversion efficiency since they are extremely selective and effective catalysts [4]. Furthermore, they use renewable and biocompatible resources, making them environmentally beneficial. However, there are drawbacks to enzymatic biofuel cells as well. Enzyme stability problems can result in a comparatively short operational lifetime for them due to the enzymes are susceptible to environmental changes, and it may be necessary to carefully regulate variables like pH and temperature [45]. Furthermore, performance can be affected by enzyme selection, and not all enzymes are easily accessible for widespread use. However, research is still being done to overcome these issues, which makes enzymatic biofuel cells an intriguing option for producing sustainable electricity in a variety of specialized applications.

1.1.2 Microbial Desalination Cell

The microbial desalination cell (MDC) is a bio-electrochemical device that has been specifically developed to tackle the dual challenges of wastewater treatment and desalination in a combined manner. MDCs employ a combination of microorganisms and electrochemical methods to effectively eliminate salt ions from saline water, simultaneously generating electrical power. The operation of MDC relies on the metabolic activities of microorganisms, which enable the transport of ions across ion-exchange membranes, thereby creating a gradient of salinity [37]. The presence of a gradient induces the movement of salt ions from the saline water toward the brine solution, resulting in the desalination of the water [24]. In the interim, the electron flow that arises from this process can be effectively utilized for the generation of electrical energy. MDC operates through a distinctive and versatile technique that integrates microbial activity with desalination. The MDC technique is a pioneering innovation that effectively tackles the pressing challenges associated with wastewater treatment and desalination in a concurrent manner. This mechanism offers a novel approach to address the issue of water shortage by leveraging the metabolic processes of microorganisms and the principles of electrochemistry to effectively manage two significant difficulties within a single integrated system.

Nevertheless, MDC also encounters other obstacles. The efficiency of the desalination process may be somewhat lower when compared to alternative desalination technologies such as reverse osmosis. Technology remains relatively novel, and the continuous investigation of enhancing its performance and scalability is a prevailing study domain. Furthermore, there is a need for enhanced energy efficiency in the context of MDC to enhance their competitiveness in comparison with currently available desalination techniques. However, with the progression of research, MDC shows great potential in tackling the challenges posed by water scarcity and supporting the adoption of sustainable water treatment methods.

1.1.3 Microbial Fuel Cell

A microbial fuel cell (MFC) is a bio-electrochemical apparatus that utilizes the metabolic processes of microorganisms, commonly bacteria, to transform organic substances into electrical energy. MFCs enable the efficient transmission of electrons generated by the microbial oxidation process of organic substrates at the anode, hence facilitating the production of power in an external circuit [21, 38]. Subsequently, the process of electron transfer is concomitantly linked with the reduction of an electron acceptor, commonly oxygen, occurring at the cathode. The process leads to the liberation of electrical energy, becoming MFC a highly promising technology for the sustainable generation of power and the treatment of wastewater. The fundamental concept of MFC is rooted in the exceptional metabolic capacities exhibited by microorganisms, specifically bacteria. MFCs rely on the pivotal involvement of microorganisms in the degradation of organic substances, such as wastewater or organic waste, under anaerobic conditions at the anode [26, 40].

The MFC process possesses the remarkable capability to efficiently convert the chemical energy contained inside organic matter into electrical energy, mostly facilitated by the metabolic activities of microorganisms. This technique exhibits environmental friendliness and possesses the capacity to fulfill a dual function by not only generating renewable energy but also efficiently treating wastewater, thereby concurrently tackling two significant concerns [11], [15]. MFCs also have demonstrated considerable efficacy in the removal of organic contaminants from sewage or industrial wastewater, while concurrently generating electricity, offering a twofold advantage [41].

However, MFCs face distinct challenges, such as somewhat lower energy conversion efficiency when compared to alternative energy sources [5], [17]. This constraint has the potential to impede their ability to scale and prevent their widespread acceptance. However, current research and advancements are actively working to overcome these constraints, thereby establishing MFCs as a highly promising technology to produce sustainable energy and the treatment of wastewater [9], [31].

1.2 Microbial Electrolysis Cell as a BES Device

Besides EBFC, MDC, and MFC, the microbial electrolysis cell (MEC) is one of the sophisticated bio-electrochemical systems that utilizes microorganisms to generate hydrogen gas or other useful compounds utilizing a synergistic interaction between microbial processes and electrolysis [10, 42].

In MEC, microorganisms, predominantly bacteria, located at the anode degrade organic matter within an oxygen-deprived setting, resulting in the liberation of electrons and protons as metabolic by-products. Subsequently, the electrons are conveyed via an external circuit toward the cathode. Concurrently, the electrons and protons are engaged in an electrochemical process at the cathode, which facilitates the electrolysis of water and subsequently leads to the production of H_2 gas with high purity, [16], [18].

The utilization of MECs exhibits considerable significance, particularly within the realm of sustainable hydrogen generation. Hydrogen is regarded as an ecologically sustainable and adaptable energy carrier, whereas MECs present a bio-inspired and energy-efficient approach to its production. One of the primary benefits associated with MECs is their capacity to facilitate the creation of renewable hydrogen by utilizing organic substrates that are easily accessible [27]. This capability serves to diminish the reliance on fossil fuels. Furthermore, they can effectively tackle the dual issues associated with wastewater treatment and nutrient removal concurrently [22]. Furthermore, the potential of MEC is being investigated in the context of nutrient recovery from wastewater [22]. This is due to their ability to effectively capture and extract valuable resources from wastewater streams that are abundant in organic matter.

The MECs demonstrate a diverse array of practical uses, hence exemplifying their multifunctionality and ecological soundness. The notable utilization of sustainable hydrogen production is regarded as one of their most prominent applications. It can also utilize microorganisms in the process of converting organic matter into H_2 gas, providing an environmentally friendly and sustainable energy source [28]. Hydrogen exhibits versatile applications, encompassing fuel cell vehicles, energy storage, and industrial operations, fostering an environmentally conscious and sustainable energy milieu.

Other than hydrogen, MEC also has been regarded as one of the potential approaches for methane production. The major advantage of producing methane through MEC is based on energy efficiency where no additional heating is required during the process. Moreover, compared to hydrogen production, the required applied potential for methane production is much lower [39]. There are two common routes for methane production in MEC which are the electrochemical route and biological route. The biological route is conducted by employing hydrogenotrophic methanogens that use the electrochemically generated CO_2 and H_2 to produce methane. Meanwhile, the electrochemical method is based on the CO_2 reduction in the cathode chamber [44].

Besides, wastewater treatment also represents a significant and vital use of MEC technology. MECs have demonstrated exceptional proficiency in the degradation of organic contaminants that are commonly found in sewage or industrial wastewater. This characteristic renders MECs a highly effective and environmentally sustainable method for the treatment of such effluents. For instance, research has directed their attention toward augmenting the anode efficiency of an MEC integrated with an anaerobic reactor to treat high-concentration industrial wastewater containing the reactive brilliant red X-3B dye [47]. In addition to its environmental benefits in reducing the impact of wastewater discharge, this method also yields electrical power as a useful secondary output.

Furthermore, there is ongoing research and investigation into the potential use of MECs to recover nutrients from wastewater. The microbes implicated in the process are responsible for the acquisition of vital nutrients, namely phosphorus and nitrogen, from streams of wastewater. Subsequently, these essential nutrients can be utilized for agricultural purposes, thus mitigating the reliance on synthetic fertilizers and effectively tackling concerns related to nutrient contamination. For instance, a study was conducted to examine the utilization of a microbial electrolysis cell in a two-step procedure for the recovery of nitrogen and subsequent chemical precipitation using Ca^{2+} or Mg^{2+} salts. The objective of the research was to explore the feasibility of sustainable nutrient recovery in municipal wastewater treatment plants by recovering nitrogen and phosphorus from digested sludge [1]. The nutrient recovery component of MEC technology offers a novel and environmentally conscious approach to improve resource efficiency and mitigate ecological consequences.

In brief, the practical applications of MEC technology encompass a wide range of areas, including sustainable hydrogen production, methane production, and nutrient recovery. The various applications of MECs highlight its promise as a versatile and environmentally sustainable technology that may effectively tackle urgent issues including energy, water management, and resource recovery in a manner that is both sustainable and environmentally responsible.

1.3 Mechanism and Working Principle of MEC

1.3.1 Basic Components and Materials

The MEC system comprises cathode, anode, membrane, and other extra equipment for the tubing and gas collection. The anode chamber in MEC consists of microbes that can be obtained from the organic material. Based on the previous studies, the *Shewanella* spp. and *Pseudomonas* spp. are the dominant species that usually exist on the anode of MEC [44]. It is worth mentioning as well that the microbial community in the MEC system is unlikely to be like MFC since the anaerobic conditions of MEC assist the growth of anaerobic bacteria such as the exoelectrogenic *Geobacter* spp. and fermentative or methanogenic microbes. Conventionally, the metal catalysts such as stainless steel, platinum, and nickel are used in the cathode to accelerate the H_2 generation purpose but due to several problems associated with cost, the biocathode implementing microbes as the biocatalyst was later introduced. However, there is conflicting result regarding the application of biocathode in MEC since there is study obtaining lower productivity of H_2 generation when using biocathode compared to metal catalysts [2].

Most MEC setup also consists of a membrane which is employed to create a chamber that contains the microbes with the function of degrading the substrate. The membrane serves as a separator to reduce the loss of H_2 sand and the mixing problem of the gas with CO_2. Besides that, the membrane is important in preventing any short circuits as the internal strength needs to be lowered by reducing the length between the electrodes [43]. The ion-exchange membranes (anion-exchange membrane (AEM) and cation-exchange membrane (CEM)) and composite membranes are the types of membranes often used in MEC. The ion-exchange membranes are usually employed due to their advantages. The cation exchange can help in proton transport leading to enhanced H_2 productivity based on the higher pH at the cathode. Meanwhile, an anion-exchange membrane has been also associated with the increased performance of MEC systems due to their permeability to chemical buffers including bicarbonate and phosphate buffers which ensure the pH consistency in the MEC [23]. They also recognized their low cost and stability to produce hydrogen except in basic conditions. Even though the membranes play significant roles in the MEC system, the membrane-less or single-chambered MEC system is currently developed as well. As the charge of the membrane accounts for about 30–40% of the whole MEC setup, removing the membrane can significantly reduce the cost of the MEC system [29].

1.3.2 Types of MEC Setup

The microbial electrolysis cell setup can be classified into two which are double chamber and single chamber. The double-chamber microbial electrolysis cell is usually a setup equipped with a membrane, functioning as a separator to reduce bacterial and organic material cross-over from the anodic chamber to the cathodic chamber. Nevertheless, there are several disadvantages of double-chamber MEC particularly in the aspect of pH stability throughout the system. Since the transport rate of hydroxyl and hydrogen ions over the membrane is restricted in the neutral condition, the pH declination all over the membrane usually occurs [25]. The pH variation condition will disrupt the H_2 generation due to the inhibition of the activity of the electrochemically active bacteria which therefore reduces the input strength. In regards, to the double-chamber MEC is not efficient for the H_2 retrieval from the unbuffered wastewater resource based on the major pH drop issue. In consequence, for less buffered wastewater treatment, occasional polarity inversion is introduced as a new mode of operation [20].

Different innovative forms of double-chamber MEC have also been developed including barrel-shaped, cruse-typed, concentric tube-designed, and disk-like types [36]. Besides that, a flat-kind MEC system was also designed to reduce the electrode distance (to minimize the resistance of the input) by implementing a membrane from the porous geotextile as the separator. Figure 1 displays the common double-chamber MEC while Fig. 2 further consists of different types of double-chamber MEC.

Meanwhile, the single-chamber MEC system on the other hand was created by eliminating the application of membrane from the two-chamber setup and immersing both the cathode and anode into a similar solution of one chamber. The advantage of the single-chamber MEC over double-chambered MEC is mainly its ability to minimize the efficiency loss related to membrane resistance [45]. Moreover, other associated drawbacks linked with the membrane part including deterioration, fouling, and high cost can also be eliminated. The manufacturing and assembling process is also much simpler for the single-chamber MEC compared to the double-chamber MEC [20]. Commonly, the single-chamber MEC is constructed in a diverse design

Fig. 1 Schematic illustrations of double-chamber MEC. Reproduced from [25] with permission from Taylor & Francis

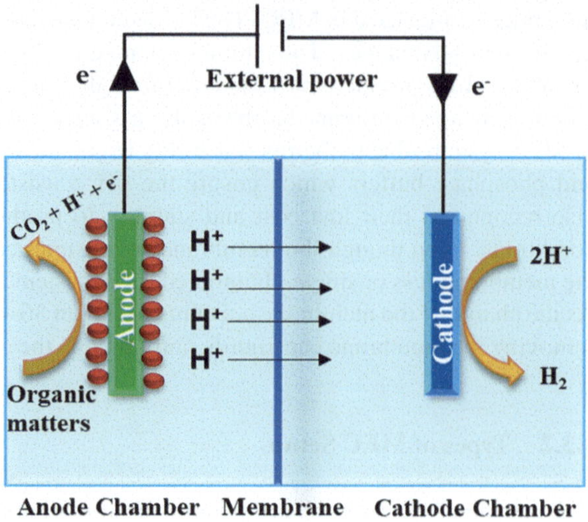

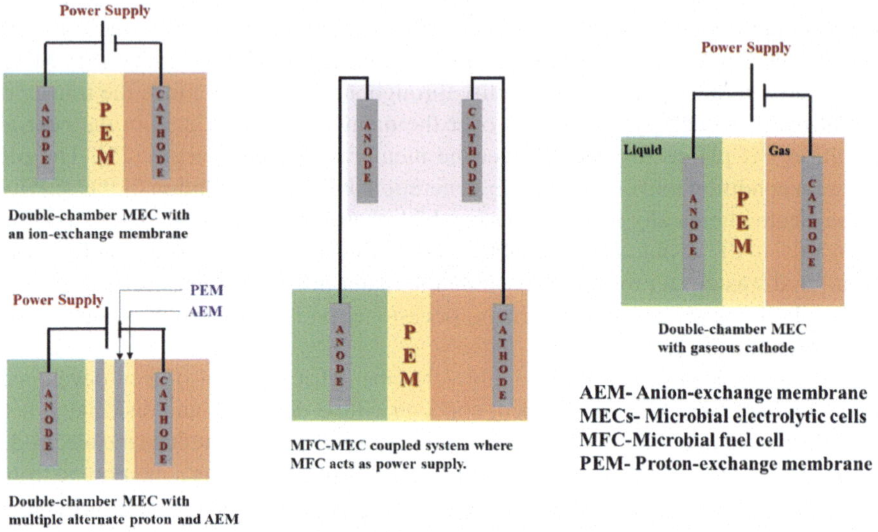

Fig. 2 Different forms of double-chamber MEC. Reproduced from [25] with permission from Taylor & Francis

consisting of material like plastic block or glass cruse. Figure 3 illustrates the schematic diagram of the single-chamber MEC setup.

However, single-chamber MEC displays tough challenges regarding the re-oxidation of H_2 by the electrochemically active bacteria and the consumption of the yielded H_2 by the methanogenic bacteria. To abate the challenges, UV radiation is implemented to hinder methanogenesis [7]. The most current solution is by

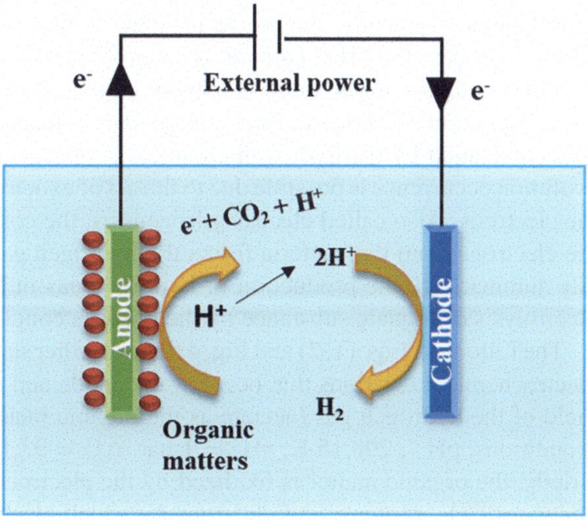

Fig. 3 Diagrammatic representation of single-chamber MEC. Reproduced from [25] with permission from Taylor & Francis

introducing an active gas method using a gas-permeable membrane and vacuum to enable rapid H_2 removal before being consumed by the bacteria. This bubbleless hydrogen diffusion approach offers several advantages especially in terms of the utilization of hydrogen productivity, hydrogen transfer, and the prevention of the explosion risks [19]. Besides, the gas-permeable membrane also can be integrated with either connected or suspended growth systems. Via the membrane walls, hydrogen disseminates into the bulk liquid as it passes through the lumen.

1.3.3 Brief Electrochemistry of MEC for Hydrogen Production

In an MEC, bio-waste exploitation offers large profits as the renewable and abundant cellulosic biomass can supply sufficient hydrogen for various uses. Compared to other production methods, MEC is considered a renewable and sustainable hydrogen generation strategy from diverse sources and types of wastewaters [28]. The brief concept originated from the chemical fuel cell where at the anode, the H_2 gas is oxidized, and the electron travels to the cathode and reduces the O_2 to H_2O. Then, the potential difference resulting from the reactions at the anode and cathode can be exploited to produce electrical energy. Even though the MEC exhibits a common feature with the chemical fuel cells, there are main differences that significantly distinguish the two electrochemical systems. While a chemical fuel cell demands the incorporation of highly pure H_2, with the requirement of a costly catalyst at the anode to enable the oxidation process, the MEC in the opposite much less expensive by the exploitation of organic chemicals from the wastewater BOD as the fuel [49]. Besides, instead of the metal catalyst, the MEC anode's catalyst is the biofilm of bacteria able to oxidize the organic matter into H_2O and CO_2 and provide electrons to the anode. Typically, the anode in MEC is like the MFC anode but instead of aerobic cathode, the MEC cathode is anaerobic [14].

The resultant current value is usually equivalent to the hydrogen gas produced since the organic compound catalyzed oxidation reaction in the anode compartment is accompanied by the hydrogen chemical evolution at the cathode. The hydrogen evolution occurrence is possible due to the act of exoelectrogenic bacteria that release the electrons. Also called electrohydrogenesis, the combination of the protons with the electrons from the bacteria forms the hydrogen gas as the product [20]. Generally summarized, the production of hydrogen gas in MEC is contributed from the electrolysis of organic substance by the bacteria coupled with the voltage addition.

The following Eqs. (1, 2) and Fig. 4 below further state and illustrate the respective electrochemical reactions that occur at the anode and cathode that contribute to the yield of the hydrogen gas (acetate as the organic matter and H_2 gas as the product (conditions: pH 7, 298.15 K, $pH_2 = 1$ bar, $\Delta G = 93.14$ kJ/mol; emf = − 0.12 V)). Briefly, the organic matter is oxidized by the electrochemically active bacteria and generates CO_2, protons, and electrons by which electrons are transferred to anode while protons are deliberated into the solution. Then, the electrons are then combined with the free protons in the solution to form hydrogen gas. The ΔG is positive indicating the non-spontaneous nature of these reactions, hence requiring the external voltage supply to induce the H_2 generation at the cathode.

$$C_2H_4O_2 + 2H_2O \rightarrow 2CO_2 + 8H^+ + 8e^- (\text{at anode}) \tag{1}$$

$$8H^+ + 8e^- \rightarrow 4H_2 (\text{ at cathode}) \tag{2}$$

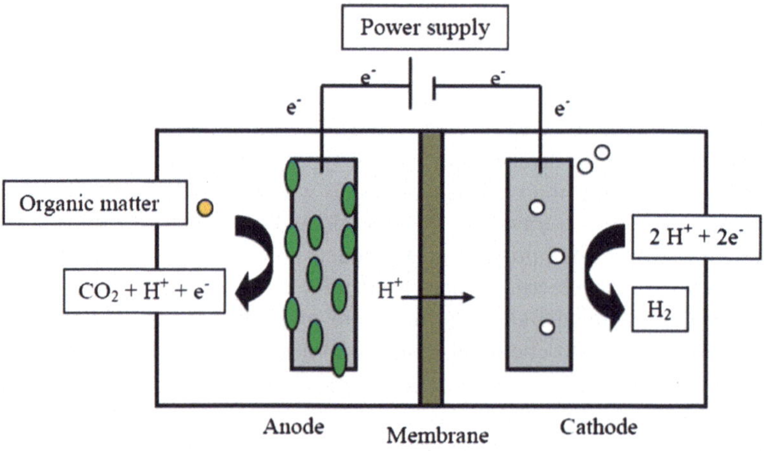

Fig. 4 Schematic representative of MEC system for hydrogen production. Reproduced from [20] with permission from Elsevier

The Nernst equations are applied to estimate the amount of the voltage needed to trigger the H_2 production. Depending on the organic substances used and the organism's types at the anode, the MEC application is not limited to hydrogen production but can extend to wastewater treatment applications as well as methane production [32].

1.3.4 Brief Electrochemistry of MEC for Methane Production

Conventionally, methane is harvested from the bio-waste anaerobic digestion, but it is a long process that requires days to complete. Through the microbial electrochemical system technology, the methane harvesting approach is developed by the presence of methanogens at the MEC cathode via CO_2 electromethanogenesis [6]. During the hydrogen production phase, the growth of methanogens usually occurs, which results in diminished hydrogen yield. Previously, many techniques have been utilized to abate the methanogen growth in MEC but mostly are ineffective and a waste of energy. Consequently, instead of focusing on eliminating the methanogens, the researchers found that harvesting the methane directly from the MEC offers a lot of advantages in comparison with the anaerobic digestion approach. For instance, the production of methane in MEC does not necessitate the heating process and can be conducted in ambient condition and hence is much more energy efficient [32]. Despite that, the MECs can develop methane in less concentrated organic compounds in contrast to anaerobic digestion which often needs the high organic content to produce methane.

There are two pathways of electromethanogenesis in MEC which are direct electromethanogenesis (direct electron uptake from the electrode Eq. 3) and indirect electromethanogenesis (hydrogen-mediated or other compounds such as acetate or formate Eq. 4, 5). The schematic illustrations of the electromethanogenesis process in MEC are further portrayed in Fig. 5.

$$CO_2 + 8H^+ + 8e^- \rightarrow CH_4 + 2H_2O \tag{3}$$

$$2H^+ + 2e^- \rightarrow H_2 \tag{4}$$

$$CO_2{+}4H_2 \rightarrow\rightarrow CH_4 + 2H_2O \tag{5}$$

To produce methane, four major components are required including a biocathode, anode, electricity source, and separator. The oxidation stage occurs at the anode which is significant to supply electrons for the CO_2 reduction at the biocathode [34]. The microorganism then assists in methane production by the utilization of the electrons from the anode. The ion-exchange membrane is important for the migration of the positive charge ions, i.e., K^+, H^+, Na^+ from the anodic chamber to the cathode chamber and ensuring the electroneutral condition of the solution. Finally, the additional voltage supply (0.2–0.8 V) is compulsory to thermodynamic drive the reaction to produce the methane.

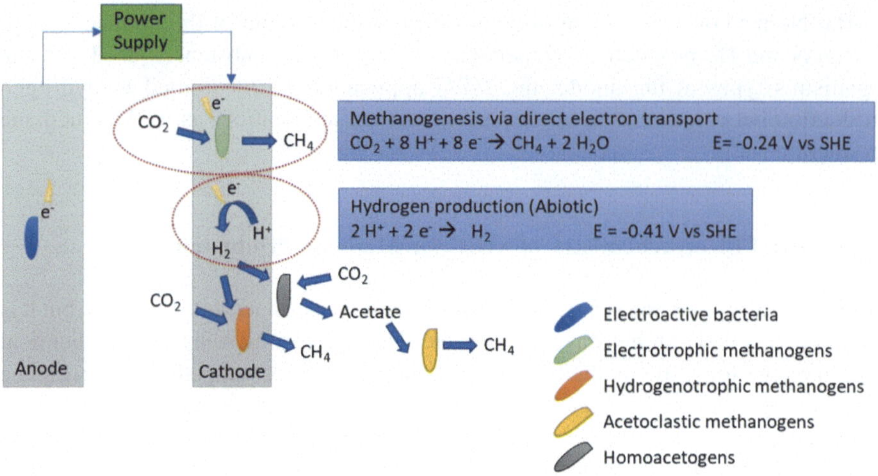

Fig. 5 Schematic illustrations of electromethanogenesis in MEC. Reproduced from Pawar et al., (2020) with permission from the Korean Society of Environmental Engineers

1.3.5 Brief Electrochemistry of MEC for Wastewater Treatment

As aforementioned, the MEC reactions occurring in the MEC system allow the production of hydrogen gas and methane. Simultaneous to this process, wastewater treatment also happens. After the electrochemical reactions, the electroactive bacteria reduced the substrates contained in the wastewater such as SO_4, NO_2, $C_2H_2Cl_4$, or NO_3 by utilizing the cathode as the electron donor [8].

Even though the substrate can be similar to wastewater, the configuration of MEC in terms of reactor architecture design particularly the number of chambers or types of the membrane may vary depending on the targeted outputs (Fig. 6). For instance, MEC specified for H_2 production is commonly equipped with a separator between the cathode and anode to stop the evolution of H_2 to CH_4 (Fig. 6) [6]. Different from the H_2 production where the microorganism only adheres to the anode, the MEC for the CH_4 production also has a biocathode, housing the methanogenic bacteria that are responsible for the conversion of H_2 to CH_4 and CO_2 [34]. When supplied with electrical power and the wastewater as the substrate, the biogas is generated reducing the organic pollutants concentration in the wastewater.

1.4 Advantages and Sustainability of MEC

MECs possess numerous inherent benefits, hence underscoring their capacity as viable and sustainable technological solutions for the generation of renewable energy and the retrieval of valuable resources. To begin with, it is worth noting that MECs

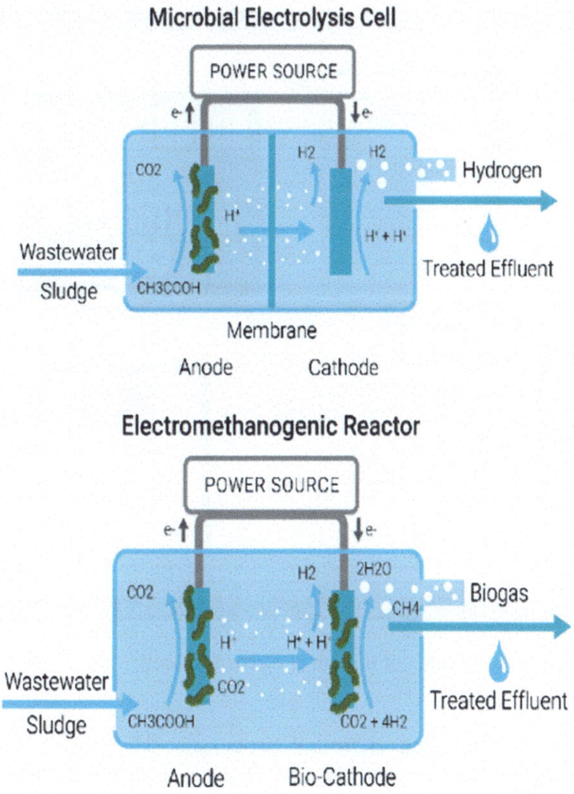

Fig. 6 Different MEC setups implement wastewater as the substrate. Reproduced from [8] with permission from Multidisciplinary Digital Publishing Institute

offer a viable and environmentally friendly means of producing hydrogen, a renewable energy source. This clean energy carrier holds great potential in facilitating the shift toward a sustainable energy landscape. The generation of sustainable hydrogen answers the imperative for ecologically conscious alternatives to conventional fossil fuels, thereby contributing to the mitigation of greenhouse gas emissions [33].

MECs demonstrate exceptional proficiency in the field of wastewater treatment by effectively degrading organic contaminants present in both sewage and industrial wastewater. The utilization of this dual-purpose strategy serves to alleviate the environmental consequences associated with the outflow of wastewater, while concurrently producing electrical energy. This not only contributes to the improvement of the eco-friendliness of the treatment process but also promotes the recovery of energy and the efficiency of resource utilization. In addition, MECs play a crucial role in the extraction of essential nutrients, including phosphate and nitrogen, from wastewater. The process of nitrogen recovery plays a crucial role in promoting sustainable agriculture by reducing reliance on chemical fertilizers [13]. This, in turn, contributes to the conservation of resources and helps mitigate nutrient contamination in aquatic ecosystems. Moreover, MECs have a distinctive attribute of biological selectivity

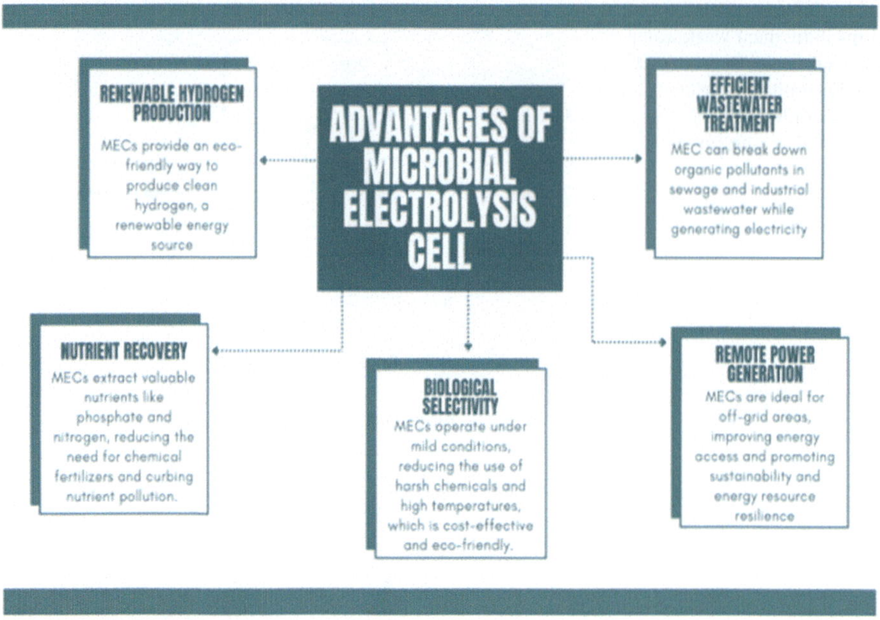

Fig. 7 Advantages of microbial electrolysis cell

when facilitating chemical processes. The ability to function under mild conditions enables them to minimize the use of harsh chemicals and high temperatures, resulting in economic and environmental benefits [48]. The utilization of MEC technology enables the generation of power in remote locations, rendering it a highly suitable option for off-grid scenarios and decentralized energy systems. The implementation of decentralization has the potential to improve energy accessibility in geographically isolated or underserved regions, hence fostering sustainability and resilience in the provision of energy resources. MECs possess a range of benefits that underscore their capacity as a viable and sustainable technology for the generation of renewable energy and the retrieval of valuable resources. The capacity of these systems to generate hydrogen with minimal pollution, treat wastewater, reclaim valuable nutrients, function under moderate operating circumstances, and enable remote power generation highlights their significance in tackling environmental and energy-related issues, while also encouraging the preservation of resources and the concept of sustainability. Figure 7 summarizes the advantages of MEC.

1.5 Challenges and Limitations of MEC

MECs have a range of advantages; yet they also include distinct challenges and limitations that necessitate careful consideration. The operational parameters of

MECs exhibit a high degree of sensitivity, necessitating meticulous management of variables such as pH, temperature, and feedstock concentration to achieve optimal performance. The requirements have the potential to introduce complications in the actual execution of MEC systems and contribute to an escalation in their operational intricacy.

MECs face the additional obstacle of scalability. Although laboratory-scale research has demonstrated the potential of the technology, the process of scaling up MECs for large-scale applications might present significant challenges. The attainment of consistent performance in larger systems and the resolution of technical and design challenges can provide intricate and expensive endeavors.

The issue of cost is of considerable importance. The utilization of MEC technology may require the utilization of expensive materials for electrodes and ion-exchange membranes, resulting in significant costs related to maintenance and monitoring. To enhance the economic feasibility of MECs, it is imperative to prioritize research and innovation efforts aimed at cost reduction and enhancing overall technological efficiency. The operational sensitivity, scalability concerns, and cost considerations are among the obstacles and negatives associated with MECs. The resolution of these problems will be of utmost importance to fully harness the potential of MECs as sustainable technologies for renewable energy and resource recovery on a broader scope.

References

1. Barua S, Zakaria BS, Chung T et al (2019) Microbial electrolysis followed by chemical precipitation for effective nutrients recovery from digested sludge centrate in WWTPs. Chem Eng J 361:256–265. https://doi.org/10.1016/j.cej.2018.12.067
2. Batlle-Vilanova P, Puig S, Gonzalez-Olmos R et al (2014) Assessment of biotic and abiotic graphite cathodes for hydrogen production in microbial electrolysis cells. Int J Hydrog Energy 39:1297–1305. https://doi.org/10.1016/J.IJHYDENE.2013.11.017
3. Clauwaert P, Aelterman P, Pham TH et al (2008) Minimizing losses in bio-electrochemical systems: the road to applications. Appl Microbiol Biotechnol 79:901–913
4. Cooney MJ, Svoboda V, Lau C et al (2008) Enzyme catalysed biofuel cells. Energy Environ Sci 1:320–337
5. Daud NNM, Hussin MH (2023) Evaluating the electrode materials to improve electricity generation with the removal of multiple pollutants through microbial fuel cells. Biomass Convers Biorefin. https://doi.org/10.1007/s13399-023-05256-9
6. Dange P, Pandit S, Jadhav D et al (2021) Recent developments in microbial electrolysis cell-based biohydrogen production utilizing wastewater as a feedstock. Sustainability (Switzerland) 13:1–37. https://doi.org/10.3390/su13168796
7. Escapa A, Mateos R, Martínez EJ, Blanes J (2016) Microbial electrolysis cells: an emerging technology for wastewater treatment and energy recovery from laboratory to pilot plant and beyond. Renew Sustain Energy Rev 55:942–956. https://doi.org/10.1016/j.rser.2015.11.029
8. Fudge T, Bulmer I, Bowman K et al (2021) Microbial electrolysis cells for decentralised-wastewater treatment: the next steps. Water (Switzerland) 13(14):45. https://doi.org/10.3390/w13040445

9. Ghada MA, Alshammari AS, Alomari AD, Almukhlifi AH (2023) Dual role of sugarcane waste in benthic microbial fuel to produce energy with degradation of metals and chemical oxygen demand. Processes 11(4):1060. https://doi.org/10.3390/pr11041060
10. Gautam R, Nayak JK, Ress NV, Steinberger-Wilckens R, Ghosh UK (2023) Bio-hydrogen production through microbial electrolysis cell: structural components and influencing factors. Chem Eng J 455:140535
11. Guerrero-Barajas C (2022) Local fruit wastes driven benthic microbial fuel cell: a sustainable approach to toxic metal removal and bioelectricity generation. Environ Sci Pollut Res 29(22):32913–32928. https://doi.org/10.1007/s11356-021-17444-z
12. Gul MM, Ahmad KS (2019) Bioelectrochemical systems: sustainable bio-energy powerhouses. Biosens Bioelectron 142:111576
13. Harindintwali JD, Zhou J, Muhoza B, Wang F, Herzberger A, Yu X (2021) Integrated eco-strategies towards sustainable carbon and nitrogen cycling in agriculture. J Environ Manag 293:112856
14. Hassan RYA, Febbraio F, Andreescu S (2021) Microbial electrochemical systems: principles, construction and biosensing applications. Sensors (Switzerland) 21:1–19. https://doi.org/10.3390/s21041279
15. He L, Du P, Chen Y et al (2017) Advances in microbial fuel cells for wastewater treatment. Renew Sustain Energy Rev 71:388–403
16. Idris MO, Kim HC (2022) Exploring the effectiveness of microbial fuel cell for the degradation of organic pollutants coupled with bio-energy generation. Sustain Energy Technol Assess 52:102183. https://doi.org/10.1016/j.seta.2022.102183
17. Jadhav DA, Das I, Ghangrekar MM, Pant D (2020) Moving towards practical applications of microbial fuel cells for sanitation and resource recovery. J Water Process Eng. https://doi.org/10.1016/j.jwpe.2020.101566
18. Jafary T, Daud WRW, Ghasemi M et al (2015) Biocathode in microbial electrolysis cell; present status and future prospects. Renew Sustain Energy Rev 47:23–33
19. Kadier A, Jiang Y, Lai B et al (2019) Biohydrogen production in microbial electrolysis cells from renewable resources. Bioenergy Biofuels. https://doi.org/10.1201/9781351228138-12
20. Kadier A, Simayi Y, Abdeshahian P et al (2016) A comprehensive review of microbial electrolysis cells (MEC) reactor designs and configurations for sustainable hydrogen gas production. Alex Eng J 55:427–443. https://doi.org/10.1016/j.aej.2015.10.008
21. Khatoon A, Mohd Setapar SH, Rafatullah M (2020) Outlook on the role of microbial fuel cells in remediation of environmental pollutants with electricity generation. Catalysts 10(8):819. https://doi.org/10.3390/catal10080819
22. Katuri KP, Ali M, Saikaly PE (2019) The role of microbial electrolysis cell in urban wastewater treatment: integration options, challenges, and prospects. Curr Opin Biotechnol 57:101–110
23. Ki D, Parameswaran P, Popat SC et al (2017) Maximizing coulombic recovery and solids reduction from primary sludge by controlling retention time and pH in a flat-plate microbial electrolysis cell. Environ Sci (Camb) 3:333–339. https://doi.org/10.1039/C6EW00305B
24. Kim HC, Ahmad A, Alshammari MB, Yaakop AS (2022) Oxidation of food waste as an organic substrate in a single chamber microbial fuel cell to remove the pollutant with energy generation. Sustain Energy Technol Assess 52:102282. https://doi.org/10.1016/j.seta.2022.102282
25. Koul Y, Devda V, Varjani S et al (2022) Microbial electrolysis: a promising approach for treatment and resource recovery from industrial wastewater. Bioengineered 13:8115–8134. https://doi.org/10.1080/21655979.2022.2051842
26. Kumar SS, Kumar V, Malyan SK et al (2019) Microbial fuel cells (MFCs) for bioelectrochemical treatment of different wastewater streams. Fuel. https://doi.org/10.1016/j.fuel.2019.05.109
27. Logan BE, Call D, Cheng S et al (2008) Microbial electrolysis cells for high yield hydrogen gas production from organic matter. Environ Sci Technol 42:8630–8640
28. Lu L, Ren ZJ (2016) Microbial electrolysis cells for waste biorefinery: a state of the art review. Bioresour Technol 215:254–264. https://doi.org/10.1016/j.biortech.2016.03.034

29. Muoio R, Palli L, Ducci I et al (2019) Optimization of a large industrial wastewater treatment plant using a modeling approach: a case study. J Environ Manag. https://doi.org/10.1016/J.JENVMAN.2019.109436
30. Nabila AK, Yang H (2021) Mini-review: recent technologies of electrode and system in the enzymatic biofuel cell (EBFC). Appl Sci 11(11):5197. https://doi.org/10.3390/app11115197
31. Nabil AZ (2023) A pilot trial in the remediation of pollutants simultaneously with bioenergy generation through microbial fuel cell. J Environ Chem Eng 11(5):110643. https://doi.org/10.1016/j.jece.2023.110643
32. Pawar AA, Karthic A, Lee S et al (2022) Microbial electrolysis cells for electromethanogenesis: materials, configurations and operations. Environ Eng Res. https://doi.org/10.4491/eer.2020.484
33. Qureshi F, Yusuf M, Kamyab H et al (2022) Latest eco-friendly avenues on hydrogen production towards a circular bioeconomy: currents challenges, innovative insights, and future perspectives. Renew Sustain Energy Rev. https://doi.org/10.1016/j.rser.2022.112916
34. Ramírez-Vargas CA, Prado A, Arias CA et al (2018) Microbial electrochemical technologies for wastewater treatment: principles and evolution from microbial fuel cells to bioelectrochemical-based constructed wetlands. Water (Switzerland) 10:1–30. https://doi.org/10.3390/w10091128
35. Rodríguez-Couto S (2021) Preparation, characterization, and application of modified carbonized lignin as an anode for sustainable microbial fuel cell. Process Saf Environ Prot 155:49–60. https://doi.org/10.1016/j.psep.2021.09.006
36. Rousseau R, Etcheverry L, Roubaud E et al (2020) Microbial electrolysis cell (MEC): strengths, weaknesses and research needs from electrochemical engineering standpoint. Appl Energy 257:113938. https://doi.org/10.1016/j.apenergy.2019.113938
37. Saeed HM, Husseini GA, Yousef S et al (2015) Microbial desalination cell technology: a review and a case study. Desalination 359:1–13
38. Schröder U (2007) Anodic electron transfer mechanisms in microbial fuel cells and their energy efficiency. Phys Chem Chem Phys 9:2619–2629. https://doi.org/10.1039/b703627m
39. Seelajaroen H, Spiess S, Haberbauer M et al (2020) Enhanced methane producing microbial electrolysis cells for wastewater treatment using poly(neutral red) and chitosan modified electrodes. Sustain Energy Fuels 4:4238–4248. https://doi.org/10.1039/D0SE00770F
40. Tabassum P, Akil A, Lokhat D (2021) A glimpse into the microbial fuel cells for wastewater treatment with energy generation. Desalin Water Treat 214:379–389. https://doi.org/10.5004/dwt.2021.26737
41. Torlaema TAM, Ibrahim MNM, Ahmad A et al (2022) Degradation of hydroquinone coupled with energy generation through microbial fuel cells energized by organic waste. Processes. https://doi.org/10.3390/pr10102099
42. Tian JH, Lacroix R, Bureau C, Midoux C, Desmond-Le Quéméner E, Bouchez T (2023) Study of a pilot scale microbial electrosynthesis reactor for organic waste biorefinery. Energies 16(2):591. https://doi.org/10.3390/en16020591
43. Varanasi JL, Veerubhotla R, Pandit S, Das D (2018) Biohydrogen production using microbial electrolysis cell: recent advances and future prospects. Elsevier B.V.
44. Waqas M, Rehan M, Aburiazaiza AS, Nizami AS (2018a) Wastewater biorefinery based on the microbial electrolysis cell: opportunities and challenges. In: Progress and recent trends in microbial fuel cells, pp 347–374https://doi.org/10.1016/B978-0-444-64017-8.00017-8
45. Xiao X, Xia HQ, Wu R et al (2019) Tackling the challenges of enzymatic (bio)fuel cells. Chem Rev 119:9509–9558
46. Yu EH, Scott K (2010) Enzymatic biofuel cells-fabrication of enzyme electrodes. Energies (Basel) 3:23–42
47. Zhang J, Zhang Y, Quan X et al (2013) Enhanced anaerobic digestion of organic contaminants containing diverse microbial population by combined microbial electrolysis cell (MEC) and anaerobic reactor under Fe(III) reducing conditions. Bioresour Technol 136:273–280. https://doi.org/10.1016/j.biortech.2013.02.103
48. Zhen G, Lu X, Kumar G et al (2017) Microbial electrolysis cell platform for simultaneous waste biorefinery and clean electrofuels generation: current situation, challenges and future perspectives. Prog Energy Combust Sci 63:119–145

49. Ziara RMM, Dvorak BI, Subbiah J (2018) Sustainable waste-to-energy technologies: Bioelectrochemical systems. Elsevier Inc

Principal, Mechanism, and Basic Thermodynamics of Microbial Electrolysis Cell

Luis Alberto Estudillo-Wong and Claudia Guerrero-Barajas

Abstract Microbial electrolysis cell is a technology that was initially designed to produce hydrogen. This technology requires a biologically assisted anodic chamber, while hydrogen is produced in the cathodic chamber under an applied external voltage. In the microbial colonized anodic chamber, in which electroactive bacteria (EAB) promote electron transfer, a wide variety of organic compounds can be oxidized, which makes this technology promising in wastewater treatment with the advantage of producing hydrogen at the same time. However, scaling up of this technology has not been fully successful yet and production of hydrogen is still low in comparison with water electrolysis. In this chapter, the basic thermodynamics and fundamentals of MEC are presented first initiating with the thermodynamics of an abiotic cell which is the basis of the water electrolysis and a brief description of the mechanism occurring in these biological assisted systems.

Keywords Microbial electrolysis cell (MEC) · Thermodynamics of MEC · Hydrogen production

1 Fundamental Thermodynamics for the Understanding of MEC

In chemical thermodynamics, we define an exergonic process where the Gibbs energy is negative. This criterion applies to a spontaneous reaction, such as a Daniel cell, as illustrated in Fig. 1.

L. A. Estudillo-Wong
Departamento de Biociencias E Ingeniería, CIIEMAD, Instituto Politécnico Nacional, Calle 30 de junio de 1520 s/n, Barrio La Laguna Ticomán, Alcaldía GAM, 07340 Mexico City, Mexico
e-mail: laestudillo-wong@ipn.mx

C. Guerrero-Barajas (✉)
Departamento de Bioprocesos, Laboratorio de Biotecnología Ambiental, Unidad Profesional Interdisciplinaria de Biotecnología, Instituto Politécnico Nacional, Av. Acueducto s/n, Barrio la Laguna Ticomán, Alcaldía GAM, 07340 Mexico City, Mexico
e-mail: cguerrerob@ipn.mx

A. A. Yaqoob and A. Ahmad (eds.), *Microbial Electrolysis Cell Technology*, Sustainable Materials and Technology, https://doi.org/10.1007/978-981-97-3356-9_2

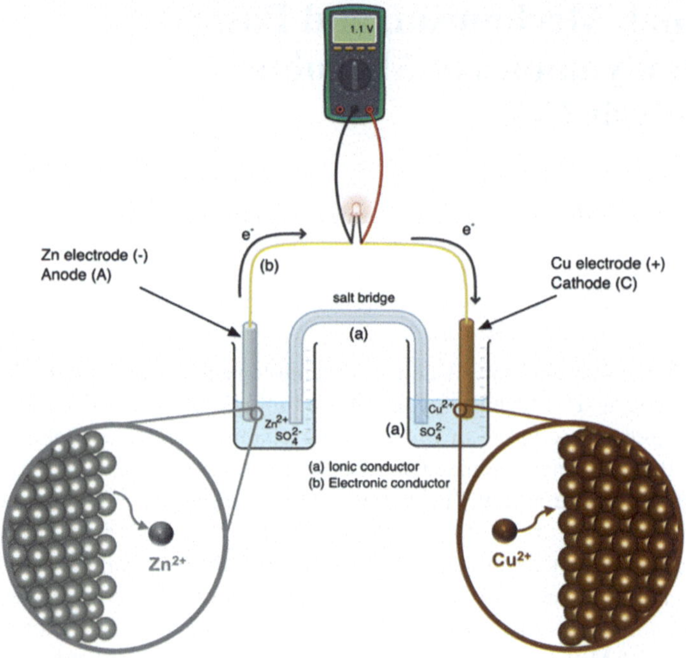

Fig. 1 Schematic of Daniell cell for an exergonic process in electrochemistry system

In electrochemistry, an exergonic reaction is related to galvanic electrochemical system. Here, the Daniell cell produces energy, which can be measured with a multimeter. First, we introduce the half-reaction in the left and right side of Fig. 1.

Anode (left side):

$$Zn_{(s)} \leftrightarrow Zn^{2+}_{(aq)} + 2e^{-} \tag{1}$$

Cathode (right side):

$$Cu^{2+}_{(aq)} + 2e^{-} \leftrightarrow Cu_{(s)} \tag{2}$$

Adding both half-reactions, we obtain the overall reaction, based on the classical definition.

$$Cu^{2+}_{(aq)} + Zn_{(s)} \leftrightarrow Cu_{(s)} + Zn^{2+}_{(aq)}\ \Delta E^{0}_{eq,cell} = 1.1\ V \tag{3}$$

Therefore, a diagram cell defines all the electrochemical processes.

$$Zn_{(s)}\left|Zn^{2+}_{(aq)}\right|\left|Cu^{2+}_{(aq)}\right|Cu_{(s)} \tag{3a}$$

The Daniell cell can produce energy with a voltage equal to 1.1 V at standard-state, which means that is an exergonic process [1]. For a hypothetical reaction, we define a standard reference electrode as a half-reaction and the Gibbs energy based on Lavoisier-Laplace's law, which occurs in the anode, at standard-state.

$$\nu_{e^-} H_{2(g)} \rightarrow 2\nu_{e^-} H^+(aq) + 2\nu_{e^-} e^- \quad \Delta_R G_A^0 = -\nu_{e^-} \Delta_R G^0_{H^+/H_2} \tag{4}$$

In the cathode, we define a reduction reaction, with the same conditions.

$$2\nu_O O^{z_O+} + 2\nu_{e^-} e^- \leftrightarrow 2\nu_R R^{z_R+} \quad \Delta_R G_C^0 = 2\Delta_R G^0_{O^{z_O+}/R^{z_R+}} \tag{5}$$

Adding both half-reactions, we obtain the overall reaction and the total Gibbs energy for the cell.

$$\nu_{e^-} H_{2(g)} + 2\nu_O O^{z_O+} + 2\nu_{e^-} e^- \rightarrow 2\nu_{e^-} H^+(aq) + \nu_R R^{z_R+} + 2\nu_{e^-} e^- \tag{6}$$

$$\Delta_R G^0_{Cell} = \Delta_R G_c^0 + \Delta_R G_a^0 \tag{6a}$$

To balance the overall equation, the total number of electrons must be $2\nu_{e^-}$, like 2 electrons in the reaction (3). Here, the Gibbs energy is a state function, dG is an exact differential; then, we can choose an electric charge q, and ξ as the degree of advancement.

$$G = G(q, \xi) \tag{7}$$

We can write the total differential as

$$dG = \left(\frac{\partial G}{\partial q}\right)_\xi dq + \left(\frac{\partial G}{\partial \xi}\right)_q d\xi \tag{8}$$

First, we can define $\left(\frac{\partial G}{\partial q}\right)_\xi$ as the electrical work, which is produced by electrons.

$$\Delta\varphi = \left(\frac{\partial W}{\partial q}\right)_\xi = -\left(\frac{\partial G}{\partial q}\right)_\xi \tag{9}$$

Considering the Faraday's constant and the degree of advancement definition by the electron, Eqs. (10) and (11), respectively.

$$\frac{dq}{dn_{e^-}} = F \tag{10}$$

$$dn_{e^-} = \nu_{e^-} d\xi \tag{11}$$

And combining those equations, we obtain Faraday's law [2].

$$dq = \nu_{e^-} F d\xi \tag{12}$$

We introduce Eq. (12) and (9) into Eq. (8), we obtain.

$$dG = -\nu_{e^-} F \Delta\varphi d\xi + \left(\frac{\partial G}{\partial \xi}\right)_q d\xi \tag{13}$$

Second, it is possible to define $\left(\frac{\partial G}{\partial \xi}\right)_q$ as chemical potential of the electron. Introducing this definition into Eq. (13), we obtain.

$$dG = [\mu_{e^-} - \nu_{e^-} F \Delta\varphi] d\xi \tag{14}$$

That is finally the electrochemical potential for the electron.

$$\frac{dG}{d\xi} = \mu_{e^-} - \nu_{e^-} F \Delta\varphi \equiv \tilde{\mu}_{e^-} = \mu_{e^-} + z_{e^-} F \Delta\varphi \tag{15}$$

where z_{e^-} is the electric charge of the electron. For a similar argument, we can define that for any charged specie in a phase.

$$\tilde{\mu}_i^\alpha = \mu_i^\alpha + z_i F \varphi^\alpha \tag{16}$$

Here, it is easy to show that for every metal, $z_{e^-} = -1$. Therefore, for electrons in every metal (Eq. 16) becomes $\tilde{\mu}_{e^-}^\alpha = \mu_{e^-}^\alpha - F\varphi^\alpha$ (Eq. 15) [3]. Besides, the electric potential difference between two metals is called a contact potential, which is not possible to measure a different potential at a single interface [3]. However, equilibrium is established for all electrochemical systems. Therefore, we can write the Gibbs–Duhem equation and introduce the definition of the electrochemical potential [4].

$$dG = \left(\frac{\partial G}{\partial T}\right)_{p,n_i} dT + \left(\frac{\partial G}{\partial p}\right)_{T,n_i} dp + \sum_i^n \tilde{\mu}_i dn_i \tag{17}$$

$$dG = \left(\frac{\partial G}{\partial T}\right)_{p,n_i} dT + \left(\frac{\partial G}{\partial p}\right)_{T,n_i} dp + \sum_i^n \mu_i dn_i + F \sum_j^m \varphi^j \sum_i^n z_i dn_i + \sum_i^n \tilde{\mu}_{i,e^-} dn_{e^-} \tag{17a}$$

Based on this equation, the electric potential (φ^j) is related to the phase, where we can define two phases, in aqueous solution (S) or metal (M). At T and P constant, and the definition of $dn_i = \nu_i d\xi$, where ν_i is the stoichiometric coefficient, we obtain.

$$dG = \left\{ \sum_{i}^{n} \nu_i \mu_i + F \sum_{j}^{m} \varphi^j \sum_{i}^{n} \nu_i z_i + \sum_{i}^{n} \nu_i \tilde{\mu}_{i,e^-} \right\} d\xi \tag{18}$$

The overall electrons transferred are equal in the right and left, for the overall reaction (Eq. 6); therefore, we consider that $\nu_{R,e^-} = \nu_{L,e^-} = 2\nu_{e^-} = z$. This implies that the third term of Eq. (18) is the electrochemical potential of the electron, in Cu phase, as metal (*c.f.* Fig. 1). Here, we define the electric potential difference of any cell (the cell potential), Eq. (18b).

$$\begin{aligned} \sum_{i}^{n} \nu_{i,e^-} \tilde{\mu}_{i,e^-} &= -\nu_{R,e^-} \tilde{\mu}_{R,e^-} + \nu_{L,e^-} \tilde{\mu}_{L,e^-} \\ &= -\nu_{R,e^-} \left(\mu_{R,e^-}^{Cu} - F \varphi_R^{Cu} \right) + \nu_{l,e^-} \left(\mu_{L,e^-}^{Cu} - F \varphi_L^{Cu} \right) \\ &= zF \left(\varphi_R^{Cu} - \varphi_L^{Cu} \right) \end{aligned} \tag{18a}$$

$$\sum_{i}^{n} \nu_{i,e^-} \tilde{\mu}_{i,e^-} = zF \Delta\varphi \tag{18b}$$

The second term is 0 because the sum of the total electric charge of each species is equal to 0, which means that the overall equation is balanced, see Eqs. (3) or (6).

$$F \sum_{j}^{m} \varphi^j \sum_{i}^{n} \nu_i z_i = 0 \tag{18c}$$

The differential of $\left(\frac{\partial G}{\partial \xi}\right)_{T,p}$, usually defines a criterion for equilibrium and spontaneity as follows; $\left(\frac{\partial G}{\partial \xi}\right)_{T,p} = 0$, the Gibbs energy has a minimum value, and the reaction is at equilibrium. However, $\left(\frac{\partial G}{\partial \xi}\right)_{T,p,} < 0$ and $\left(\frac{\partial G}{\partial \xi}\right)_{T,p,} > 0$ correspond to galvanic and electrolytic cell, respectively [5]. Therefore, we obtain Eq. (19), in equilibrium.

$$\left(\frac{\partial G}{\partial \xi}\right)_{T,p,\text{eq}} = \sum_{i}^{n} \nu_i \mu_i + zF \Delta\varphi = 0 \tag{19}$$

It is easy to demonstrate kinetically that the limit value of $\Delta\varphi$ measured as the current goes to zero is called the electromotive force of the cell [6].

$$\lim_{i \to 0} \Delta\varphi = \Delta E_{\text{eq,cell}} \tag{19a}$$

Then, Eq. (19) becomes.

$$\left(\frac{\partial G}{\partial \xi}\right)_{T,p,\mathrm{eq}} = -zF\Delta E_{\mathrm{eq,cell}} = \sum_{i}^{n} \nu_i \mu_i \tag{20}$$

At standard-state, the chemical potential is defined by Hess law.

$$\sum_{i}^{n} \nu_i \mu_i^0 = \Delta_R G_{298.15}^0 = \Delta\mu^0 = \Delta\tilde{\mu}^0 = -zF\Delta E_{\mathrm{eq,cell}}^0 \tag{21}$$

For the overall reaction (Eq. 6), we define

$$-2\nu_{\mathrm{e^-}} F\Delta E_{\mathrm{eq,cell}}^0 = -2\nu_{\mathrm{e^-}} F\Delta E_{O^{z_O+}/R^{z_R+}}^0 + 2\nu_{\mathrm{e^-}} F\Delta E_{\mathrm{H^+}/\mathrm{H_2}}^0 \tag{22}$$

we obtain

$$-2\nu_{\mathrm{e^-}} F\Delta E_{\mathrm{eq,cell}}^0 = -2\nu_{\mathrm{e^-}} F\left[\Delta E_{O^{z_O+}/R^{z_R+}}^0 - \Delta E_{\mathrm{H^+}/\mathrm{H_2}}^0\right] \tag{23}$$

For convention, $\Delta E_{\mathrm{H^+}/\mathrm{H_2}}^0$ is zero and if we have $\Delta E_{O^{z_O+}/R^{z_R+}}^0 > \Delta E_{\mathrm{H^+}/\mathrm{H_2}}^0$, Eq. (23) becomes.

$$\Delta E_{\mathrm{eq,cell}}^0 = E_{O^{z_O+}/R^{z_R+}}^0 \text{ versus standard hydrogen electrode(SHE)} \tag{24}$$

This equation defines the electrochemical reaction cell $\left(E_{\mathrm{cell}}^0\right)$, using a standard hydrogen electrode. The E_{cell}^0 is purely thermodynamic, where Lavoisier-Laplace and Hess law can be applied. If we define that all reactions go in the direction of reduction (Eq. 5), we can define the emf for the left and right reaction, at standard-state, Eqs. (25–27).

$$\Delta E_{\mathrm{eq},L}^0 = \Delta E_A^0 - \Delta E_{\mathrm{H^+}/\mathrm{H_2}}^0 \tag{25}$$

$$\Delta E_{\mathrm{eq},R}^0 = \Delta E_c^0 - \Delta E_{\mathrm{H^+}/\mathrm{H_2}}^0 \tag{26}$$

$$\Delta E_{\mathrm{eq,Cell}}^0 = \Delta E_{\mathrm{eq},R}^0 - \Delta E_{\mathrm{eq},L}^0 = \Delta E_c^0 - \Delta E_A^0 \tag{27}$$

Equation (27) defines the electric potential difference, when we measure the electric potential alone for the anode and the cathode with a standard hydrogen electrode. Instead of this, we can now use the definition of chemical potential.

$$\left(\frac{\partial G}{\partial \xi}\right)_{T,p} = \sum_{i}^{n} \nu_i \mu_i \tag{28}$$

The equilibrium constant (k) defines when the differential $\left(\frac{\partial G}{\partial \xi}\right)_{T,p}$ is zero for any reaction at the equilibrium [5]. For an electrochemical reaction, Eq. (20) confirms the definition of the left side of Eq. (28). Therefore, the right side of Eq. (28) is defined by the chemical potential of each species, who is participating in the overall reaction.

$$\begin{aligned} -zF\Delta E_{\text{eq,cell}} &= \Delta_R G^0_{298.15} + RT \ln k_p k_c \\ &= \Delta_R G^0_{298.15} + RT \ln\left\{\prod_i \left(\frac{f_i}{p^0}\right)^{\nu_i} \prod_i \left(\frac{a_i}{a^0}\right)^{\nu_i}\right\} \end{aligned} \tag{29}$$

where k_p and k_c are related to the gas and liquid phase in the reaction. The parameter of f_i and a_i is the fugacity and the activity for gas and ion in aqueous solution, respectively.

$$-zF\Delta E_{\text{eq,cell}} = \Delta_R G^0_{298.15} + RT \ln\left\{\prod_i \left(\frac{y_i \phi_i P}{p^0}\right)^{\nu_i} \prod_i (\gamma_i x_i)^{\nu_i}\right\} \tag{30}$$

where y_i, x_i, ϕ_i, γ_i, are mol fraction, fugacity coefficient, activity coefficient in gas and liquid phase, respectively. The P and ν_i parameters are the total pressure and the stoichiometric coefficient. In the case of gas phase, we can use the Lewis' fugacity rule as a first approximation if we have a mixture gas in the reaction system. In many electrochemical systems, we can use the second approximation, in which ideal gas can be applied at lower temperature and pressure. That means the ϕ_i is equal to 1 and we only consider the partial pressure of the gas. In liquid phase, we must consider the unsymmetrical convention (Convention II) for standard-state. If the mol fraction (x_A) and the activity coefficient (γ_A) of the solvent tends to 1, we define an ideal-dilute solution, an electrolyte solution. Based on molality definition, we can obtain the mol fraction of the species i.

$$m_i = \frac{n_i}{n_A \text{PM}_A} \frac{n_T^{-1}}{n_T^{-1}} = \frac{x_i}{x_A \text{PM}_A} \therefore x_i = m_i x_A \text{PM}_A \tag{31}$$

where n_T and PM_A are total mol of the system and the mass weight of the solvent. Then, we introduce the concentration at standard-state.

$$x_i = m_i x_A \text{PM}_A \frac{m^0}{m^0} \tag{31a}$$

Introducing the definition of mol fraction and partial pressure, in liquid and gas phase, we obtain.

$$-zF\,\Delta E_{\text{eq,cell}} = \Delta_R G^0_{298.15} + RT\ln\left(\text{PM}_A m^0\right) \prod_i^{\text{liquid phase}} \nu_i + RT\ln\left\{\prod_i\left(\frac{p_i}{p^0}\right)^{\nu_i}\prod_i\left(\gamma_i x_A\frac{m_i}{m^0}\right)^{\nu_i}\right\} \tag{32}$$

Based on this equation, we include the solvent into the Gibbs energy of reaction. That means that Gibbs energy of reaction includes the standard Gibbs energy of formation as ion and follows the Hess law. Therefore, Eq. (32) becomes.

$$-zF\,\Delta E_{\text{eq,cell}} = \Delta_R G^0_{298.15} + RT\ln\left\{\prod_i\left(\frac{p_i}{p^0}\right)^{\nu_i}\prod_i\left(\gamma_i x_A\frac{m_i}{m^0}\right)^{\nu_i}\right\} \tag{33}$$

Using water as a solvent, $c_i = \frac{\rho_w}{1000}m_i$, we can change from molality to molarity. Then, we obtain.

$$-zF\,\Delta E_{\text{eq,cell}} = \Delta_R G^0_{298.15} + RT\ln\left\{\prod_i\left(\frac{p_i}{p^0}\right)^{\nu_i}\prod_i\left(\gamma_{c,i}\frac{c_i}{c^0}\right)^{\nu_i}\right\} \tag{34}$$

where p^0 and c^0 parameters are pressure and molarity at standard-state, respectively. Rearranging Eq. (34), we obtain the Nernst's equation which relates the cell potential away from standard-state, where $\Delta E^0_{\text{eq,Cell}}$ is defined by Eq. (21).

$$\Delta E_{\text{eq,cell}} = \Delta E^0_{\text{eq,Cell}} - \frac{RT}{zF}\ln\left\{\prod_i\left(\frac{p_i}{p^0}\right)^{\nu_i}\prod_i\left(\gamma_i\frac{c_i}{c^0}\right)^{\nu_i}\right\} \tag{35}$$

2 Temperature Dependence of Voltage Cell

Based on the definition of Eq. (28) and taking the first derivate with respect to the temperature.

$$\frac{\partial}{\partial T}\left(\frac{\partial G}{\partial \xi}\right)_p = \sum_i \nu_i\left(\frac{\partial \mu_i}{\partial T}\right)_p \tag{36}$$

Now, we use definition of entropy in terms of the differential of Gibbs energy.

$$\left(\frac{\partial G}{\partial T}\right)_{p,n_i} = \left(\frac{\partial \mu_i}{\partial T}\right)_p = -S_i \tag{37}$$

Combining Eqs. (36) and (37), we obtain.

$$-zF\left(\frac{\partial \Delta E_{\text{eq,cell}}}{\partial T}\right)_p = -\sum_i \nu_i S_i = -\Delta_R S \tag{38}$$

Here, it is easy to demonstrate the entropy of reaction as a function of rate of change of $\Delta E_{\text{eq,cell}}$ with respect to temperature.

$$\Delta_R S = zF\left(\frac{\partial \Delta E_{\text{eq,cell}}}{\partial T}\right)_p \tag{39}$$

Now, we start taking the rate of change of Gibbs energy with respect to degree of advancement with the temperature and pressure constant, in the common definition.

$$\left(\frac{\partial G}{\partial \xi}\right)_{T,p} = \left(\frac{\partial H}{\partial \xi}\right)_{T,p} - T\left(\frac{\partial S}{\partial \xi}\right)_{T,p} \tag{40}$$

Then, the second derivate with respect to temperature and rearranging the position of the rate of change with respect to degree of advancement.

$$\begin{aligned} \frac{\partial}{\partial T}\left(\frac{\partial G}{\partial \xi}\right)_p &= \frac{\partial}{\partial T}\left(\frac{\partial H}{\partial \xi}\right)_p - \frac{\partial}{\partial T}T\left(\frac{\partial S}{\partial \xi}\right)_{T,p} \\ \frac{\partial}{\partial T}\left(\frac{\partial G}{\partial \xi}\right)_p &= \frac{\partial}{\partial \xi}\left(\frac{\partial H}{\partial T}\right)_p - \left[T\frac{\partial}{\partial \xi}\left(\frac{\partial S}{\partial T}\right)_p + \left(\frac{\partial S}{\partial \xi}\right)_{T,p}\right] \end{aligned} \tag{41}$$

We can take the definition of $\left(\frac{\partial S}{\partial T}\right)_p = \frac{C_p}{T}$ and $\left(\frac{\partial H}{\partial T}\right)_p = C_p$ and introducing into Eq. (41), we obtain.

$$\frac{\partial}{\partial T}\left(\frac{\partial G}{\partial \xi}\right)_p = -\left(\frac{\partial S}{\partial \xi}\right)_{T,p} \tag{42}$$

Equation (42) shows a definition of rate of change of Gibbs energy with respect to the degree of advancement and its relationship with the entropy of reaction. Using Eq. (20), Eq. (42) becomes Eq. (43)

$$\left(\frac{\partial S}{\partial \xi}\right)_{T,p} = -\frac{\partial}{\partial T}\left(\frac{\partial G}{\partial \xi}\right)_p = zF\left(\frac{\partial \Delta E_{\text{eq,cell}}}{\partial T}\right)_p \tag{43}$$

With this definition, we can confirm that the definition of entropy of energy is equal to rate of change of the entropy with respect to degree of advancement at temperature and pressure constant, Eq. (44).

$$\Delta_R S = \left(\frac{\partial S}{\partial \xi}\right)_{T,p} \tag{44}$$

Now, it is possible to demonstrate by using Eq. (40), the enthalpy of reaction as a function of electric differential potential and its dependence of the temperature.

$$\begin{aligned}\Delta_R H = \left(\frac{\partial H}{\partial \xi}\right)_{T,p} &= -zF\Delta E_{\text{eq,cell}} + zFT\left(\frac{\partial \Delta E_{\text{eq,cell}}}{\partial T}\right)_p \\ &= -zF\left[\Delta E_{\text{eq,cell}} - T\left(\frac{\partial \Delta E_{\text{eq,cell}}}{\partial T}\right)_p\right]\end{aligned} \tag{45}$$

Finally, the heat capacity at constant pressure can be obtained by taking the second derivate of Eq. (43) with respect to the temperature, Eq. (46).

$$\frac{\partial}{\partial T}\left(\frac{\partial S}{\partial \xi}\right)_p = zF\frac{\partial}{\partial T}\left(\frac{\partial \Delta E_{\text{eq,cell}}}{\partial T}\right)_p \tag{46}$$

By the definition of $\left(\frac{\partial S}{\partial T}\right)_p = \frac{C_p}{T}$, Eq. (46) becomes:

$$\left(\frac{\partial}{\partial T}\sum_i \nu_i S_i\right)_p = \left(\frac{\partial \Delta_R S}{\partial T}\right)_p = \frac{\partial}{\partial T}\left(\frac{\partial S}{\partial \xi}\right)_p = \frac{1}{T}\sum_i \nu_i C_{p,i} = \frac{\Delta C_p}{T} \tag{46a}$$

And finally, we obtain the heat of capacity of electrochemical reaction, Eq. (47).

$$\Delta C_p = zFT\left(\frac{\partial^2 \Delta E_{eq,cell}}{\partial T^2}\right)_p \tag{47}$$

If we consider only the pressure at 1 bar and observe the change in thermodynamic properties as a function of temperature, these variations can be considered a standard-state. However, when changing the pressure to a value other than 1 bar, even if it remains constant, the reference point changes, and the thermodynamic properties are not considered at the standard-state. These conditions are common in electrochemical systems, where it cannot always be measured at standard-state conditions, basically at 1 bar. A scenario that is also present in biological systems (microbial electrolysis cells) in which the operational conditions are far from standard conditions.

3 Applying the Nernst Equation to Microbial Electrolysis Cells (MEC)

Once the basics of the "abiotic" MEC (Daniel cell) have been presented, the basics for the biological MEC will be described. Logan B.E, [7, 8], within his research group back in 2005 was the first in presenting the MEC for the production of hydrogen, since then, MEC became a research area of upmost interest in environmental biotechnology. For the purpose of this chapter, in Fig. 2 a scheme of a MEC along with an MFC is shown to compare the functioning of each one. In an MEC the anode compartment may contain a wide variety of organic compounds which are oxidized by the microorganisms present in the anolyte and also on the anode electrode surface forming an electroactive biofilm, the electrons flow through an external circuit to the cathode. The protons are transferred directly to the cathodic chamber in which they combine with electrons to form hydrogen gas according to (48).

Cathodic compartment reaction:

In terms of protons,

$$8H^+ + 8e^- \rightarrow 4H_2 \tag{48}$$

or

$$2H_2O + 2e^- \rightarrow H_2 + 2OH^- \tag{48a}$$

Then the cathode equilibrium potential calculated from Nernst equation arranged to consider the partial pressure of hydrogen produced taking (48),

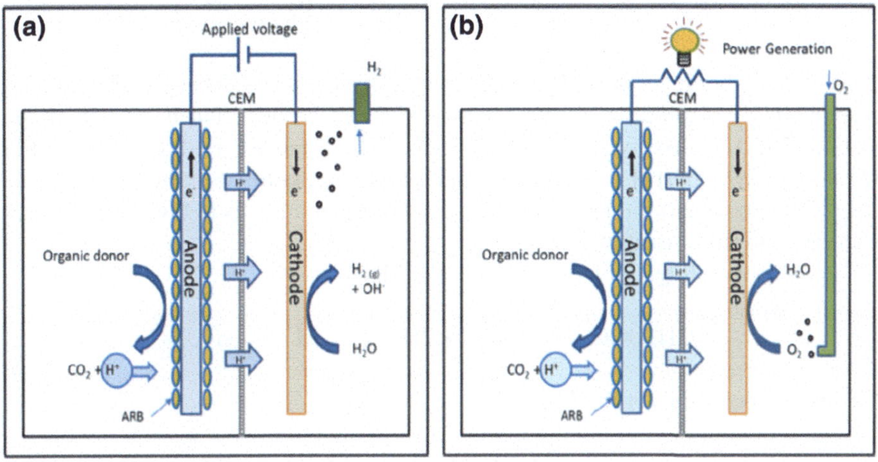

Fig. 2 **a** MEC and **b** MFC, in an MEC an applied voltage is required, whereas in an MFC, there is power generation. (Yasri et al. [9], reproduced with Elsevier permission)

$$E_{cat} = E^0_{cat} - \frac{2.3RT}{F}\text{pH} - \frac{2.3RT}{2\text{F}}\log(p\text{H}_2)$$
$$= -0.414\text{ V versus NHE at pH 7} \quad (49)$$

Taking into account the oxidation of organic compounds occurring in the anode (50), the oxidation of acetate will be taken as a model of oxidation of an organic compound to couple its oxidation to (48).

Anodic compartment reaction:

$$CH_3COO^- + 4H_2O \rightarrow 2HCO_3^- + 9H^+ + 8e^- \quad (50)$$

From Nernst equation:

$$E_{an} = E^0_{an} - \frac{RT}{8F}\ln\frac{[CH_3COO^-]}{[H^+]9[HCO_3^-]^2}$$
$$= -0.3\text{ V(versus NHE at pH 7)} \quad (51)$$

where: E^0_{an} (standard potential) = 0.187 V (at 298.15 K and $HCO_3{}^-$, CH_3COO^- concentrations of 1 M at pH = 0), $T = 298.15$ K, $R = 8.314$ J/mol K, $F = 96{,}500$ C/mol at 0.0169 M (equivalent to 1 g/L of acetate, CH_3COO^-) in a 5 mM (pH 7) bicarbonate ($HCO_3{}^-$) solution.

It is important to consider that (50) can be written with CO_2 (dissolved CO_2 or CO_2 gas) as the oxidation product of acetate instead of bicarbonate. When the product of the reaction is CO_2 gas, Nernst equation is as follows:

$$E_{an} = E^0_{an} - \frac{2.3RT}{8F}7pH + \frac{2.3RT}{8F}\log\left(\frac{p^2_{CO_2}}{C_{CH_3COO^-}}\right) \quad (52)$$

As recently pointed out by Rousseau et al. [10], there is a difference in standard potentials (E^0_{an}) depending on the acetate oxidation product that is written in the reaction ($HCO_3{}^-$, dissolved CO_2 or CO_2 gas). The standard conditions implied in (52) are $T = 298.5$ K, pH = 0 and pCO_2 (CO_2 partial pressure) = 1 atm for an acetate (CH_3COO^-) concentration of 1 M, the partial pressure of CO_2 in this case is closer to actual conditions assayed in MECs.

Then, the overall reaction occurring in the MEC is:

$$CH_3COO^- + 4H_2O \rightarrow 2HCO_3^- + H^+ + 4H_2 \quad (53)$$

Thus, the potential for the MEC is

$$E_{cell} = E_{cat} - E_{an} = (-0.414\text{ V}) - (-0.3\text{ V}) = -0.114\text{ V} \quad (54)$$

This means that theoretically, 0.114 V is required to drive the MEC. In an actual MEC, higher voltages may be required to overcome the ohmic and activation losses and mass transport limitations [11]; therefore, the actual potential applied to the MEC would be as follows:

$$E_{app} = E_{thermodynamic} - (\eta_{cathode} + \eta_{anode} + \eta_{ohmic}) \tag{55}$$

where $E_{thermodynamic}$ is the theoretical thermodynamic voltage, $\eta_{cathode}$ and η_{anode} are the cathode and anode overpotentials, respectively, and η_{ohmic} is the overall ohmic loss of the MEC (Fig. 3). In the case of oxidation of acetate to produce hydrogen, theoretically only 4 mol of H_2/mol acetate can be formed. This yield can be higher with glucose ($C_6H_{12}O_6$) in which case up to 12 mol H_2/mol glucose can be theoretically obtained ($\Delta G^0 = 3.2$ kJ/mol).

$$C_6H_{12}O_6 + 12H_2O \rightarrow 12H_2 + 6CO_2 \tag{56}$$

However, glucose fermentation in the anodic chamber of MEC proceeds to acetate and butyrate as the main fermentation products rather than to hydrogen production as follows:

$$C_6H_{12}O_6 + 4H_2O \rightarrow 2CH_3COO^- + 2HCO_3^- + 4H^+ + 4H_2 \quad (\Delta G^0 = -206.3\ \text{kJ/mol}) \tag{57}$$

$$C_6H_{12}O_6 + 2H_2O \rightarrow 2CH_3CH_2CH_2COO^- + 2HCO_3^- + 3H^+ + 2H_2 \quad (\Delta G^0 = -256.8\ \text{kJ/mol}) \tag{58}$$

$$CH_3COO^- + 4H_2O \rightarrow 2HCO_3^- + H^+ + 4H_2 (\Delta G^0 = 104.6\ \text{kJ/mol}) \tag{59}$$

$$CH_3CH_2CH_2COO^- + 10H_2O \rightarrow 2HCO_3^- + H^+ + 4H_2 \quad (\Delta G^0 = 257.3\ \text{kJ/mol}) \tag{60}$$

Calculations for hydrogen efficiency yield are the mol of hydrogen production normalized to the mol of substrate consumption and the coulombic efficiency (CE) is

$$\text{CE} = nE/nH_2\text{theoretical} \tag{61}$$

$$nE(\text{electric current generated in the MEC}) = \int_{t=0}^{t} I\Delta t/2F \tag{62}$$

nH_2 theoretical is the theoretical amount of hydrogen mol obtained from substrate

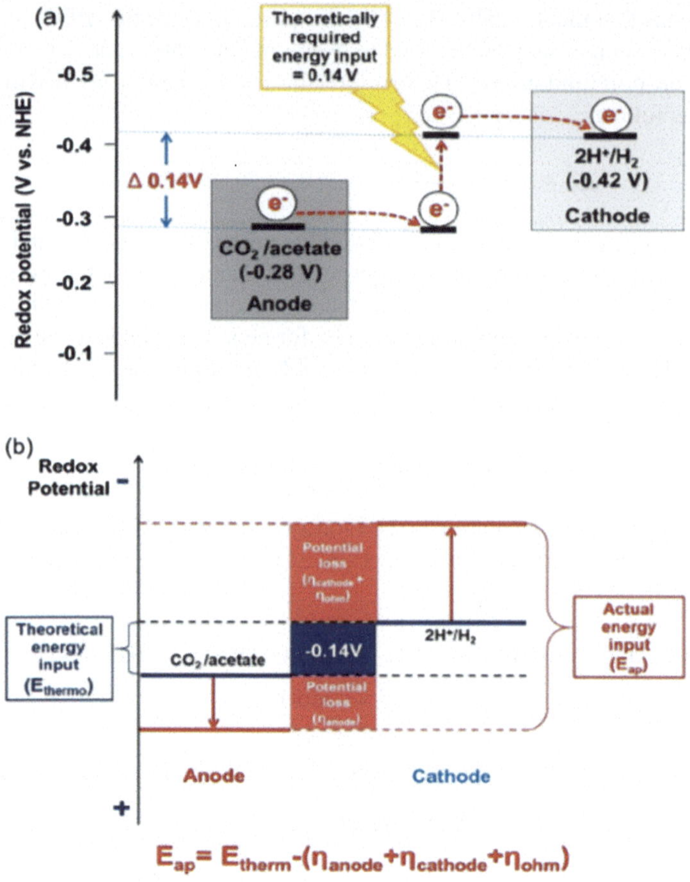

Fig. 3 **a** Theoretical required energy input for a microbial electrolysis cell (MEC) for hydrogen production, and **b** actually required energy input for an MEC. The practically required energy input is much larger owing to the potential losses attributed to the different conditions of an MEC (Yang et al. [11] reproduced with Elsevier permission)

I current.

Δt current sampling interval (s).

F Faraday´s constant 96,500 coulombs and number 2 corresponds to $2e^-$ from $2H^+ + 2e^- \rightarrow H_2$.

Thus, the energy efficiency (ηS %) related to the electron donors (substrates) used in the MEC system can be obtained as follows [12],

$$\eta S = W_{H2}/W_S \tag{63}$$

The energy efficiency based on input electricity (η_E %)

$$\eta_E = W_{H2}/W_E \tag{64}$$

The overall energy efficiency (η_{E+S} %)

$$\eta_{E+S} = W_{H2}/W_{E+S} \tag{65}$$

where:

$$W_{H2} = nH_2\Delta H_{H2}(\text{kJ})\text{is the hydrogen energy in terms of calorific value, } nH_2\text{mol of produced hydrogen and } \Delta H_{H2}\text{heat of combustion of hydrogen}(\text{kJ/mol}) \tag{66}$$

$$W_S = n_S DH_S(\text{kJ})\text{is the energy based on the substrates in terms of calorific value, } n_S\text{mol of substrate consumed and } DH_S\text{heat of combustion of substrate}(\text{kJ/mol}) \tag{67}$$

$$W_E = \sum_{i=1}^{n}\left(IE_{ap} - I^2R_{ex}\right)t.\ (\text{kJ})\text{input of electrical energy to the MEC} \tag{68}$$

where

I = current (mA).
Δt is the current sampling interval (s).
E_{ap} is the applied voltage (V).
R_{ex} is the external resistance (Ω).

Fermentation of glucose is a known process that can be easily implemented in MEC and the overall performance of the system will depend on the proper design including the microorganisms, electrodes materials and operational conditions (i.e., pH and T). For example, in a study carried out to compare the hydrogen production in an MEC using in separate acetate, propionate and butyrate, it was reported that the highest hydrogen production was achieved with acetate, obtaining an hydrogen production rate of 0.53 m^3/m^3 d (current density of 6.0 A/m^2) and a coulombic efficiency of 87%. Butyrate as substrate yielded hydrogen as 0.18 m^3/m^3 d at 2.5 A/m^2 of current density and a coulombic efficiency of 72%. Propionate yielded the lowest hydrogen production of 0.072 m^3/m^3 d at 1.6 A/m^2 and 51% of coulombic efficiency [13]. In this case, the report suggests that the acetate was the most suitable substrate for the microorganisms over propionate and butyrate which points out to the importance of the consortium in the MEC depending on the type of substrates that will be utilized. Shao et al. [14] also compared the performance of an MEC (applied voltage 1 V at 35 °C) supported with acetate, propionate and butyrate and they reported that the highest hydrogen production (25.85% v/v) and 23.4 mA was obtained. They also reported that the EAB population on the anode biofilm using acetate was richer than the population of these microorganisms using glucose as the electron donor. The genus *Desulfuromonas* was suggested as the most active in regard to hydrogen production in the MEC.

In regard to the microorganisms that are used in MEC, these must be capable of oxidizing the organic compounds and present affinity to the electrode (anode electrode). These microorganisms are known as electroactive or exoelectrogenic bacteria because they can transfer electrons to the anode without any external mediator. The electroactive bacteria (EAB) transfer the electrons to the anode (electron acceptor) by several mechanisms. These mechanisms are: direct electron transfer through direct contact of the membrane with the electrode surface, direct electron transfer through a pili (or nanowires), which is a conductive protein appendage that serves to attach the microorganism to the anode electrode and transfer the electrons and by means of electron shuttles or redox mediators like flavins produced by the microorganisms in the biofilm. Figure 4 presents the different electron transfer mechanisms that EAB utilize in MEC. Recently, the study of syntrophic associations among microorganisms to produce hydrogen in MEC has produced interesting results as in the case, in which the hydrogen producer *Ethanoligenens harbinence* was coupled to the EAB *Geobacter sulfurreducens* in an MEC supplemented with glucose [12]. In this work, it was demonstrated that *Geobacter sulfurreducens* could oxidize acetate produced from glucose fermentation by *Ethanoligenens harbinence* while transferring electrons to the anode of the MEC (applied voltage 0.7 V). The current densities (2.03 $\pm$ 0.04 A/m^2), coulombic efficiencies (CE, 30.89% $\pm$ 3.46) and energy efficiencies relative to the substrates (ηS, 20.70 $\pm$ 1.99%) were higher when both, *Ethanoligenens harbinence* and *Geobacter sulfurreducens* were inoculated in the systems against those MEC in which they were inoculated individually.

4 Final Considerations

In bioelectrochemical systems (i.e., MEC), it is important to consider the different factors that may affect the performance of the biological catalyzer in the anodic chamber regarding the production of hydrogen, which was the example presented here. The pH, temperature, type of substrates, anode material and structure, applied voltage and cell design may lead to low hydrogen productions despite of being thermodynamically feasible (in the theoretical calculations). The microorganisms (biological catalyzer) and the biofilm formed on the anode are the core of the MEC, nowadays literature offers a vast amount of reports of work related to the anode materials and the improvements in design, efficiency and cost achieved so far, taking into consideration the biomass-based carbon materials that have been developed in attempts to diminish the ohmic loss and improve the electron transfer process. It is also important to take into consideration the culturing of the microorganisms and the maintenance of the biofilm over time.

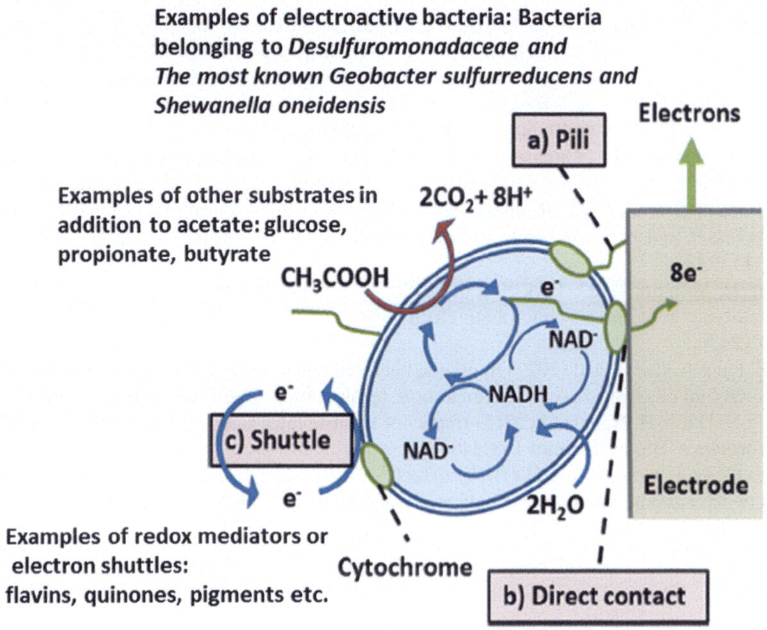

Fig. 4 Electron transfer mechanisms that are exhibited in EAB in MEC during the oxidation of acetate (Modified from Yasri et al. [9])

The development of the Nernst equation from the abiotic process presented at the beginning of this chapter provides the basis for understanding the process in the MEC and serves to clarify the contribution of the substrate (electron donor) and the microorganisms (i.e., EAB) to the production of hydrogen. The information presented in this chapter is also the basis for the understanding of systems in which could be a biological catalyzer in the cathode (biocathode of the MEC) in replacement of the hydrogen production.

References

1. Petrovic S (2021) Electrochemistry crash course for engineers, 1st edn. Springer Cham, Switzerland
2. Koretsky MD (2013) Engineering and chemical thermodynamics, 2nd edn. Wiley, USA
3. Bard AJ, Faulkner LR, White HS (2022) Electrochemical methods: fundamentals and applications, 3rd edn
4. Bard AJ, Inzelt G, Scholz F (2012) Electrochemical dictionary, 2nd edn. Springer, Berlin, Heidelberg, USA
5. Raff LM (2014) Spontaneity and equilibrium: why $\Delta G < 0$ denotes a spontaneous process and $\Delta G = 0$ means the system is at equilibrium are incorrect. J Chem Educ 91:386–395
6. Wang J (2006) Analytical electrochemistry, 3rd edn. Wiley, Canada

7. Liu H, Grot S, Logan BE (2005) Electrochemically assisted microbial production of hydrogen from acetate. Environ Sci Technol 39:4317–4320. https://doi.org/10.1021/es050244p
8. Logan BE, Call D, Cheng S, Hamelers HVM, Sleutels THJA, Jeremiasse AW et al (2008) Microbial electrolysis cells for high yield hydrogen gas production from organic matter. Environ Sci Technol 42:8630–8640. https://doi.org/10.1021/es801553z
9. Yasri N, Roberts EPL, Gunasekaran S (2019) The electrochemical perspective of bioelectrocatalytic activities in microbial electrolysis and microbial fuel cells. Energy Rep 5:1116–1136
10. Rousseau R, Etcheverry L, Roubaud E et al (2020) Microbial electrolysis cell (MEC): strengths, weaknesses and research needs from electrochemical engineering standpoint. Appl Energy 257:113938
11. Yang E, Mohamed HO, Park S-G et al (2021) The electrochemical perspective of bioelectrocatalytic activities in microbial electrolysis and microbial fuel cells. Biores Technol 320:124363
12. Li Z, Fang A, Cui H et al (2021) Synthetic bacterial consortium enhances hydrogen production in microbial electrolysis cells and anaerobic fermentation. Chem Eng J 417:127986
13. Yang N, Hafez H, Nakhla G (2015) Impact of volatile fatty acids on microbial electrolysis cell performance. Biores Technol 193:449–455
14. Saho Q, Li J, Yang S, Sun H (2019) Effects of different substrates on microbial electrolysis cell (MEC) anodic membrane: biodiversity and hydrogen production performance. Water Sci Technol 1123–1133

Material for Microbial Electrolysis Cell: Design to Reactor Configuration

Humaira Rashid Khan, Abdul Latif Ahmad, Ahmed Shuja Syed, Imran Murtaza, Bushra Rafique, and Asadullah Dawood

Abstract Microbial electrolysis cells (MECs) are potential for wastewater management and renewable energy generation. This chapter discusses MEC materials and reactor architecture. While detoxifying effluent, MECs transform organic materials into hydrogen gas or energy using electrochemically active microorganisms. Performance, efficiency, and durability of microbial electrolytic cells depend on material selection. Proton exchange membranes, reactor designs, and anode and cathode materials are critical. Conductive anodes like carbon cloth or graphite help exo-electrogenic bacteria oxidize organic molecules. Cathode materials, usually platinum or other noble metals, allow proton reduction, generating hydrogen gas or other compounds. Proton exchange membranes separate anode and cathode compartments, allowing ion transfer while inhibiting vapor or substrate exchange. Microbial electrolysis cells vary in design dependent on goals. Single-chambered electrode assemblies

H. R. Khan (✉) · A. L. Ahmad
School of Chemical Engineering, Universiti Sains Malaysia, Engineering Campus, 14300 Nibong Tebal, Pulau Pinang, Malaysia
e-mail: humairar_khan@usm.my

A. L. Ahmad
e-mail: chlatif@usm.my

H. R. Khan · A. Shuja Syed · I. Murtaza
Center for Advanced Electronics and Photovoltaic Engineering (CAEPE), International Islamic University, Islamabad 44000, Pakistan
e-mail: ahmed.shuja@iiu.edu.pk

I. Murtaza
e-mail: imran.murtaza@iiu.edu.pk

I. Murtaza
Flexible Electronics Laboratory (FEL), Department of Physics, International Islamic University, Islamabad 44000, Pakistan

B. Rafique
School of Chemistry and Chemical Engineering, Beijing Institute of Technology, Beijing 100081, China

A. Dawood
Department of Physics, National Excellence Institute (University), Islamabad 04524, Pakistan
e-mail: asaddawood@gmail.com

A. A. Yaqoob and A. Ahmad (eds.), *Microbial Electrolysis Cell Technology*, Sustainable Materials and Technology, https://doi.org/10.1007/978-981-97-3356-9_3

are simple and inexpensive. Dual-chamber membrane electrode assemblies reduce gas crossing and increase efficiency by separating the anode and cathode compartments. This chapter emphasizes material selection and reactor design in microbial electrolysis cells for bio-hydrogen production and wastewater treatment. To optimize microbial electrolysis cells for sustainable energy generation and environmental cleanup, one must understand how materials and configurations affect performance. To improve efficiency and versatility of microbial electrolysis cell technology, future research should focus on innovative materials and reactor topologies.

Keywords Microbial electrolysis cells (MECs) · Reactor configuration · Pilot-scale MEC · Anode · Membrane separator · Cathode

1 Introduction

Recent years have seen a surge in demand for sustainable and ecologically friendly energy sources due to climate change concerns. Solar, wind, and hydroelectric power have made progress, but their intermittent nature requires effective energy storage and conversion systems. Microbial electrolysis cell (MEC) technology uses microorganisms to generate clean electricity, treat wastewater, and reduce greenhouse gas emissions, offering a potential alternative. The microbial electrolysis cell is a concept that is based on the fundamental concepts of microbial fuel cells (MFCs). MFCs are devices that utilize the metabolic activity of microorganisms to produce energy from organic substances. Nevertheless, MECs possess a unique purpose and design that differentiates them from other technologies. MFCs have the ability of generating electrical energy through direct means, but microbial electrolysis cells (MECs) primarily emphasize the creation of hydrogen gas (H_2). Therefore, MECs enable the treatment of wastewater and recovery of energy simultaneously through hydrogen synthesis.

The fundamental principle underlying microbial electrochemical cells (MECs) is the utilization of electroactive bacteria, also known as exo-electrogens, which exhibit the exceptional capability of extracellular electron transfer to solid electrodes. Microbial electrochemical cells (MECs) enable the conversion of organic matter into hydrogen gas and clean water, while simultaneously generating electricity. This is achieved by incorporating microorganisms into a system that consists of an anode, a cathode, and an ion-conductive membrane. This innovative method has a multitude of benefits, encompassing effective energy recuperation, wastewater management, and diminished emissions of greenhouse gases.

One of the foremost benefits associated with microbial electrolysis cells (MECs) is in its capacity to produce hydrogen gas, which serves as a sustainable energy carrier. Hydrogen is widely acknowledged as an environmentally friendly and adaptable energy resource, owing to its ability to be utilized in several domains, ranging from the propulsion of fuel cell automobiles to the facilitation of industrial operations. MECs present a viable and sustainable alternative to traditional techniques, such as steam methane reforming, for hydrogen production. Unlike the latter, which heavily relies on fossil fuels and results in the release of carbon dioxide (CO_2) emissions, MECs utilize microbial processes to generate hydrogen. Furthermore, it should be noted that MECs have the capability to function effectively within normal temperature and pressure ranges, hence exhibiting energy efficiency and a positive impact on the environment. The significance of MEC technology extends to sewage treatment as well. MECs play a vital role in eliminating organic contaminants and nutrients from wastewater during the electricity generation process. The microorganisms present in the system degrade organic substances, thereby reducing the chemical oxygen demand (COD) of the wastewater and promoting water purification. This integrated strategy effectively tackles two urgent environmental concerns, namely energy production and wastewater treatment, in a unified system [1].

Furthermore, MECs possess the capacity to effectively reduce the release of greenhouse gas emissions. The anaerobic digestion of organic matter in wastewater treatment plants has the potential to result in the emission of methane (CH_4), which is recognized as a highly potent greenhouse gas. Microbial electrochemical cells (MECs) offer a novel approach by effectively collecting the electrons produced during microbial electrolysis, hence mitigating the generation of methane. Through the redirection of electrons toward the synthesis of hydrogen, microbial electrolysis cells (MECs) not only facilitate the generation of environmentally friendly energy but also yield a substantial reduction in the carbon emissions associated with wastewater treatment techniques [2].

This introduction only scratches the surface of microbial electrolysis cells' potential and applications. As the need for sustainable energy, wastewater treatment, and greenhouse gas reduction grows, MEC technology leads the way. Engineers and researchers are constantly developing new MEC uses, making them a promising instrument for a greener future. This review will explore MEC technology's working principles, important components, recent advances, and obstacles, highlighting its transformative significance in energy and the environment.

2 Microbial Electrolysis Cell (MEC)

Microbial electrolysis cells (MEC) are an intriguing method to produce hydrogen from organic compounds like effluent and other natural resources [1, 2]. MECs were initially discovered in 2005 by two separate groups of researchers. One of them was affiliated with Penn State University, while the other was associated with Wageningen University, Netherlands [3, 4]. Bacteria that are capable of electrochemical activity oxidize organic materials in a MEC, producing carbon dioxide (CO_2), electrons (e^-), and protons (H^+). Bacteria emit protons into the solution while transferring electrons to the anode. The electrons then mix with the free protons in solution after passing along a wire to a cathode. This, however, does not happen naturally. MEC reactors require an externally supplied voltage (> 0.22 V) in a bio-assisted conditions to create hydrogen at the cathode from the combination of these charge carriers [5]. This is accomplished by applying voltage from an external source. While conventional water electrolysis typically requires 1.23–1.8 V, MECs only need 0.2–0.8 V.

A microbial electrolysis cell (MEC) is a type of electrolyzer that facilitates the oxidation of organic materials at the anode, while the cathode performs the abiotic reduction of water in a conventional manner [6]. In electrolysis systems, microorganisms' lower over-potential, accelerate electron transfers, and reduce power use. The most crucial thing is that MECs use renewable substrates and convert efficiently. That makes MEC a good technology for local power plants, wastewater treatment, and hydrogen generation. An intriguing new area of environmental biotechnological study is microbial electrolysis cells (MECs), which use wastewater to create hydrogen. The Oakville-based Napa Wine Company conducted the first microbial electrolysis demonstration of renewable hydrogen production from wastewater [7]. The MEC experienced extensive investigation and development over the past five years. The widespread dissemination of accomplishments holds significant importance for the advancement of this groundbreaking innovation.

3 The Basics of MEC Engineering

3.1 Working Principle of MEC

The basic components of a MEC are the anode, power source, cathode, electrochemically active bacteria (EAB), and a barrier or membrane. Figure 1 is a graphical illustration of a MEC and its operating principle. In a microbial electrolysis cell, electroactive bacteria (EAB) attach to the surface of anode and catalyze the exchange of substrates into carbon dioxide (CO_2), electrons (e^-), and protons (H^+), as an integral portion of their microbial breakdown. The electroactive biofilm (EAB) facilitates the transport of generated electrons to the superficial layer of the anode. Simultaneously, the protons created are dispersed openly into the electrolyte solution within the microbial electrolysis cell to engage in the reduction reactions. Subsequently, the electrons are subsequently conveyed in a continuous manner via an external circuit, facilitated by an electrical source, to a cathode. At the cathode, these electrons amalgamate with the available H^+ ions within the solution, resulting in the production of H_2. Plain carbon electrodes produce H_2 slowly through the hydrogen evolution reaction (HER), which requires a large over-potential. An electrocatalyst has been mounted on the cathodes to lower the potential difference [8]. The electrochemical reactions involved in the process can be elucidated through the use of acetate as a substrate.

At Anode:

$$CH_3COO^- + 4H_2O \rightarrow 2HCO_3^- + 9H^+8e^-$$

At Cathode:

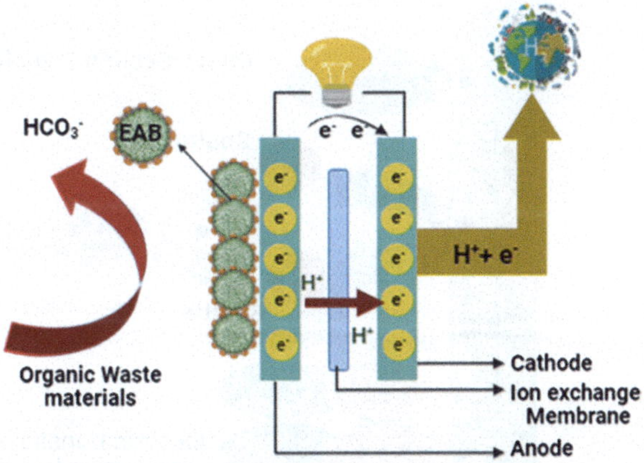

Fig. 1 Schematic depiction of a conventional two-chambered microbial electrolysis cell (MEC) and its functioning [11] (Adapted from reference [11] and redrawn in the text.)

$$8H^+ + 8e^- \rightarrow 4H_2(E_{H+/H2} = 0.414 \text{ vs NHE})$$

Overall Reaction:

$$CH_3COO^- + 4H_2O \rightarrow 2HCO_3^- + H^+ + 4H_2$$

However, it is important to acknowledge that the reduction of H^+ to H_2, also known as HER, is a thermodynamically unfavorable reaction. As a result, this process only requires a low current in a range of about 0.2–0.8 V, to facilitate the formation of H_2 in the electric circuit. The voltage need for this process is significantly lower compared to that required for water electrolysis, which normally falls within the range of 1.23 to 1.8 V [9, 10].

3.2 *MEC Electron Transfer Mechanism*

The operation of MEC systems requires the transfer of electrons from an organic substrate to the electrodes. Exoelectrogens and electrogens, collectively referred to as EAB, possess the ability to transport electrons from a substrate to extracellular electron acceptors (EEAs) in the absence of extraneous mediators. This capability enables the facilitation of essential extracellular electron transfer (EET) [12]. Figure 2 provides a comprehensive overview of the often-reported EET processes. The basic mechanism involves the passage of electrons across the outer protective layer of the bacterium to the rigid ion acceptor, the electrode. This transfer is facilitated by proteins, such as cytochromes [13]. The second approach is the transportation of electrons in the presence of a soluble electron shuttle. This shuttle, that can be a substance such as melanin, phenazines, flavins, or quinines, facilitates the transfer of electrons between bacteria and the electrode through diffusive transportation [14, 15]. The third approach involves a rigid component known as cellular pili, which

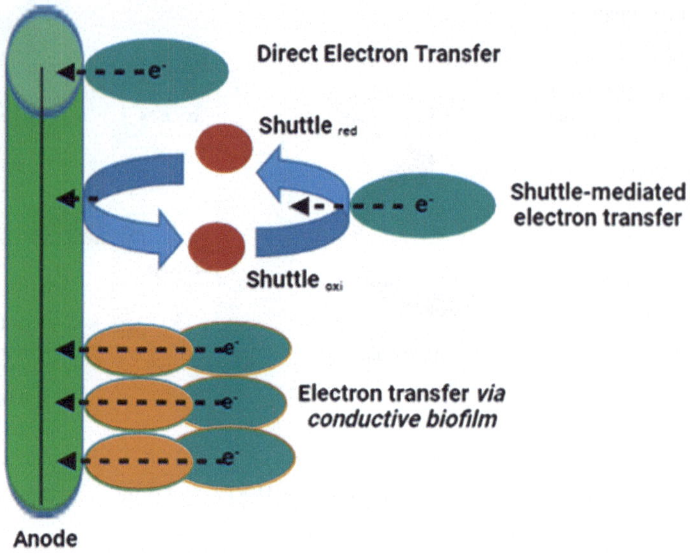

Fig. 2 MEC anode schematic for extracellular electron transfer (EET) [16] (Adapted from reference [16] and redrawn in the text.)

functions as nanowires inside the extracellular biofilm matrix. The pili in question play a crucial role in enabling the transfer of electrons from the bacteria to the electrode, thereby functioning as electron acceptors [10].

4 Reactor Configuration

The configuration and arrangement of microbial electrolysis cell (MEC) reactors have experienced substantial enhancements, resulting in a considerable augmentation of the overall efficiency of MECs to a considerable degree [17–19]. The performance of microbial electrochemical cells (MECs) can be influenced by a range of physicochemical parameters shown by the electrodes. These properties include surface area, substrate shape, separation between electrodes, electrode position, activation resistance, and conductance, overall effectiveness of MECs [20, 21]. In their study, Call and Logan (2008) reported an increase in hydrogen production rate (HPR) through the utilization of graphite granules to improve the topography of an anode. Additionally, the reduction in internal resistance within a microbial electrolytic cell was achieved by decreasing the separation distance between the anode and cathode electrodes from 14 to 4 cm [22, 23]. Additionally, Liang et al. (2019) provided evidence to support the notion that the internal resistance of the microbial electrolysis cell (MEC) system can be reduced through the optimization of the anode configuration [24]. In their research, the authors of the study conducted an experiment using a microbial electrolytic cell to investigate the effects of positioning two anodes on opposite sides of the cathode in a parallel configuration.

4.1 MECs Electrode Materials

The primary objective of MEC is to facilitate hydrogen synthesis through an efficient approach. The anode of a microbial electrolytic cell serves as the origin of both electrons and protons. The presence of an extracellular electron acceptor in the anode can facilitate the oxidation of the source material within the anode chamber. This process involves the passage of electrons from the cell membrane to the anode, hence establishing extracellular electron pathways. The cathodic reaction pertains to the generation of hydrogen, specifically known as the hydrogen evolution reaction (HER). Hydrogen generation can be facilitated through the combination of electrons on the cathode with free protons in solution, requiring a comparatively low energy input in a particular range of about 0.2–0.8 V. The optimization of these two-electrode reactions is of utmost importance for enhancing the performance of a MEC. As a result, numerous techniques are employed to enhance the speeds of these two-electrode reactions, while simultaneously boosting current density and hydrogen recoveries.

4.2 Anodic Materials of MEC

Electrons and protons, which are fundamental constituents involved in the process of hydrogen synthesis, are produced at the outer layer of the anode. The aforementioned phenomenon takes place via the oxidation of substrates, which is helped by electrochemically active bacteria (EAB) residing on the exterior of the anode. The generation of hydrogen is intricately connected to the current operational efficiency of the essential electrode. Hence, it is imperative for the anode to possess certain distinct characteristics in order to fulfill the criteria for satisfactory microbial electrochemical cell (MEC) performance. The aforementioned characteristics contain (i) a substantial relative dimension, (ii) elevated electrical conductance, (iii) biocompatibility and stability, and (iv) appropriate mechanical strength [25]. The substantial relative dimension can supply EAB enriched adhesion sites, increasing current densities [26]. The Ohmic standard affects anodic conductance, which can limit the flow of current and supply of energy. The formation of EAB biofilm and the transfer of electrons between the anode and EAB are reliant upon biocompatibility. Moreover, the economical anode exhibits compatibility with the process of scaling up microbial electrolysis cells. Most MEC research studies use carbon-based anodes due to their superior conductivity, varied morphological characteristics, and biocompatibility. In addition, carbon-based materials are known for their chemical stability and cost-effectiveness [2]. Examples of such materials are carbon fabric, carbon papers, graphite felt, graphite brushes, and graphite flakes. These materials have been found to facilitate a strong connection between the EAB biofilm and the anode, resulting in a reduction in the internal resistance of the electrode distance. However, the particular surface area may be a limiting parameter when compared to other carbon compounds. Graphite brushes and graphite granules are very notable anode materials due to their substantial surface area, which facilitates an increased number of electrochemically active bacteria (EAB) adhering to the anode. This, in turn, leads to an augmented current production. Both of these carbon-based anodes, however, have drawbacks in terms of electron resistance for the break in the circuit between the electrode and the external circuit [27]. The carbon paper possesses a distinct surface area; yet, its restricted mechanical strength serves as the limiting factor for its utilization as a MEC anode throughout prolonged periods of operation. There exist numerous approaches to improve the performance of the anode using various conceptual frameworks. Certain intriguing and innovative nano-carbon materials exhibit significant promise as anodes for microbial electrochemical cells (MECs). The hollow structures of carbon nanotubes (CNTs) possess significant characteristics such as a high specific surface area and conductivity. These properties have garnered substantial attention as materials for supporting anodes [28]. Recent studies on CNT have shown that they can be hazardous to cells, leading to both increased cell growth and cell death. This has prevented their effective use as a MEC anode. Gold nanoparticles, polypyrrole, and polyaniline are just a few examples of the many biologically compatible substances that could be utilized to make these modifications.

Graphene, with its amazing electrical characteristics, is another potential material for the MEC anode [29]. However, the biodegradability of graphene is suboptimal, hence restricting the microbial attachment capacity and efficiency of extracellular electron transfer (EET). When studying the properties of polyaniline, it is seen that it possesses favorable biocompatibility but exhibits inadequate conductivity. However, by combining polyaniline with graphene, a composite material may be created that effectively harnesses the advantageous characteristics of both components. In addition, certain metallic elements have the potential to be employed for the purpose of enhancing the performance of the anode surface. Mn(IV), Fe_3O_4, Ni(II), and an anode coated with a semiconductor (TiO_2) are examples of electron mediators that have better properties, including increased surface area and electrical conductivity[30]. Several pretreatment techniques are being explored in order to enhance the efficiency of the anode. These methods include heat treatment, heat treatment with N_2 or NH_3 gas, acidic, and surfactants treatment [31]. It has been observed that these pretreatment techniques can effectively increase the anode surface area while improving its biological compatibility [32]. Some metals and materials, including as stainless steel, Ag, Cu, and Ni, are potentially employed as MEC anode due to their higher conductivity compared to carbon-based anodes.

4.3 Cathodic Materials of MEC

The cathode of a microbial electrolysis cell functions as the focal point for the hydrogen evolution reaction (HER), which has a substantial impact on the efficiency of hydrogen generation. The HER serves as the rate-determining step in the context of microbial electrolytic cells for facilitating mass transfer across solid–liquid–gas interfaces. These reactions are consisted of two main processes: the electrochemical Volmer and the chemical Tafel reactions, or alternatively, electrochemical Heyrovsky reaction. The hydrogen evolution reaction is influenced by the presence of over-potentials, which can be mitigated through the application of additional voltage and the use of certain catalysts. Carbon-based materials are widely acknowledged as the predominant choice for cathode-based materials in microbial electrolysis cells. The primary configuration of the double-chamber microbial electrolysis cell entails the use of carbon papers as the electrode at the cathode, which serves to facilitate H_2 production [33]. Stainless steel possesses favorable aspects such as stability, affordability, and less toxicity toward microbes, rendering it a viable candidate for implementation as a cathode material in microbial electrolytic cells. Similar hydrogen production rates can be achieved using either a nano-Pt coated carbon cathode or a large area of coated stainless-steel brushes [23]. The 3D macroporous stainless-steel fiber felt exhibited remarkable catalytic performance in hydrogen production due to its substantial active surface area. Ni demonstrates a favorable cost-to-benefit ratio as a metallic element, exhibiting a notable level of catalytic performance in the hydrogen production. The effectiveness of Ni and Ni-based metal cathodes in the production of hydrogen using microbial electrolysis cells (MEC) has been confirmed, with nickel 625 demonstrating greater performance in comparison with alternative Ni alloys [34]. The potential application of Ni foams with a significant specific area of contact makes them a promising candidate for use as cathodes in microbial electrolysis cells. The production of hydrogen on a bare carbon-based cathode exhibits very poor efficiency, necessitating the application of a high over-potential for effective hydrogen production. Hence, it is imperative to introduce catalysts to modify cathode-based materials in order to optimize their application and enhance their performance as cathodes in microbial electrolysis cells (MECs).

4.4 Different Substrates, Concentrations, and Loads

MECs have the capacity to utilize a diverse array of substrates, encompassing not only individual components like sugars, carboxylic acids, alcohols, and proteins, but also intricate combinations such as biomass hydrolysate and wastewaters derived from residential, animal, and food-processing sources [35, 36]. The rates of organic decay and H^+ release, as well as the resulting HPR, vary to a greater extent depending on the specific types of substrates utilized. Acetate is often employed as the primary substrate in microbial electrochemical cells (MECs). The highest recorded hydrogen production rate (HPR) achieved in MECs thus far is 50 m^3 H_2/m^3 d, which was accomplished by utilizing acetate as the substrate and applying a voltage of 1 V [37]. Furthermore, Cheng and Logan (2007) conducted a study in which they investigated several biodegradable substrates, such as glucose and cellulose, as well as non-fermentable substrates, namely volatile fatty acids (VFAs) including acetic, butyric, lactic, propionic, and valeric acids, within the MECs.

In the same way, Yang and coworkers conducted an experiment to examine the viability of utilizing the prevalent by-products of dark fermentation (acetate, butyrate, and propionate) as potential fuel sources in microbial electrolysis cells. The findings indicated that the highest current density observed for acetate, butyrate, and propionate were 6.0 ± 0.25 A/m^2, 2.5 ± 0.06 A/m^2, and 1.5 ± 0.14 A/m^2, respectively. The study conducted by Lu et al. (2009) provided evidence of the feasibility of generating extra hydrogen gas from the effluent of a reactor utilizing ethanol as the feedstock. In a study conducted by Lu et al., it was shown that hydrogen (H_2) may be produced from glucose at reduced temperatures (4 °C) via utilizing microbial electrolysis cells [38]. Nevertheless, the findings of this study diverged from the outcomes reported in a prior investigation conducted by the MEC [39]. In their investigation, Sakai and Yagishita employed glycerol (B-glycerol) as a feed in a two-chamber microbial electrolysis cell (MEC). They achieved a maximum hydrogen yield (Y_{H2}) of 0.76 mol-H_2/mol.-glycerol [40]. Lignocellulose, abundant, and renewable, has garnered attention as a possible MEC H_2 source. To convert lignocellulosic materials into H_2, DF and MEC were integrated in a two-stage process. Yan and coworkers studied power and H_2 generation in a dual-chambered MFC-MEC using Morella thermoacetica to perform hemicellulose DF [41]. Lewis et al. developed a switch grass H_2 generation method using integrated pyrolysis and MEC [42]. A recent study by Rivera and coworkers used a MEC for the treatment of cheese whey (CW) and recover industrial by-products and H_2 gas [43].

5 Pilot-Scale MEC

The initial implementation of a microbial electrolysis cell for the purpose of biohydrogen production is currently taking place in the United States [44]. The architecture of the reactor is founded upon the methodology of submerging brush anodes and flat cathodes, both constructed from stainless steel, into a tank. The reactor is comprised of a total of 24 modules, whereby each module consists of six pairs of electrodes. The overall volume of the reactor is roughly 1 m^3. Cusick et al. presented a comprehensive analysis of the outcomes obtained from their pilot-scale continuous flow microbial electrolysis cell with a volume of 1000 L, which was supplied with wine effluent as the feedstock [45]. The reactor performed utilizing a total of 144 electrode pairs, distributed over 24 modules. An externally applied voltage of 0.9 V was used during the operation. The gas generation rate observed in this study was rather low in comparison with laboratory investigations conducted under continuous flow or batch flow conditions, suggesting the necessity for additional optimization efforts. Despite the intended production of hydrogen gas, a significant portion of the gas generated was transformed into methane, as evidenced by the volume analysis. Furthermore, the process of biofilm enrichment required a significantly longer duration of up to 60 days, which is considerably greater than the timeframe typically observed in controlled research environments. Strong biofilms are needed for practical applications due to these issues. Over the course of 12 months, a 100 L microbial electrolysis cell (MEC) was supplied with untreated home sewage, resulting in the production of 0.6 L per day of hydrogen. The experimental setup involved the utilization of a plant with a total volume of 130 L. The plant was initiated in a batch mode, employing acetate and glucose as the primary substrates. The operation of the plant spanned a duration of five months. The production of hydrogen exhibited a rise, surpassing 4 L per day, accompanied by a gas purity of 95%. Furthermore, the cathodic gas recovery reached 82%, and the energy recovery achieved a remarkable 121% in relation to the power supply [46]. In a recent study, a team of researchers employed a microbial electrochemical cell (MEC) scale-up strategy to transition from a 50 ml to a 10 L cell for the purpose of treating domestic wastewater. The researchers observed that the reduction of chemical oxygen demand (COD) amounted to 5.7 g, while the creation of hydrogen ranged from 1.0 to 2.6 L [47]. The use of the pilot MEC exhibited the absence of membrane partitioning, which, while advantageous, may have facilitated the flow of hydrogen into the anode zone, hence potentially promoting methanogenesis. Therefore, it is imperative to minimize methanogenesis during the process of hydrogen production. Moreover, the potential applications of MECs, such as chemical synthesis, pollution removal, and metal recovery, have not yet transitioned beyond the laboratory setting. Scaling-up presents significant challenges in terms of electrode materials and reactor architecture.

6 Biocatalysts in MEC

The bio-cathode in a microbial electrolysis cell consists of electroactive microorganisms that serve as catalysts on the cathode, facilitating the reduction of protons or other chemical species. The utilization of a bio-cathode in microbial electrolysis cells (MECs) has great potential in enhancing the application of MECs by achieving similar electrochemical performance and reducing the resistance of bio-electrochemical systems (BES) [48]. In addition, it is worth noting that the bio-cathode exhibits characteristics such as affordability, renewability, and self-regeneration. Initially, microorganisms were employed as electrocatalysts in microbial fuel cells (MFCs) as a substitute for noble metal catalysts. In recent times, the utilization of bio-cathodes in microbial electrolysis cells (MECs) has gained prominence as an effective means for the generation of hydrogen evolution reaction (HER), methane, and other valuable compounds. The initial HER-catalyzing bio-cathode was utilized through the reversal of polarity of a bio-anode that oxidizes acetate and hydrogen [49]. This particular microbial electrolysis cell (MEC) utilizes a heterotrophic electroactive biofilm, which is composed of a naturally chosen consortium of microorganisms with electrochemical activity. The bio-cathode microbial electrolysis cell (MEC) is also employed in the direct synthesis of biohythane [50]. Both the electron acceptor and donor are required for bacteria respiration and reproduction. In a bio-cathode, the cathode serves as an electron acceptor while the applied voltage provides the electrons for the circuit. Biofilms on cathodes require the electrons. As a result, the ideal voltage would promote microbial development and serve various functions in the cathode chamber. For instance, the MEC bio-cathode chamber could be used to remove the heavy metals (Cd, Ni, Zn, Co, and Sb). MEC cathode may also be able to remove certain additional organic impurities, such as azo dye and chlorine contamination.

7 Two-Chambered MECs

In each of the MEC experiments described below, a fundamental element consistently incorporated is the presence of a membrane. This membrane is probably employed to enhance the purity of the generated hydrogen and to impede the microbial consumption of this hydrogen. The addition of a membrane and the migration of hydrogen across it and into the anode have been linked to significant potential losses. A membrane prevents short circuits, reduces fuel and bacteria crossover from the anode to the cathode chamber, and maintains hydrogen gas purity. Microbial electrolysis cells have employed a range of different membranes, with the proton exchange membrane (PEM) being the most prevalent. The PEM is specifically engineered with $-SO^{3-}$ functional groups to selectively facilitate the passage of unbound protons (H^+) [35, 51]. Additional types of membranes have been evaluated in microbial electrolysis cells (MECs) [35], including anion-exchange membranes (AEM) like AMI-7001, bipolar membranes, and charge-mosaic membranes (CMM) [52].

7.1 Bio-Electrochemically Aided Microbial Reactor (BEAMR)

In order to demonstrate the effectiveness of the bio-electrochemically assisted method, Liu et al. [5] conducted a study. Two hydrogen generating reactors were constructed by modifying the designs of microbial fuel cell (MFC) reactors. Both reactors were designed as two-chamber microbial electrolysis cells (MECs), where the anode and cathode were placed in separate chambers, with a proton exchange membrane separating them (Fig. 3). The anode utilized in the experiment consisted of a carbon cloth devoid of any other modifications. On the other hand, the cathode was constructed using carbon paper, which was impregnated with platinum at a density of 0.5 mg/cm^2. The initial experimental setup consisted of a dual-bottle reactor, with

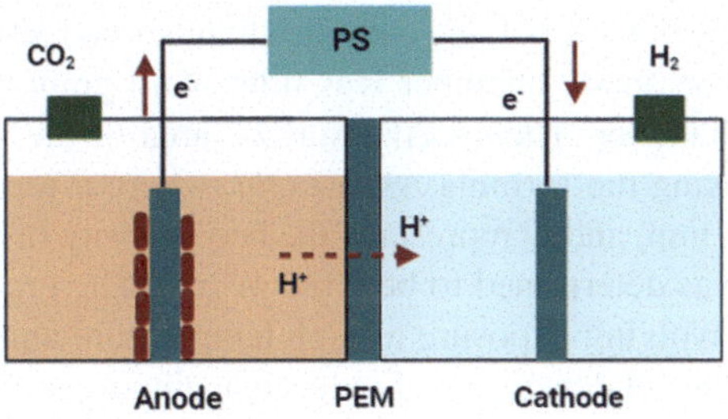

Fig. 3 Schematic layout of first bio-electrochemically aided microbial reactor (BEAMR) [5] (Adapted from reference [5] and redrawn in the text.)

each bottle having a capacity of 0.31 L. The proton exchange membrane (PEM) was securely fastened within the reactor using a clamp, effectively separating the two chambers. The electrodes were positioned at a distance of 15 cm from each other. Instead of introducing air into the cathode chamber through sparging, the chamber was hermetically sealed and subjected to periodic analysis to determine the creation of hydrogen gas. The surface area of each electrode utilized in the experiment was 12 cm^2, whereas the proton exchange membrane (PEM) had a surface area of 3.5 cm^2. Additionally, the bottles were filled with a volume of 0.2 L (Fig. 3).

In order to gather samples of hydrogen gas, the MEC reactors' highest portions were outfitted with sampling ports. There is currently a limit of 4 mol of $H_2/C_6H_{12}O_6$, for hydrogen production by bacterial fermentation. However, this process generates an end product (acetate; 2 mol/mol glucose) that bacteria are incapable of further converting into hydrogen. The present study demonstrates that the aforementioned biochemical barrier can be effectively bypassed through the utilization of a fully anaerobic microbial fuel cell (MFC) for the production of hydrogen gas from acetate. By enhancing the electrochemical potential attained by bacteria within the MFC with the application of an external voltage of 0.25 V or higher, it became feasible to directly generate hydrogen at the cathode through the oxidation of organic substrates. The hydrogen gas recovery from the oxidation of acetate by bacteria resulted in the retrieval of over 90% of H^+ and electrons, yielding an overall Coulombic efficiency (CE) ranging from 60 to 78%. The aforementioned value corresponds to a total production of 2.9 mol. H_2/CH_3COO^- ions. The anaerobic microbial fuel cell (MFC) technique has the potential to produce hydrogen gas from a wide range of biodegradable dissolved organic substrates, not solely confined to carbohydrates as seen in traditional fermentation processes.

7.2 An Innovative, High-Efficiency MEC

Cheng and Logan have developed a novel MEC reactor architecture [53, 54]. The construction of the MEC reactor involved the utilization of an AEM (AMI-7001) that was securely fastened between the anode and cathode chambers. The anode chamber had a diameter of 30 mm and a length of 20 mm, with a total volume of 14 mL. On the other hand, the cathode chamber had a length of 40 mm and a volume of 28 mL. The anode chamber was filled with graphite granules ranging from 2 to 6 mm in diameter. The specific surface area of the granules, denoted as As, was estimated using the formula As $= 6Q/d$, where d represents the average particle diameter of 4 mm, and Q represents the bed porosity of 53%. The resulting specific surface area was determined to be 1320 m^2/m^3. The granules underwent a pretreatment process involving exposure to high temperature ammonia gas (NH_3), resulting in enhanced current densities and reduced acclimation times of the reactor [55]. The cathode was constructed using carbon cloth and a platinum (Pt) catalyst with a loading of 0.5 mg/ cm^2 Pt, following the procedure described in a previous study. It was positioned in close proximity to the membrane within the cathode chamber and linked to the external circuit using a titanium wire.

Hydrogen gas (H_2) was obtained through the attachment of a glass tube, including a crimp cap and a thick rubber stopper. This assembly was affixed to a hole that was cut into the top of the cathode chamber. Through enhancements in both the materials utilized and the reactor architecture, the production of hydrogen gas was achieved with yields ranging from 2.01 to 3.95 mol. This was accomplished by applying voltages within the range of 0.2–0.8 V, with acetic acid as the substrate. Acetic acid is commonly generated as a byproduct in the fermentation processes of glucose or cellulose. At an applied voltage of 0.6 V, the process exhibited an overall energy efficiency of 288% purely based on the electricity delivered. Additionally, the hydrogen production rate (HPR) was measured to be 1.1 m^3 H_2/m^3 d.

8 Single-Chambered MECs

Due to its comparatively poor solubility in water, hydrogen exhibits limited miscibility. Consequently, in scenarios where production rates are sufficiently elevated, the microbial conversion of hydrogen to methane is expected to proceed at a slow rate. Furthermore, it should be noted that unlike microbial fuel cells (MFCs), microbial electrolysis cells (MECs) operate under anaerobic conditions. Consequently, the elimination of the membrane in MECs does not result in the introduction of oxygen to the anode. As a result, the removal of the membrane is not expected to have a detrimental effect on the efficiency of MECs. In order to mitigate potential losses and enhance energy recovery in the membrane-based process, an experimental investigation was conducted on a novel design of microbial electrolysis cells (MECs) that eliminates the use of a membrane. This alternative design incorporates many features, including ammonia-treated anodes, high surface area graphite brush anodes, and reduced electrode spacing. Single-chamber membrane-less MECs eliminate the need for membranes, optimizing the design and decreasing the initial investment. However, microbial hydrogen losses to methanogens are a serious problem in MECs because of the membrane's absence. Both CH_3COONa and H_2 are substrates for methanogens, which compete with electrochemically active bacteria [56].

8.1 Single-Chambered MEC with Flat-Carbon Cathode Configuration

Call and Logan [17] proposed a novel membrane-less MEC design. The MEC was fabricated using a cylindrical chamber made of polycarbonate, with dimensions of 4 cm in length and 3 cm in diameter. The anode utilized in the experiment was a graphite brush that had been treated with ammonia. It had a diameter of 25 mm and a length of 25 mm, resulting in a surface area of 0.22 m^2. The anode had a specific surface area of 18,200 m^2/m^3 and a porosity of 95%. It was positioned at the center of the chamber. The cathode utilized in the experiment was a wet-proofed carbon cloth with a surface area of 7 cm^2. It was coated with a platinum (Pt) catalyst at a density of 0.5 mg/cm^2. The cathode was positioned on the opposing side of the chamber. The gas generated at the cathode in the microbial electrochemical cell (MEC) was introduced into the reactor solution. The gas was subsequently collected by means of an anaerobic tube, which had a headspace volume of 0.015 L. This tube was securely attached to the upper part of the reactor, positioned above an opening with a diameter of 1.6 cm. The uppermost part of the tube was hermetically sealed using a stopper made of butyl rubber and an aluminum crimp cap. In order to prevent the entry of light, aluminum foil was used to cover all the MEC reactors (Fig. 4). In a MEC system, cathodic hydrogen recoveries ranging from 77 ± 1.5 to 97 ± 0.5% were successfully attained, even in the lack of an inter-electrode membrane. These results were obtained at applied voltages ranging from 0.3 to 0.8 V and a solution conductivity of 7.5 mS/cm. By employing a membrane-less configuration, employing a graphite fiber brush anode, and maintaining a narrow electrode gap, the high-pressure reactor (HPR) achieved a peak hydrogen production rate (HPR) of 3.13 ± 0.03 m^3 H_2/m^3 reactor per day when subjected to an applied voltage of 0.8 V. The current production rate exceeds the rate achieved in prior MEC experiments by more than double.

8.2 Single-Chambered MEC with Bottle Configuration

Two distinct types of single-chambered membrane-free microbial electrolysis cells were fabricated in order to estimate the performance of hydrogen production in these systems [6]. The initial setup, denoted as setup (a), was constructed utilizing wide mouth glass bottles with a capacity of 0.5 L. This system was employed to examine the process of hydrogen synthesis facilitated by a hybrid culture. The anode, with a surface area of 3.5 × 4 cm^2, and the cathode, with a surface area of 4 × 5 cm^2, were securely fastened using plastic screws, ensuring that the electrodes were positioned 2 cm apart. The second setup (b) was fabricated utilizing translucent borosilicate ceramic serum tubes having a volume of 0.1 L. This system was employed to examine the process of hydrogen generation by S. oneidensis. The selection of serum vials for the pure culture test was primarily based on their ease of sealing and

Fig. 4 Single-chambered MEC with glass collection tube [17]. (Adapted and permission granted from reference [17])

autoclaving, as well as their ability to sustain anaerobic conditions following autoclave treatment. In this experimental setup, the anode and cathode, each measuring 3 × 3 cm, were spatially separated by a J-Cloth layer to prevent any potential short circuit. In both systems, the anode consisted of carbon cloth of type A, while the cathode was composed of carbon cloth of type B, which incorporated a Pt catalyst with a density of 0.5 mg/cm^2. The electrodes were connected to the circuit using TiO_2 wire (TW). At an applied voltage of 0.6 V, the MEC system utilizing a mixed culture demonstrated per day rate of hydrogen production of 0.53 m^3 at neutral pH. The rate of hydrogen production increased to about 0.69 m^3 at pH 5.8, respectively.

8.3 Single-Chambered MEC with Top-Mounted Cathode

The reactor is comprised of a primary chamber and an upper cover, constructed from glass material, possessing a vacant capacity of 0.4 L (as depicted in Fig. 5) [57]. The anode consisted of graphite granules of 3–5 mm in diameter that were set up at the base of the 0.32 L chamber. The cathode was a titanium tube with little structure with specific dimensions of 20 mm and a minor diameter of 10 μm. This titanium tube was coated with a layer of platinum and positioned at the uppermost part of the chamber. The minimum distance separating the anode and cathode was measured to be 30 mm. The circuit was supplied with voltage using a power source, while the voltage along an external resistor was monitored using a data collection board in order to determine the current. The electrolyte and substrate were introduced into the

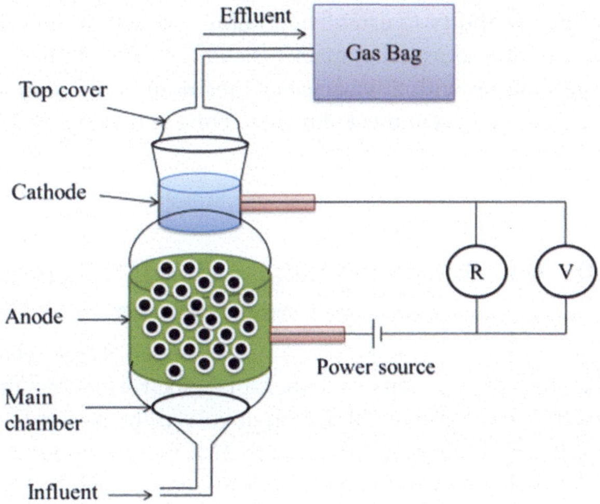

Fig. 5 Schematic representation of single-chambered MEC with top-mounted cathode [2]. (Adapted and permission granted from reference [2])

chamber via the lower intake, while the gas generated at the cathode was captured using a gas bag. The highest achieved energy recovery rate was 88% when an applied voltage of 0.5 V was utilized.

9 The Small-Scale MEC

A basic microbial electrolysis cell (MEC) system was constructed utilizing readily accessible materials for the purpose of conducting bio-electrochemical investigations with large-scale potential. The system was operated by employing a sole energy source [58]. The MECs were fabricated utilizing clear glass reagent vials with a volume of 0.005 L. The anodes utilized in the experiment were graphite plates that required iso-molding, resulting in a uniform thickness of 0.32 cm. These plates were then precisely chopped to measurements of 1.5 cm in length and 1 cm in width. The anodes underwent a polishing process using sandpaper, followed by a cleaning step including immersion in a 1N HCl solution for an extended period of time, and subsequently rinsed three times with Milli-Q water. A titanium wire (0.08 cm in diameter) was subjected to abrasion using sandpaper. Subsequently, the wire was divided into segments measuring 5 cm in length and manipulated to form a J-shaped configuration at one end. Upon inserting the linear segment of the wire, which remained unaltered in shape, through a drilled aperture located in close proximity to the upper central region of the graphite plate, utilizing a drill bit with a diameter of 0.08 cm, the curved extremity of the wire was subsequently introduced into a separate aperture and secured by means of crimping, therefore establishing a secure and taut linkage between the wires and the graphite plate. Additional components, such as carbon cloth and carbon paper, were found to be unsuitable for usage due to their inability to establish reliable contacts when crimped with wire, resulting in excessively elevated contact resistances. The stainless-steel (SS) mesh cathode, in combination with a wastewater inoculum containing acetate as the electron donor, yielded the maximum volumetric current density of 240 A/m^3 at a voltage input of 0.7 V.

10 Dye-Sensitized Solar Cell (DSSC)-Powered MECs

The utilization of a dye-sensitized solar cell (DSSC) has been employed to augment the reductive capacity of a microbial electrolysis cell through the conversion of light energy. The experimental setup involved the use of H-shaped two-chambered glass bottle microbial electrolysis cells (MECs) in conjunction with dye-sensitized solar cells (DSSCs) as an external power supply. The integrated system included testing to evaluate its performance in the production of hydrogen (shown in Fig. 6).

The dye-sensitized solar cell (DSSC) generated an open circuit voltage of 0.6 V, which was subsequently delivered to the microbial electrolytic cell (MEC). The entire system generated a total of 400 mmol of H_2 over a period of 5 h, with a cathode recovery capacity of 78% [60]. The platinum catalyst-free MEC cathode was designed to lower the cost of this linked system. When the voltage exceeded 0.7 V, the hydrogen production levels were comparable between the system utilizing plain cathode and the system utilizing Pt-loaded carbon felt electrodes. Moreover, a notable improvement in the production of hydrogen occurred while employing a carbon nano-powder-coated electrode in the absence of platinum (Pt) [59]. The solar cell-MEC-coupled system facilitates the conversion of sunlight into liquid or gaseous

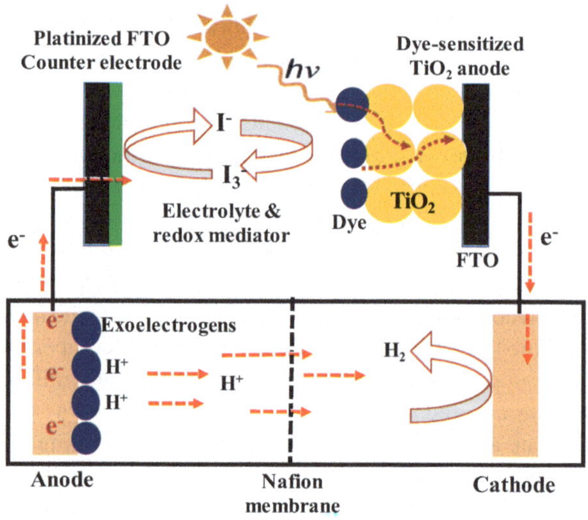

Fig. 6 Schematic diagram of dye-sensitized solar cell (DSSC)-powered MEC [59] (Adapted from reference [59] and redrawn in the text)

fuels for transport, such as H_2, CH_4, and C_2H_5OH. These fuels can be effectively stored for subsequent utilization. In order to enhance the overall performance of the system, it is imperative to consider the integration of many solar cells in a series configuration as a potential avenue for future research and development.

11 MEC Layout and Management

As an initial observation in this chapter, it is worth noting that the potential to employ elevated pressure MEC is an intriguing route for exploration that, to the best of our understanding, has not yet been explored. A potentially significant area of interest lies in the development of a unique configuration that effectively separates the bio-anode from the hydrogen evolution cathode. In this particular scenario, individual electrodes are placed within distinct electrochemical reactors. The connection between the two reactors is established through a hydraulic cycle that incorporates a redox mediator. The mediator experiences a process of reduction within the reactor that is fitted with the bio-anode and then undergoes oxidation within the reactor responsible for the production of hydrogen. Hydrogen and oxygen production reactors can be kept completely separate using this detached method, which is being effectively adapted to water electrolysis [61]. Different electrolytes can be utilized at the bio-anode and the hydrogen-evolving cathode in microbial electrolyte cells (MECs). While this may help with pH control in the two independent reactors and eliminate the issue of cathode deposit, it comes at the expense of some thermodynamic efficiency. MECs have yet to begin to use this method, which should be of significant importance. In order to design basic conceptual instruments to manage MEC operation, it should be advantageous to include local pH calculations and then combine the model with experimental pH readings. Thus, electrode operating conditions may be determined, and MEC operation instructions established. A proper control strategy can overcome the severe constraint caused by microbial anode acidification. To consume biofilm-accumulated protons, frequently reverse the polarity so the anode becomes the cathode. The subject under investigation is a developing field of study, and there is still a need to evaluate the extent and occurrence of polarity inversion as well as its effects on MEC development. Recent developments in reversible electrodes [62], or bidirectional electrodes, may help establish new MEC control techniques that minimize microbial acidification by periodical polarity flipping. Finally, MECs are part of increasingly complicated production units. MECs can synthesize methane and other chemicals. Hydrogen is the intermediate reactant. Large-scale electrolyzers require operation at high current density to be cost-effective. This implies that the evolution of hydrogen should occur rapidly, resulting in the formation of large bubbles characterized by brief periods of residency. On the other hand, the present state of the art reveals that microbially catalyzed hydrogenation processes exhibit relatively low reaction rates. This can be attributed, in part, to the restrictions imposed by gas–liquid mass transfer, necessitating the utilization of specialized gas–liquid contactors. To achieve the best results, it is necessary to physically separate the MEC from the synthesis reactor. Hydrogen (either pure or blended with CO_2 and maybe CH_4) would be produced by the MEC, and its gas outlet would be fed into a gas–liquid contactor where the microbially catalyzed synthesis would take place. The implementation of a two-step approach could enable the optimization of gas–liquid transfer within the contactor, while simultaneously ensuring a high rate of hydrogen evolution in the microbial electrolysis cell. As a result, optimizing the synthesis reaction rate within the MEC would become superfluous. This limitation is removed from MEC optimization. In addition, the MEC's CO_2 byproduct production is of little significance if the last step is to hydrogenate CO_2 in the downstream synthetic process. By using a two-stage process, CO_2 may be safely introduced into the cathode compartment to regulate pH without contaminating the resultant flow.

12 Conclusions

When designing a microbial electrolytic system, we aim to achieve a dual objective. In simple terms, a MEC is required to attain the integration of two opposing systems. One perspective to consider is that electrolyzer designed for large-scale manufacturing is optimally utilized with clean electrolytes that possess a chemically simplest composition, exhibit strong ionic mobility, and possess extremely high pH values. In contrast, microbial processes necessitate culture media that are chemically abundant and have a neutral pH. Additionally, these processes typically do not accommodate solutions with high conductivity. The best conditions for the two types of systems do not overlap. The aforementioned factor does not provide a barrier when considering modest current densities that are adequate for sustaining metabolic processes. On the contrary, when the engineer attempts to compel the process toward industrial output at elevated current densities, the incongruity in the ideal conditions disrupts the system. As the current density rises, the MEC is hampered by some sort of adverse feedback mechanism. Would MECs be viable enough to enter the manufacturing sector? This is still a captivating task and a difficult question. This crucial section suggests avenues of inquiry that could lead to a solution for this problem.

In this chapter, we will review the several MEC reactors that have been designed for high-yield, environmentally friendly H_2 production and waste management. Since the invention of MEC rectors in 2005, numerous variations have been created. The majority of the investigations were conducted using two-chamber reactors containing membranes. The primary motivation for this action was to prevent the escape of hydrogen into the anode chamber, where it could potentially serve as a substrate for hydrogenotrophic microbes. The utilization of a membrane successfully assisted the separation of H_2 production from the carbon dioxide generated at the anode, hence ensuring improved purity of H_2 at the cathode. But Rozendal et al. [63] have demonstrated that membranes cause a significant pH differential along their surfaces. Another important drawback of the MEC with membranes is the high price of the membrane itself.

Recent research has demonstrated that membrane-free MEC designs can increase hydrogen recoveries and output. Several investigations have produced membrane-free reactors since then. The absence of a membrane in the functioning of the system results in the elimination of the pH gradient, any potential losses, and internal resistance. This, in turn, enables the application of a larger voltage to drive the electrode reactions. When using a mixed culture, the gas collected will contain CO_2 and CH_4 (<1%). This phenomenon arises as a result of the existence of microorganisms in multiple cultures that metabolize H_2 and convert CO_2 into methane through reduction reactions. Methanogens, organisms known for their slow growth rates and susceptibility to oxygen, can be effectively managed in terms of methane generation by employing periodic airflow and implementing short retention intervals. However, it is important to note that these measures do not entirely eradicate methane formation. The creation of continuous flow microbial electrolysis cells on a pilot size has advanced recently. The high internal resistances of these designs were attributed to the distance between electrodes and the resistance encountered by ions during diffusion across the membranes. Higher hydrogen production rate is possible if membrane separators are removed and the solution conductivity is increased. There is an optimistic outlook for the future, as advancements in MEC reactor design and cost reduction are anticipated to facilitate the use of more scalable MEC designs. Consequently, it is anticipated that this will make a significant contribution to the advancement of renewable and economically feasible hydrogen energy systems. The advanced systems will exhibit the capacity to produce hydrogen gas from a diverse array of sustainable sources, including different types of waste and biofuels obtained from plant matter.

References

1. Logan BE, Call D, Cheng S, Hamelers HV, Sleutels TH, Jeremiasse AW, Rozendal R (2008) Microbial electrolysis cells for high yield hydrogen gas production from organic matter. Environ Sci Technol 42:8630–8640
2. Dange P, Pandit S, Jadhav D, Shanmugam P, Gupta PK, Kumar S, Kumar M, Yang Y-H, Bhatia SKJS (2021) Recent developments in microbial electrolysis cell-based biohydrogen production utilizing wastewater as a feedstock. Sustainability 13:8796
3. Zawieja I (2019) Enhanced sludge degradation process using a microbial electrolysis cell. In: Industrial and municipal sludge. Elsevier, pp 385–397
4. Kadier A, Chaurasia AK, Sapuan S, Ilyas R, Ma PC, Alabbosh KFS, Rai PK, Logroño W, Hamid AA, Hasan H, et al (2020) Essential factors for performance improvement and the implementation of microbial electrolysis cells (MECs), pp 139–168
5. Liu H, Grot S, Logan B (2005) Electrochemically assisted microbial production of hydrogen from acetate. Environ Sci Technol 39:4317–4320
6. Hu H, Fan Y, Liu H (2008) Hydrogen production using single-chamber membrane-free microbial electrolysis cells. Water Res 42:4172–4178
7. Chorbadzhiyska E, Hristov G, Mitov M, Hubenova Y (2011) Microbial electrolysis cells as innovative technology for hydrogen production

8. Kundu A, Sahu JN, Redzwan G, Hashim MA (2013) An overview of cathode material and catalysts suitable for generating hydrogen in microbial electrolysis cell. Int J Hydrog Energy 38(4):1745–1757
9. Sun M, Mu ZX, Sheng GP, Shen N, Tong ZH, Wang HL, Yu HQ (2010) Hydrogen production from propionate in a biocatalyzed system with in-situ utilization of the electricity generated from a microbial fuel cell. Int Biodeterior Biodegrad 64(5):378–382
10. Kadier A, Jiang Y, Lai B, Rai PK, Chandrasekhar K, Mohamed A, Kalil M (2018) Biofuels. Biohydrogen production in microbial electrolysis cells from renewable resources, pp 331–356
11. Kadier A, Simayi Y, Abdeshahian P, Azman NF, Chandrasekhar K, Kalil MS (2016) A comprehensive review of microbial electrolysis cells (MEC) reactor designs and configurations for sustainable hydrogen gas production. Alex Eng J 55(1):427–443
12. Kadier A, Kalil MS, Mohamed A, Hamid AA (2017) A new design enhances hydrogen production by G. sulfurreducens PCA strain in a single-chamber microbial electrolysis cell (MEC). J Teknol 79:71–79
13. Torres CI, Marcus AK, Lee HS, Parameswaran P, Krajmalnik-Brown R, Rittmann B (2010) A kinetic perspective on extracellular electron transfer by anode-respiring bacteria. FEMS Microbiol Rev 34:3–17
14. Newman DK, Kolter R (2000) A role for excreted quinones in extracellular electron transfer. Nature 405:94–97
15. Von Canstein H, Ogawa J, Shimizu S, Lloyd JR (2008) Secretion of flavins by Shewanella species and their role in extracellular electron transfer. Appl Environ Microbiol 74:615–623
16. Kadier A, Kalil MS, Rai PK, Kumar SS, Abdeshahian P, Sivagurunathan P, Hasan HA, Hamid AA, Mohamed A (2019) Microbial electrolysis cells (MECs) a promising and green approach for bioenergy and biochemical production from waste resources, pp 209–234
17. Call D, Logan B (2008) Hydrogen production in a single chamber microbial electrolysis cell lacking a membrane. Environ Sci Technol 42(9):3401–3406
18. Cusick RD, Ullery ML, Dempsey BA, Logan B (2014) Electrochemical struvite precipitation from digestate with a fluidized bed cathode microbial electrolysis cell. Water Res 54:297–306
19. Wang L, Gu K, Zhang Y, Sun J, Gu Z, Zhao B, Hu C (2022) Enhanced struvite generation and separation by magnesium anode electrolysis coupled with cathode electrodeposition. Sci Total Environ 804:150101
20. Wang L, Linowski K, Liu H (2023) Scalable membrane-less microbial electrolysis cell with multiple compact electrode assemblies for high performance hydrogen production. Chem Eng J 454:140257
21. Gautam R, Nayak JK, Ress NV, Steinberger-Wilckens R, Ghosh UK (2023) Bio-hydrogen production through microbial electrolysis cell: structural components and influencing factors. Chem Eng J 455:140535
22. Kadier A, Simayi Y, Logrono W, Kalil MS (2015) The significance of key operational variables to the enhancement of hydrogen production in a single-chamber microbial electrolysis cell (MEC). Iran J Hydrog Fuel Cell 2(2):85–97
23. Cheng S, Logan B (2011) High hydrogen production rate of microbial electrolysis cell (MEC) with reduced electrode spacing. Bioresour Technol 102:3571–3574
24. Yasri N, Roberts EP, Gunasekaran S (2019) The electrochemical perspective of bioelectrocatalytic activities in microbial electrolysis and microbial fuel cells. Energy Rep 5:1116–1136
25. Santoro C, Guilizzoni M, Baena JC, Pasaogullari U, Casalegno A, Li B, Babanova S, Artyushkova K, Atanassov PJC (2014) The effects of carbon electrode surface properties on bacteria attachment and start up time of microbial fuel cells. Carbon 67:128–139
26. Chatterjee P, Dessì P, Kokko M, Lakaniemi AM, Lens P (2019) Selective enrichment of biocatalysts for bioelectrochemical systems: a critical review. Renew Sustain Energy Rev 109:10–23
27. Bora A, Mohanrasu K, Swetha TA, Ananthi V, Sindhu R, Chi NTL, Mathimani T (2022) Microbial electrolysis cell (MEC): reactor configurations, recent advances and strategies in biohydrogen production. Fuel 328:125269

28. Mehdinia A, Ziaei E, Jabbari A (2014) Multi-walled carbon nanotube/SnO_2 nanocomposite: a novel anode material for microbial fuel cells. Electrochim Acta 130:512–518
29. Mier AA, Olvera-Vargas H, Mejía-López M, Longoria A, Verea L, Sebastian PJ, Arias DM (2021) A review of recent advances in electrode materials for emerging bioelectrochemical systems: from biofilm-bearing anodes to specialized cathodes. Chemosphere 283:131138
30. Wei K, Faraj Y, Yao G, Xie R, Lai B (2021) Strategies for improving perovskite photocatalysts reactivity for organic pollutants degradation: a review on recent progress. Chem Eng J 414:128783
31. Bhatnagar A, Hogland W, Marques M, Sillanpää M (2013) An overview of the modification methods of activated carbon for its water treatment applications. Chem Eng J 219:499–511
32. Hua T, Li S, Li F, Zhou Q, Ondon BS (2019) Microbial electrolysis cell as an emerging versatile technology: a review on its potential application, advance and challenge. J Chem Technol Biotechnol 94:1697–1711
33. Xing D, Yang Y, Li Z, Cui H, Ma D, Cai X, Gu J, Products T (2020) Hydrogen production from waste stream with microbial electrolysis cells, pp 39–70
34. Selembo PA, Merrill MD, Logan B (2009) The use of stainless steel and nickel alloys as low-cost cathodes in microbial electrolysis cells. J Power Sour 190:271–278
35. Varanasi JL, Veerubhotla R, Pandit S, Das D (2019) Biohydrogen production using microbial electrolysis cell: recent advances and future prospects, pp 843–869
36. Sivagurunathan P, Kuppam C, Mudhoo A, Saratale GD, Kadier A, Zhen G, Trably E, Kumar G (2018) A comprehensive review on two-stage integrative schemes for the valorization of dark fermentative effluents. Crit Rev Biotechnol 38:868–882
37. Jeremiasse AW, Hamelers HV, Saakes M, Buisman CJ (2010) Ni foam cathode enables high volumetric H2 production in a microbial electrolysis cell. Int J Hydrog Energy 35:12716–12723
38. Lu L, Xing D, Liu B, Ren N (2012) Enhanced hydrogen production from waste activated sludge by cascade utilization of organic matter in microbial electrolysis cells. Water Res 46(4):1015–1026
39. Selembo PA, Perez JM, Lloyd WA, Logan B (2009) High hydrogen production from glycerol or glucose by electrohydrogenesis using microbial electrolysis cells. Int J Hydrog Energy 34(13):5373–5381
40. Sakai S, Yagishita T (2007) Microbial production of hydrogen and ethanol from glycerol-containing wastes discharged from a biodiesel fuel production plant in a bioelectrochemical reactor with thionine. Biotechnol Bioeng 98(2):340–348
41. Yan D, Yang X, Yuan W (2015) Electricity and H_2 generation from hemicellulose by sequential fermentation and microbial fuel/electrolysis cell. J Power Sour 289:26–33
42. Lewis AJ, Ren S, Ye X, Kim P, Labbe N, Borole AP (2015) Hydrogen production from switchgrass via an integrated pyrolysis–microbial electrolysis process. Bioresour Technol 195:231–241
43. Rivera I, Bakonyi P, Cuautle-Marín MA, Buitrón G (2017) Evaluation of various cheese whey treatment scenarios in single-chamber microbial electrolysis cells for improved biohydrogen production. Chemosphere 174:253–259
44. Gajda I, Greenman J, Ieropoulos IA (2018) Recent advancements in real-world microbial fuel cell applications. Curr Opin Electrochem 11:78–83
45. Cusick RD, Bryan B, Parker DS, Merrill MD, Mehanna M, Kiely PD, Liu G, Logan B (2011) Biotechnology performance of a pilot-scale continuous flow microbial electrolysis cell fed winery wastewater. Appl Microbiol Biotechnol 89:2053–2063
46. Samsudeen N, Spurgeon J, Matheswaran M, Satyavolu J (2020) Simultaneous biohydrogen production with distillery wastewater treatment using modified microbial electrolysis cell. Int J Hydrog Energy 45:18266–18274
47. Escapa A, San-Martín MI, Mateos R, Morán A (2015) Scaling-up of membraneless microbial electrolysis cells (MECs) for domestic wastewater treatment: bottlenecks and limitations. Bioresour Technol 180:72–78
48. Wang YZ, Wang AJ, Liu WZ, Kong DY, Tan WB, Liu C (2013) Accelerated azo dye removal by biocathode formation in single-chamber biocatalyzed electrolysis systems. Bioresour Technol 146:740–743

49. Rozendal RA, Jeremiasse AW, Hamelers HV, Buisman CJ (2008) Hydrogen production with a microbial biocathode. Environ Sci Technol 42:629–634
50. Noori MT, Min B (2022) Fundamentals and recent progress in bioelectrochemical system-assisted biohythane production. Bioresour Technol 361:127641
51. Rozendal RA, Hamelers HV, Euverink GJ, Metz SJ, Buisman CJ (2006) Principle and perspectives of hydrogen production through biocatalyzed electrolysis. Int J Hydrog Energy 31:1632–1640
52. Liu H, Hu H, Chignell J, Fan Y (2010) Microbial electrolysis: novel technology for hydrogen production from biomass. 1: 129–142
53. Cheng S, Logan B (2007) Sustainable and efficient biohydrogen production via electrohydrogenesis. Proc Natl Acad Sci 104:18871–18873
54. Jung DH, Lee KD, Lee JK (2023) A study on design method of synchronous reluctance motor based on nonlinear MEC using Newton-Raphson method. IEEE Access 11:10723–10729
55. Cheng S, Logan B (2007) Ammonia treatment of carbon cloth anodes to enhance power generation of microbial fuel cells. Electrochem Commun 9:492–496
56. Zhang Y, Angelidaki I (2014) Microbial electrolysis cells turning to be versatile technology: recent advances and future challenges. Water Res 56:11–25
57. Guo K, Tang X, Du Z, Li H (2010) Hydrogen production from acetate in a cathode-on-top single-chamber microbial electrolysis cell with a mipor cathode. Biochem Eng J 51:48–52
58. Call DF, Logan B (2011) A method for high throughput bioelectrochemical research based on small scale microbial electrolysis cells. Biosens Bioelectron 26:4526–4531
59. Chae KJ, Choi MJ, Kim KY, Ajayi FF, Chang IS, Kim IS (2009) A solar-powered microbial electrolysis cell with a platinum catalyst-free cathode to produce hydrogen. Environ Sci Technol 43(24):9525–9530
60. Ajayi FF, Kim KY, Chae KJ, Choi MJ, Kim SY, Chang IS, Kim IS (2009) Study of hydrogen production in light assisted microbial electrolysis cell operated with dye sensitized solar cell. Int J Hydrog Energy 34(23):9297–9304
61. Wang A, Sun D, Cao G, Wang H, Ren N, Wu WM, Logan BE (2011) Integrated hydrogen production process from cellulose by combining dark fermentation, microbial fuel cells, and a microbial electrolysis cell. Bioresour Technol 102(5):4137–4143
62. Sharma M, Sarma P (2018) Microbially mediated electrosynthesis processes, pp 421–442
63. Rozendal RA, Hamelers HV, Molenkamp RJ, Buisman CJ (2007) Performance of single chamber biocatalyzed electrolysis with different types of ion exchange membranes. Water Res 41:1984–1994

Impact of Electrode Material on Microbial Electrolysis

Krishan Balachandran, Irwan Ibrahim, Ryan Yeo Yow Zhong, Mohd Nur Ikhmal Salehmin, Mimi Hani Abu Bakar, Manal Binti Ismail, Wei Lun Ang, and Swee Su Lim

Abstract Annually, the budget allocated to developing new or improving existing green technologies continues to rise, reflecting the growing demand and their potential to address ongoing environmental challenges. Among these technologies, Microbial Electrolysis Cells (MECs) play a pivotal role in concurrently treating wastewater and producing (bio)hydrogen at the anode and cathode, respectively, with the assistance of electroactive microorganisms as biocatalysts. As the anode serves as the system's driving force, selecting or adapting an appropriate electrode material with desirable characteristics like biocompatibility, chemical stability, non-corrosiveness, electrical conductivity, low resistance, affordability, and porosity is crucial. These attributes promote bacterial adhesion while facilitating electron transfer and flow. Conversely, the cathode must offer a flexible platform for hydrogen evolution reactions (HER) in MECs. Hence, researchers have explored various cathode materials as more cost-effective alternatives to metal-based catalysts, emphasizing sustainability and the reduction of HER overpotential. This chapter provides a review of diverse electrode material designs and types investigated to enhance both substrate oxidation and hydrogen production in MECs.

Keywords Bioanode · Biocathode · Photocathode · Hydrogen evolution reaction · Wastewater treatment

---

K. Balachandran · I. Ibrahim · R. Y. Y. Zhong · M. H. A. Bakar · S. S. Lim (✉)
Fuel Cell Institute, Universiti Kebangsaan Malaysia, UKM, 43600 Bangi, Selangor, Malaysia
e-mail: limss@ukm.edu.my

M. N. I. Salehmin
Institute of Sustainable Energy (ISE), Universiti Tenaga Nasional (UNITEN), Putrajaya Campus, Jalan Ikram-Uniten, 43000 Kajang, Selangor, Malaysia

M. B. Ismail · W. L. Ang
Department of Chemical and Process Engineering, Faculty of Engineering and Built Environment, Universiti Kebangsaan Malaysia, 43600 Bangi, Selangor, Malaysia

A. A. Yaqoob and A. Ahmad (eds.), *Microbial Electrolysis Cell Technology*, Sustainable Materials and Technology, https://doi.org/10.1007/978-981-97-3356-9_4

1 Anode Materials

Microorganism resides on the anode will catalyzes the oxidation of substrate via internal respiration and releases electrons. The released electron will be captured by the anode and transferred to the cathode for hydrogen production via external circuits. As a microorganism's habitat, anode should have several features such as the ability to promote biofilm formation and enhance the EET rate. Furthermore, the anode must exhibit exceptional chemical and biological stability, as well as cost-effectiveness, to support the industrialization of MECs. Previously, many conventional carbon-based materials like carbon cloth, carbon paper, carbon felt, graphite sheet, and graphite rod have been widely employed as MEC anodes due to their advantages such as excellent specific conductance, biocompatibility, affordability, and outstanding chemical stability [1]. Additionally, conventional electrode materials have played a pioneering role in advancing MEC technology. For instance, the use of graphite felt as a bioanode not only unveiled its role as a limiting factor in biocathode hydrogen production but also determined the maximum applied potential to maintain bioanode activity at 1.0 V [2]. Furthermore, Baek et al. [3] disclosed that the chloride ion in phosphate saline buffer inhibits the performance of graphite plate bioanode at high applied voltage.

However, such conventional materials often encounter issues including limited growth of bacteria and subpar electron transport efficiency, prompting the scientific community to explore alternative materials for either improving existing carbon substrates or developing entirely new electrodes. Plasma-treated carbon cloth, immobilizing bacteria with alginate, demonstrated impressive results when compared to the non-immobilized anode, achieving a 78% reduction in COD, possibly due to the prevalence of the *Geobacter* bacterial species, which accounted for up to 90% of the total population [4]. Meanwhile, a bioanode constructed from a graphene-based sponge offered biocompatible, hydrophilic microenvironments that facilitated the density of electrogenic *Shewanella oneidensis* MR-1 biomass resulting in steady-state biocurrent density within a short period [5]. Additionally, a three-dimensional (3D) nano-structured electrode, created via cultivation of a self-supported N-doped C/Fe_3O_4 nanotube composite array on carbon fiber cloth, improved electrode conductivity and serving large surface area leading to low charge transfer resistance and providing more space for bacterial growth, respectively [6]. Furthermore, a self-supported 3D porous carbon aerogel bioanode has been demonstrated to boost bacterial incubation and improve the transfer of electrons due to its extensive specific surface area, outstanding electrical conductivity, and improved bonding with microbes, ultimately yielding five times more hydrogen production compared to bio-carbon fiber [7]. Additionally, Yasri and Nakhla [8] developed a 3D electrode using granular activated carbon and compared its performance when doped with three different substances, namely conductive magnetite, semi-conductive iron(II) sulfide, and nonconductive calcium sulfide where the latter exhibited better performances than the remaining two.

To enhance the sustainability of MECs, a 3D porous cellulose sponge carbon was created by carbonizing wood pulp-derived cellulose sponges at various temperatures (ranging from 600 to 1100 °C). The intricately pleated multilayer porous structure facilitates microbial attachment and allows for the development of dense biofilm where the highest biomass recorded for the one developed at 1100 °C comprising 86% electrogenic bacteria, primarily *Geobacter* [9]. Furthermore, biochar, a product of the thermal decomposition of agricultural residues in oxygen-limited conditions, is commonly used as an electrocatalyst in MFCs and can also be employed to construct anodes due to its large surface area, cost-effectiveness, minimal energy consumption, high cation-exchange capacity, and structural stability when compared to other low-cost carbon materials [10]. Recent findings indicate that modifying an anode with biochar derived from bamboo enhances the electrode's surface area, thereby facilitating electron transfer and promoting the growth of electroactive biofilms, resulting in increased power density in MFCs [11].

Moreover, numerous other materials that have been studied in MFCs can be utilized as bioanodes in MECs, as both systems share similar foundational requirements. For instance, chitosan beads as bioanode material in MFC showcased to retrieve ammonium and phosphate ions from municipal wastewater, in the presence of impeding substances, sulfate ions [12]. Additionally, conductive polymers like polyaniline (PANI) have been extensively explored in BES systems due to their economical production, excellent electrochemical activity, and favorable biocompatibility. For instance, Wang et al. [13] demonstrated that electropolymerized PANI nanoparticles onto carbon paper loaded with sodium citrate improve hydrophilicity, facilitating microbe adhesion and leading to a greater bacterial loading compared to unmodified carbon paper. All these experiments clarified that the modification of the conventional electrode material is vital as there is a drastic change in terms of overall performance when a slight change in the composition of the material is conducted. However, the choice of material has to be done according to the ideal characteristics of bioelectrode to avoid any compensation of its vital features. More importantly, utilizing organic materials originating from food waste, agriculture can boost the sustainability of MEC besides supporting the bioeconomy.

2 Cathode Materials

The role of the cathode as a site for hydrogen evolution is well-known in MECs. Mostly, platinum (Pt) is used as a catalyst to overcome the overpotential of the cathode. However, due to several drawbacks owed by Pt researchers are keen on finding alternative catalysts to reduce the cost and the same time match the performance of the Pt catalyst. There are several types of catalysts made up of different materials such as carbon, nickel, palladium, and so on. Most of the catalysts are developed in the composite form to compensate for the low performance of the pristine material. Besides that, different configurations of cathode also has been actively researched to make MEC more sustainable such as bioanode and biocathode. When it comes to biocathode, basic requirements for the colonization of the bacteria need to be fulfilled. Besides that, the biocathode also seems to be advantageous compared to the abiotic cathode because it can produce more sustainable hydrogen and at the same time it can reduce hazardous chemicals from wastewater simultaneously. Apart from that, the photocathode uses sunlight to supply additional energy for hydrogen production was also discovered with different configurations and photocatalysts.

2.1 Carbon-Based Materials

Carbon nanoparticles (NPs) are recognized for their expansive surface area and exceptional electrical conductivity, making them favorable as cathodic catalysts. Additionally, they are more cost-effective than platinum (Pt) catalysts. However, despite these attributes, carbon NPs cannot be unequivocally considered an alternative to Pt, as scientific reports suggest that pristine carbon NPs have yet to achieve performance levels equivalent to Pt. For instance, a study revealed that multi-walled carbon nanotubes (MWCNTs) combined with Pt exhibited a higher hydrogen production rate compared to pristine MWCNTs. Another study claimed that a cathode modified with single-walled carbon nanotubes (SWCNT) resulted in low hydrogen production as compared to the electrode fabricated with transition metal NPs [14]. A comparative study by Xiao et al. [15] asserted that novel carbon/iron-based catalysts outperformed CNTs and unmodified cathodes but fell short of the efficiency of platinum-based catalysts. While not surpassing the efficacy of metal catalysts, the authors suggest that the composite catalyst could be a cost-effective alternative, compensating for its lower performance compared to Pt/C catalysts.

In a study conducted by Jayabalan et al. [16], the improved performance resulting from incorporating reduced graphene with Ni or Co metal oxide was highlighted. This combination leveraged the strong catalytic activity of Ni/Co alongside the excellent conducting capacity of graphene. These studies collectively underscore the potential of metal NPs to enhance the catalytic performance of carbon NPs. Besides metal-based materials, conducting polymers such as polyaniline (PANI) are actively utilized by the MEC community to elevate the catalytic property of carbon NPs. For example, cathode modified with 75 wt% PANI/MWCNTs performed better than Pt/C thus making it the cost-effective and efficient catalyst [17]. Furthermore, Ghasemi et al. [18] demonstrated that the synergistic effect of PANI and graphene resulted in high rate of hydrogen production compared to the pristine of each materials.

2.2 Stainless Steel-Based Material

Apart from carbon NPs, materials made up of stainless steel (SS) have garnered tremendous attention from the MEC community to be utilized as catalysts. Different forms of SS have been discovered so far namely, plates, wool, brushes, felt, mesh, and the recent one electroformed mesh. It boasts exceptional attributes, such as affordability, robust durability, and excellent electrical conductivity. In addition, the compatibility of SS with the phosphate group aids in effectively reducing overpotential and enhancing the HER, where its large surface area helps in elevating the current densities [19], making it a preferable catalyst material. Previously, several reports have highlighted the excellent performances of the SS either the modified version or the exclusively prepared one over Pt.

For instance, SS alloy known as SS A286 achieved 61% recovery of cathodic hydrogen recovery compared to 47% with platinum [20]. In contrast, a comparative study involving nickel foam (NF), SS wool, molybdenum disulfide-coated SS mesh (MoS_2), and platinum-coated stainless steel mesh (Pt) demonstrated that NF exhibited superior performance [21]. However, the authors pointed out that the final choice should rely just on the efficiencies, the rate and material cost also need to be considered thus making SS wool the best option. Besides that, 3D structured SS also has been successfully implemented to scale up the MEC performances. For example, large pores possessed by 3D SS fiber felt displayed exceptional catalytic performance for hydrogen production, matching a platinum cathode and surpassing an SSM [22]. In another investigation, Kim et al. [23] compared the performances of the 3D structure of different forms of SS, where the wool contributed to the highest production of hydrogen that is comparable to the Pt control.

2.3 *Nickel-Based Material*

Nickel (Ni) is a commonly employed catalyst for the HER due to its lower overpotential and stability in alkaline conditions when compared to other precious metals. In a study by Jeremiasse et al. [24], it was noted that Ni foam exhibited a high specific surface area and lower cathode overpotential in contrast to a Pt-based cathode. The economic evaluation indicated a 56% cost reduction when substituting Pt-coated nickel foam as the supporting base for achieving equivalent performance instead of using carbon cloth [25]. Additionally, a more cost-effective nickel-based material, referred to as electroformed Ni mesh, outperformed other cathode materials typically used in MECs [26].

Furthermore, the performance of electroformed Ni mesh was on par or comparable to that of precious metals and standard cathodes (Pt/CC) in terms of hydrogen production rate and current density. Rossi et al. [27] demonstrated that nickel-molybdenum cathode catalysts synthesized via hydrothermal means produced a similar current with a minimal overpotential of 50 mV compared to Pt thus having a higher H_2 production rate. Besides that, the incorporation of activated carbon in nickel via nitric acid oxidation results also in tremendous hydrogen production surface [28]. Kim et al. showcased that carbon-supported phase-pure nickel phosphide (Ni_2P/C) catalyst produced a similar quantity of hydrogen over a 24 h cycle as that obtained using Pt/C, therefore be a useful alternative to Pt/C or other Ni-based catalysts in MECs due to its chemical stability over time [29].

2.4 Molybdenum Sulfide-Based Materials

Molybdenum sulfides (MoS_X) represent a cost-effective and chemically stable choice with significant potential for improving the hydrogen production. They are essentially ionic compounds comprising Mo^{4+} cations, sulfide (S^{2-}) ions, and disulfide anions. Research has shown that within molybdenum sulfides, hydrogen atoms readily bond to sulfur ions located on the edges of MoS_X layers, resulting in a substantial acceleration of the HER rate. The effective application of MoS_2 as a catalyst in nitrogenase studies has inspired scientists to employ it as a catalyst in MEC [30]. The initial experimentation of the HER in MECs using MoS_2 revealed that the catalyst possessed lower hydrogen evolution overpotential at higher current densities compared to the Pt-based catalyst [31].

Previously, MoS_2 was noted for having a limited number of active sites and poor conductivity, issues that have been addressed through chemical exfoliation. Rozendfel et al. [32] compared the hydrogen evolution rate of pristine MoS_2, exfoliated MoS_2, and Pt, with the exfoliated form outperforming the other catalysts. MoS_2-EF yielded highly purified hydrogen and proved to be an electrochemically active and cost-effective alternative to Pt. Additionally, the MoS_2/CNT composite displayed promising results when compared to Pt and MoS_2, suggesting that CNT enhances electronic conductivity for MoS_2 and provides a greater number of active sites for CNTs [33]. MoS_X stands out as an affordable, efficient, and exceptionally durable electrocatalyst for HER in practical applications involving real acidic wastewater, offering a valuable contribution to energy production from wastewater in the form of hydrogen applications [34].

2.5 Palladium-Based Materials

Being among the metals most similar to platinum (Pt), palladium (Pd) holds significant promise as an effective catalyst for the hydrogen evolution reaction (HER) due to its outstanding catalytic capabilities. Research indicates that when considering the catalytic efficiency relative to the mass of Pd nanoparticles, it can be roughly fifty times greater than that of Pt black catalysts [35]. Furthermore, Pd is significantly more abundant in nature and cost-effective compared to Pt [36]. Previously, Pd was employed in air–cathode MFCs as a Pt–Pd alloy for oxygen reduction [37]. The generation of Pd nanoparticles was achieved by utilizing *S. oneidensis*, which was subsequently applied as a coating on an anode electrode to improve the bioelectricity production of MFCs [37]. A similar approach was also used to create Pd catalysts coated on graphite, which were then utilized as cathodes in MFCs to dehalogenate pharmaceuticals [38]. MEC with bioelectrochemically deposited Pd catalysts achieved a significantly higher hydrogen production rate than the chemically deposited Pd catalyst which was due to its smaller size and higher electrochemical surface area [39]. Huang and colleagues [35] developed a cathode in which

Pd nanoparticles were deposited onto carbon cloth displayed a lower overpotential when compared to a carbon paper coated with Pt black. These results indicate that Pd nanoparticles can function as an efficient catalyst for the cathode in MECs, potentially leading to reduced manufacturing costs and diminished voltage losses.

3 Biocathode

In general, expensive metal such as Pt is widely utilized to catalyze the HER at the cathode due to their ability to reduce the overpotential. However, the usage of such non-renewable catalysts exhibits negative impacts on the environment as the mining and extraction of the metal results in scarcity and difficult to dispose of. Besides that, the catalyst has a high likelihood of being poisoned by CO and sulfur and has negative effects on the environment. Furthermore, the loading of the expensive metal catalyst increases the total cost of MEC, thus challenging its commercialization. Therefore, the scientific community is keen on finding cost-effective catalysts as an alternative to reform the sustainability of MEC. The application of microorganisms as biocatalysts on the cathode, known as biocathode seems to be a promising effort as they are environmentally friendly, inexpensive, and possess self-regenerative and resistant to certain levels of impurities. Previously, the biocathodes employed to catalyze HER were mostly enzyme-based which has several drawbacks such as being unstable and not self-regenerative. In 2008, for the first time, Rozendal et al. demonstrated the production of hydrogen via the application of a biocathode which has been developed by reversing the polarity of acetate- and hydrogen-oxidizing bioanode into hydrogen-producing biocathode [40].

Numerous research works have been published on the modification of different carbon-based and metal-based electrode materials via various methods to enhance the performance of the biocathode based on the desired type of reduction reaction. Recently, Arvizu et al. demonstrated that laccase oxide-reductase enzyme-loaded titanium oxide (TiO_2) nanoparticles on glassy carbon (GCE) electrode reduce the overpotential and elevate the cathodic current which symbolizes that it could perform better as a biocathode [41]. The development of biocathode via enzyme immobilization allows the researcher to explore specific types of reduction reactions and its electron transfer as an enzyme target on a specific region known as an active site which interacts with specific substrates to form products. A novel composite cathode developed by wrapping the bimetallic MOF catalyst (Fe/Ni–NH_2BDC) on stainless steel mesh exhibited smaller charge transfer resistance thus accelerating the electron transfer which is advantageous for HER that supports hydrogenotrophic methanogenesis of MEC-assisted anaerobic digestion cathode [42]. Previously, similar modification techniques have been utilized to develop a hybrid graphite felt (GF)-biocathode, where a piece of GF was rolled into a tubular shape and physically fixed to a plain carbon cathode [43]. The inclusion of GF added an advantage to the overall performance of the biocathode as it provides a more open and conductive matrix that offers vast space for the colonization of microbes and enhances the electron transfer for methane production. Apart from physically modifying the biocathode, the performance of the biocathode can be enhanced via the direct addition of conductive minerals in the catholyte. For example, the addition of magnetite elevates the biomass on the cathode, the thickness of the biofilm, and the relative abundance of *Desulfovibrio* sp. exhibited a higher rate of sulfate reductive rate compared to the one without magnetite suggesting that the mineral improves the electron transfer interspecies directly or indirectly by stimulating the generation of pili-like EPS [44].

Apart from hydrogen and methane production, numerous research studies have reported that biocathodes can perform the reduction of metal in wastewater with simultaneous hydrogen production. For example, Dai et al. demonstrated that ZIF-67-modified sulfate-reducing bacteria resulted in a tremendous bioreduction of sulfate and antimony with the removal efficiency of 93.37% and 92%, respectively, along with a substantial amount of hydrogen production [45]. Furthermore, PANI-modified graphite felt has been demonstrated to facilitate the enrichment of perchlorate-reducing bacteria (PCRB) and thus exhibited a higher reduction rate of perchlorate compared to abiotic cathode [46]. In another study, the synergic effect of PANI and MWCNT was demonstrated to enhance the performance of the biocathode in terms of hydrogen production compared to the electrode modified with the material individually [47].

4 Photocathode

From a thermodynamic standpoint, traditional MECs often necessitate an additional energy input ranging from 0.2 to 1 V to facilitate electrohydrogenesis, primarily due to overpotential and device resistance. This aspect has made them less appealing and sustainable. To make MEC more sustainable, researchers explored renewable energy such as sunlight to catalyze HER by harnessing dye-sensitized solar cells (DSSC). Typically, a DSSC is composed of a photocathode (dye-sensitized TiO_2-coated conducting glass), a counter electrode (Pt coated glass substrate), and an electrolyte containing an iodide/tri-iodide redox couple within an organic solvent, positioned between the two electrodes. The initial experiments involving the hybrid MEC-DSSC, using a ruthenium dye (N719), have shown promising hydrogen production results under xenon lamp irradiation and in natural sunlight, promoting environmentally friendly MECs [48]. However, further optimization of critical parameters is required to determine overall efficiency and process costs. Moreover, solar-powered MECs can operate effectively without the need for a Pt catalyst on the cathode, which enhances their potential for commercialization by reducing significant expenses related to electricity and catalyst utilization [49].

Apart from using DSSC separately, electrodes made up of photo-catalytic material can be merged as the constituent of MEC either in a single- or two-chamber configuration turning it into the solar-driven microbial photoelectrochemical cell (MPC). By implementing this, the cost and mechanical difficulties can be lessened as it eliminates separate construction and interconnections of DSSCs and MECs.

For example, Qian et al. demonstrated the synergetic effect of p-type cuprous oxide nanowire-arrayed photocathode with *S. oneidensis* MR-1-colonizing anode induced substantial current generation, thus opening up exciting opportunities at the unique interface between semiconductor nanowires and bacterial systems [50]. However, the system suffered from cathode overpotential due to the requirement of semiconductors to possess more positive valance band edge than the electrochemical potentials of the bacterial outer membrane c-type cytochrome (Omc) proteins to receive microbial electrons. To resolve this issue, a new dual-chambered MEC system consisting of p-type Si nanowires (SiNWs) photocathode and a microbially catalyzed anode powered by single-chamber MFC function as extra voltage supply has been developed [51]. The principle behind the operation of the cell started with electron transfer from the MFC anode to the photocathode of MPC via an external circuit resulting in the capturing of several electrons of photogenerated hole under visible light and allowing the remaining electrons to participate in hydrogen evolution reaction.

At low applied potential and long-term operation, single chamber MEC is favorable to methanogenesis instead of electrohydrogenesis as the electrochemically formed H_2 in the cathode is utilized as an electron donor by methanogen. However, there are several methods to suppress methanogenesis such as adding methanogen inhibitors, heat treatment, exposing electrodes to air, and utilizing UV irradiation. Instead of the suggested methods, many researchers are keen on discovering distinct semiconductors to be employed as photocathodes to exclusively produce H_2 without CO_2 and CH_4. For example, Cu_2O coated with MoS_2 as cocatalyst (MoS_2/Cu_2O) demonstrated that high production of H_2 rate was reported at -0.8 V without methane and carbon dioxide suggesting fast kinetics of proton reduction is the reason behind the scene [52]. The author also claimed that the production rate of hydrogen is higher than a similar system with polyaniline nanofiber at a similar potential with negligible methane production [53]. Recently, Song et al. proved that the intuitive interplay of circuital current, light illumination, and bacterial communities driven by the $ZnFe_2O_4$/g-C_3N_4 photo-cathode resulted in a slightly similar hydrogen production rate with the one used polyaniline [54]. The same semiconductor has been utilized to treat etching terminal wastewater with simultaneous Ni(II) removal and hydrogen production proving that MEC has high potential for real practical application [54]. Another novel photocathode known as CDs/g-C_3N_4 was successfully utilized to treat ETW treatment with simultaneous H_2 production, allowing the effluent to meet national discharge standards [55].

5 Conclusion and Future Perspective

MECs are an exciting bioelectrochemical system as they can simultaneously treat wastewater and generate biohydrogen. The biohydrogen produced via MECs is considered one of the most sustainable forms of production as there is less damage to the environment. Electrode material has a great influence on the overall performances of MECs as everything begins and ends at the electrode. An anode is a site where oxidation of substrate by the residing bacteria takes place and the electrons are released post-respiration will be transferred to the cathode for hydrogen generation. Thus anode material needs to be biocompatible, conductive, and less expensive. There are three types of cathode have been employed in MEC so far, namely, abiotic cathode, biocathode, and photocathode. Different type of catalyst and materials has been discussed to enhance the understanding of different materials in the performance of the cathode. The employment of biocathode and photocathode can make the MEC more sustainable as they use live organisms and sunlight to produce hydrogen. A lot of efforts need to be taken to make MEC produce higher production of hydrogen so that MEC can become a hydrogen fuel supplier for fuel cells that operate using hydrogen as a fuel thus securing the environment.

References

1. Bian B et al (2018) 3D printed porous carbon anode for enhanced power generation in microbial fuel cell. Nano Energy 44:174–180. https://doi.org/10.1016/j.nanoen.2017.11.070
2. Lim SS, Yu EH, Daud WRW, Kim BH, Scott K (2017) Bioanode as a limiting factor to biocathode performance in microbial electrolysis cells. Bioresour Technol 238:313–324. https://doi.org/10.1016/j.biortech.2017.03.127
3. Baek G, Shi L, Rossi R, Logan BE (2021) The effect of high applied voltages on bioanodes of microbial electrolysis cells in the presence of chlorides. Chem Eng J 405:126742. https://doi.org/10.1016/j.cej.2020.126742
4. Gandu B, Rozenfeld S, Ouaknin Hirsch L, Schechter A, Cahan R (2020) Immobilization of bacterial cells on carbon-cloth anode using alginate for hydrogen generation in a microbial electrolysis cell. J Power Sources 455:227986. https://doi.org/10.1016/j.jpowsour.2020.227986
5. Omenesa IM et al (2023) Scalability of biomass-derived graphene derivative materials as viable anode electrode for a commercialized microbial fuel cell: a systematic review. Chinese J Chem Eng 2023(55):277–292. https://doi.org/10.1016/j.cjche.2022.05.009
6. Wang Y et al (2021) Hierarchical N-doped C/Fe_3O_4 nanotube composite arrays grown on the carbon fiber cloth as a bioanode for high-performance bioelectrochemical system. Chem Eng J 406:126832. https://doi.org/10.1016/j.cej.2020.126832
7. Wang W-K, Tang B, Liu J, Shi H, Xu Q, Zhao G (2019) Self-supported microbial carbon aerogel bioelectrocatalytic anode promoting extracellular electron transfer for efficient hydrogen evolution. Electrochim Acta 303:268–274. https://doi.org/10.1016/j.electacta.2019.02.099
8. Yasri NG, Nakhla G (2017) The performance of 3-D graphite doped anodes in microbial electrolysis cells. J Power Sources 342:579–588. https://doi.org/10.1016/j.jpowsour.2016.12.081
9. Yang P et al (2023) A wood pulp sponge cleaning wipe as a high-performance bioanode material in microbial electrochemical systems for its vast biomass carrying capacity, large capacitance, and small charge transfer resistance. J Mater Sci Technol. https://doi.org/10.1016/j.jmst.2023.08.065
10. Rodríguez-Couto S (2021) Preparation, characterization, and application of modified carbonized lignin as an anode for sustainable microbial fuel cell. Process Safe Environ Protect 155:49–60. https://doi.org/10.1016/j.psep.2021.09.006
11. Kumar A, Pandit S, Sharma K, Mathuriya AS, Prasad R (2023) Evaluation of bamboo derived biochar as anode catalyst in microbial fuel cell for xylan degradation utilizing microbial co-culture. Bioresour Technol 390:129857. https://doi.org/10.1016/j.biortech.2023.129857

12. Shahid K, Ramasamy DL, Sillanpää M (2021) Chitosan beads as a bioanode for simultaneous recovery of nutrients and energy from municipal wastewater using a microbial nutrient recovery cell. J Clean Prod 298:126756. https://doi.org/10.1016/j.jclepro.2021.126756
13. Wang Y-X, Li W-Q, Zong W-M, Su T-Y, Mu Y (2019) Polyaniline-decorated honeycomb-like structured macroporous carbon composite as an anode modifier for enhanced bioelectricity generation. Sci Total Environ 696:133980. https://doi.org/10.1016/j.scitotenv.2019.133980
14. Choi M-J, Yang E, Yu H-W, Kim IS, Oh S-E, Chae K-J (2019) Transition metal/carbon nanoparticle composite catalysts as platinum substitutes for bioelectrochemical hydrogen production using microbial electrolysis cells. Int J Hydrogen Energy 44(4):2258–2265. https://doi.org/10.1016/j.ijhydene.2018.07.020
15. Xiao L, Wen Z, Ci S, Chen J, He Z (2012) Carbon/iron-based nanorod catalysts for hydrogen production in microbial electrolysis cells. Nano Energy 1(5):751–756. https://doi.org/10.1016/j.nanoen.2012.06.002
16. Jayabalan T, Matheswaran M, Preethi V, Naina Mohamed S (2020) Enhancing biohydrogen production from sugar industry wastewater using metal oxide/graphene nanocomposite catalysts in microbial electrolysis cell. Int J Hydrogen Energy 45(13):7647–7655. https://doi.org/10.1016/j.ijhydene.2019.09.068
17. Yang Q et al (2015) Hydrogen production with polyaniline/multi-walled carbon nanotube cathode catalysts in microbial electrolysis cells. J Chem Technol Biotechnol 90(7):1263–1269. https://doi.org/10.1002/jctb.4425
18. Ghasemi B, Yaghmaei S, Abdi K, Mardanpour MM, Haddadi SA (2020) Introducing an affordable catalyst for biohydrogen production in microbial electrolysis cells. J Biosci Bioeng 129(1):67–76. https://doi.org/10.1016/j.jbiosc.2019.07.001
19. Tabassum P, Akil A, Lokhat D (2021) A glimpse into the microbial fuel cells for wastewater treatment with energy generation. Desalin Water Treat 214:379–389. https://doi.org/10.5004/dwt.2021.26737
20. Selembo PA, Merrill MD, Logan BE (2009) The use of stainless steel and nickel alloys as low-cost cathodes in microbial electrolysis cells. J Power Sources 190(2):271–278. https://doi.org/10.1016/j.jpowsour.2008.12.144
21. Ribot-Llobet E, Nam J-Y, Tokash JC, Guisasola A, Logan BE (2013) Assessment of four different cathode materials at different initial pHs using unbuffered catholytes in microbial electrolysis cells. Int J Hydrogen Energy 38(7):2951–2956. https://doi.org/10.1016/j.ijhydene.2012.12.037
22. Su M, Wei L, Qiu Z, Wang G, Shen J (2016) Hydrogen production in single chamber microbial electrolysis cells with stainless steel fiber felt cathodes. J Power Sources 301:29–34. https://doi.org/10.1016/j.jpowsour.2015.09.108
23. Kim K-Y, Zikmund E, Logan BE (2017) Impact of catholyte recirculation on different 3-dimensional stainless steel cathodes in microbial electrolysis cells. Int J Hydrogen Energy 42(50):29708–29715. https://doi.org/10.1016/j.ijhydene.2017.10.099
24. Jeremiasse AW, Hamelers HVM, Saakes M, Buisman CJN (2010) Ni foam cathode enables high volumetric H_2 production in a microbial electrolysis cell. Int J Hydrogen Energy 35(23):12716–12723. https://doi.org/10.1016/j.ijhydene.2010.08.131
25. Wang L, Liu W, He Z, Guo Z, Zhou A, Wang A (2017) Cathodic hydrogen recovery and methane conversion using Pt coating 3D nickel foam instead of Pt-carbon cloth in microbial electrolysis cells. Int J Hydrogen Energy 42(31):19604–19610. https://doi.org/10.1016/j.ijhydene.2017.06.019
26. Kadier A, Simayi Y, Chandrasekhar K, Ismail M, Kalil MS (2015) Hydrogen gas production with an electroformed Ni mesh cathode catalysts in a single-chamber microbial electrolysis cell (MEC). Int J Hydrogen Energy 40(41):14095–14103. https://doi.org/10.1016/j.ijhydene.2015.08.095
27. Rossi R, Nicolas J, Logan BE (2023) Using nickel-molybdenum cathode catalysts for efficient hydrogen gas production in microbial electrolysis cells. J Power Sources 560:232594. https://doi.org/10.1016/j.jpowsour.2022.232594

28. Moreno-Jimenez DA, Kim K-Y (2022) Enhanced wettability improves catalytic activity of nickel-functionalized activated carbon cathode for hydrogen production in microbial electrolysis cells. Bioresour Technol 350:126881. https://doi.org/10.1016/j.biortech.2022.126881
29. Kim K-Y, Habas SE, Schaidle JA, Logan BE (2019) Application of phase-pure nickel phosphide nanoparticles as cathode catalysts for hydrogen production in microbial electrolysis cells. Bioresour Technol 293:122067. https://doi.org/10.1016/j.biortech.2019.122067
30. Hinnemann B et al (2005) Biomimetic Hydrogen Evolution: MoS_2 nanoparticles as catalyst for hydrogen evolution. J Am Chem Soc 127(15):5308–5309. https://doi.org/10.1021/ja0504690
31. Tokash JC, Logan BE (2011) Electrochemical evaluation of molybdenum disulfide as a catalyst for hydrogen evolution in microbial electrolysis cells. Int J Hydrogen Energy 36(16):9439–9445. https://doi.org/10.1016/j.ijhydene.2011.05.080
32. Rozenfeld S et al (2018) Exfoliated molybdenum di-sulfide (MoS_2) electrode for hydrogen production in microbial electrolysis cell. Bioelectrochemistry 123:201–210. https://doi.org/10.1016/j.bioelechem.2018.05.007
33. Yuan H, Li J, Yuan C, He Z (2014) Facile synthesis of MoS_2@CNT as an effective catalyst for hydrogen production in microbial electrolysis cells. ChemElectroChem 1(11):1828–1833. https://doi.org/10.1002/celc.201402150
34. Kokko M, Bayerköhler F, Erben J, Zengerle R, Kurz P, Kerzenmacher S (2017) Molybdenum sulphides on carbon supports as electrocatalysts for hydrogen evolution in acidic industrial wastewater. Appl Energy 190:1221–1233. https://doi.org/10.1016/j.apenergy.2016.12.097
35. Huang Y-X et al (2011) A new cathodic electrode deposit with palladium nanoparticles for cost-effective hydrogen production in a microbial electrolysis cell. Int J Hydrogen Energy 36(4):2773–2776. https://doi.org/10.1016/j.ijhydene.2010.11.114
36. Savadogo O, Lee K, Oishi K, Mitsushima S, Kamiya N, Ota K-I (2004) New palladium alloys catalyst for the oxygen reduction reaction in an acid medium. Electrochem commun 6(2):105–109. https://doi.org/10.1016/j.elecom.2003.10.020
37. Quan X, Mei Y, Xu H, Sun B, Zhang X (2015) Optimization of Pt-Pd alloy catalyst and supporting materials for oxygen reduction in air-cathode microbial fuel cells. Electrochim Acta 165:72–77. https://doi.org/10.1016/j.electacta.2015.02.235
38. De Gusseme B et al (2011) Biogenic palladium enhances diatrizoate removal from hospital wastewater in a microbial electrolysis cell. Environ Sci Technol 45(13):5737–5745. https://doi.org/10.1021/es200702m
39. Wang W, Zhang B, He Z (2019) Bioelectrochemical deposition of palladium nanoparticles as catalysts by *Shewanella oneidensis* MR-1 towards enhanced hydrogen production in microbial electrolysis cells. Electrochim Acta 318:794–800. https://doi.org/10.1016/j.electacta.2019.06.038
40. Rozendal RA, Jeremiasse AW, Hamelers HVM, Buisman CJN (2008) Hydrogen production with a microbial biocathode. Environ Sci Technol 42(2):629–634. https://doi.org/10.1021/es071720+
41. Arvizu IC, García-Reyes RB, Pech-Canul MA, García-González A (2023) Evaluating the role of laccase immobilized TiO_2 for potential use as biocathode in a microbial electrolysis cell. Biochem Eng J 194:108883. https://doi.org/10.1016/j.bej.2023.108883
42. Wang C et al (2022) Enhanced anaerobic digestion for degradation of swine wastewater through a Fe/Ni-MOF modified microbial electrolysis cell. J Clean Prod 380:134773. https://doi.org/10.1016/j.jclepro.2022.134773
43. Zhen G, Lu X, Kobayashi T, Kumar G, Xu K (2016) Promoted electromethanosynthesis in a two-chamber microbial electrolysis cells (MECs) containing a hybrid biocathode covered with graphite felt (GF). Chem Eng J 284:1146–1155. https://doi.org/10.1016/j.cej.2015.09.071
44. Hu J, Zeng C, Liu G, Luo H, Qu L, Zhang R (2018) Magnetite nanoparticles accelerate the autotrophic sulfate reduction in biocathode microbial electrolysis cells. Biochem Eng J 133:96–105. https://doi.org/10.1016/j.bej.2018.01.036
45. Dai J et al (2023) Promoted Sb removal with hydrogen production in microbial electrolysis cell by ZIF-67-derived modified sulfate-reducing bacteria bio-cathode. Sci Total Environ 856:158839. https://doi.org/10.1016/j.scitotenv.2022.158839

46. Li J-J et al (2015) Perchlorate reduction in microbial electrolysis cell with polyaniline modified cathode. Bioresour Technol 177:74–79. https://doi.org/10.1016/j.biortech.2014.11.065
47. Chen Y, Xu Y, Chen L, Li P, Zhu S, Shen S (2015) Microbial electrolysis cells with polyaniline/multi-walled carbon nanotube-modified biocathodes. Energy 88:377–384. https://doi.org/10.1016/j.energy.2015.05.057
48. Ajayi FF et al (2009) Study of hydrogen production in light assisted microbial electrolysis cell operated with dye sensitized solar cell. Int J Hydrogen Energy 34(23):9297–9304. https://doi.org/10.1016/j.ijhydene.2009.09.096
49. Chae K-J, Choi M-J, Kim K-Y, Ajayi FF, Chang I-S, Kim IS (2009) A solar-powered microbial electrolysis cell with a platinum catalyst-free cathode to produce hydrogen. Environ Sci Technol 43(24):9525–9530. https://doi.org/10.1021/es9022317
50. Qian F, Wang G, Li Y (2010) Solar-driven microbial photoelectrochemical cells with a nanowire photocathode. Nano Lett 10(11):4686–4691. https://doi.org/10.1021/nl102977n
51. Wan L-L, Li X-J, Zang G-L, Wang X, Zhang Y-Y, Zhou Q-X (2015) A solar assisted microbial electrolysis cell for hydrogen production driven by a microbial fuel cell. RSC Adv 5(100):82276–82281. https://doi.org/10.1039/C5RA16919D
52. Jeon Y, Kim JH, Koo K, Kim S (2018) A photo-assisted microbial electrolysis cell for the exclusive biohydrogen production using a MoS_2-coated p-type copper oxide. J Power Sources 373:79–84. https://doi.org/10.1016/j.jpowsour.2017.11.003
53. Jeon Y, Kim S (2016) Persistent hydrogen production by the photo-assisted microbial electrolysis cell using a p-type polyaniline nanofiber cathode. Chemsuschem 9(23):3276–3279. https://doi.org/10.1002/cssc.201600906
54. Song S, Huang L, Zhou P (2023) Efficient H_2 production in a $ZnFe_2O_4$/g-C_3N_4 photo-cathode single-chamber microbial electrolysis cell. Appl Microbiol Biotechnol 107(1):391–404. https://doi.org/10.1007/s00253-022-12293-3
55. Zhang J, Wang Q, Wan H, Shi Y, Huang L (2023) Enhanced etching terminal wastewater treatment and H_2 production by in-situ deposited heavy metals on carbon dots/g-C_3N_4 photocathode microbial electrolysis cells. J Hazard Mater 459:132178. https://doi.org/10.1016/j.jhazmat.2023.132178

Effect of Substrate on the Efficiency of Microbial Electrolysis Cells

N. Hemashenpagam, S. Selvajeyanthi, and N. Gunasheela

Abstract Microbial Electrolysis Cells (MECs) present a promising avenue for sustainable bioenergy generation by harnessing microbial activity to drive electrochemical reactions. The choice of substrate plays a crucial role in dictating the efficiency and overall performance of MECs. This chapter provides a thorough investigation into the influence of various substrates on MEC efficiency, offering insights into their impact on microbial behavior, electrochemical responses and resultant bioenergy production. The substrates utilized in MECs categorizing them based on their composition and availability. Critical factors guiding substrate selection such as chemical makeup, cost considerations and biodegradability are examined to underscore their importance in optimizing MEC performance. Detailed discussions on the effects of different substrate types and concentrations elucidate the direct influence on key electrochemical parameters including current density, voltage and efficiency metrics. Moreover, the chapter explores the intricate interplay between substrates and microbial communities, highlighting shifts in microbial composition and their implications for MEC performance. Challenges related to substrate utilization are discussed, paving the way for future research and advancements in substrate engineering for MECs. In a deep understanding of how substrates affect MEC efficiency is pivotal for optimizing MEC design and operation. This knowledge lays the foundation for advancing microbial electrolysis technology, ultimately contributing to sustainable bioenergy production and waste treatment.

Keywords Microbial electrolysis cells (MECs) · Substrate diversity · Hydrogen production · Sustainable bioenergy · Environmental remediation

N. Hemashenpagam (✉)
Research Department of Microbiology, Hindusthan College of Arts & Science, Coimbatore, Tamilnadu 641028, India
e-mail: hemashenpagam.n@hicas.ac.in

S. Selvajeyanthi
Department of Microbiology, Theeran Chinnamali College of Arts and Science for Women, Tirupur, Tamilnadu 641663, India

N. Gunasheela
Department of Microbiology, Sri Ramakrishna College of Arts & Science for Women, Coimbatore, Tamilnadu 641044, India

A. A. Yaqoob and A. Ahmad (eds.), *Microbial Electrolysis Cell Technology*, Sustainable Materials and Technology, https://doi.org/10.1007/978-981-97-3356-9_5

1 Introduction

Global energy use has increased considerably in recent years. The depletion of fossil fuel reserves, considerable cost increases and the resulting environmental consequences have all contributed to the need for alternative energy resources. Green fuels are viewed as a promising sustainable energy source due to their high efficiency, ease of manufacture and pollution-free operation. Hydrogen is the most efficient and environmentally friendly bio-green fuel for autos. Previously, hydrogen gas was produced using non-renewable resources. Electrochemical, photochemical, thermochemical and biochemical processes are the most effective methods for producing hydrogen gas. Recently, biohydrogen gas was created utilizing renewable resources. Because of its low energy usage of the MEC is a significant and practical technology for biohydrogen generation [1].

In the ever-evolving landscape of sustainable energy, the quest for environment cleaner and more efficient ways to harness renewable energy sources remains paramount. Within this dynamic arena, the Microbial Electrolysis Cell (MEC) emerges as a transformative technology, offering a promising solution to the dual challenges of wastewater treatment and bioenergy production. At the heart of this innovative process lies a fundamental question this chapter will explore: How does the choice of substrate influence the efficiency and performance of Microbial Electrolysis Cells? Microbial Electrolysis Cells (MECs) represent a groundbreaking convergence of microbiology and electrochemistry, offering a compelling solution for the sustainable production of hydrogen gas and wastewater treatment. At the core of these bio-electrochemical systems lies a symbiotic relationship between microorganisms and electrodes, where organic matter is efficiently converted into hydrogen while simultaneously generating electrical energy. MECs hold promise in revolutionizing clean energy production and wastewater management, with their potential applications spanning from biohydrogen generation to ecologically responsible municipal wastewater treatment.

The concept of Microbial Electrolysis Cells (MECs) was first proposed in 2005 by Liu et al. Microbial bio-electrochemical systems are promising techniques for simultaneous wastewater treatment and hydrogen production through MECs or power production through MFCs. To produce hydrogen using MEC systems require external power at an electrical potential much lower than the theoretical one used in water electrolysis. This source of external power could be a regular power supply, photovoltaics, wind energy, thermoelectric generators or MFCs.

Microbial Electrolysis Cells have garnered widespread attention for their ability to generate hydrogen gas, a clean and versatile energy carrier from organic materials while simultaneously treating wastewater. To operate through the symbiotic interaction of microorganisms, electroactive bacteria and electrodes, providing an elegant synergy between biology and electrochemistry. The magic happens within the depths of these electroactive biofilms, where electrons flow and generate hydrogen, unlocking the potential for sustainable energy production.

The organic substrate material introduced to the MEC plays a pivotal role in determining its performance and efficiency. This chapter delves into the fascinating world of substrates to examining how various organic compounds from simple sugars to complex wastewater effluents; influence the microbial communities, electrochemical reactions and overall output of MECs and the intricate interplay between microbial life, electrochemical processes and the critical role of substrate selection. The exploration into the effect of substrate on MEC efficiency will not only deepen the understanding of this innovative technology but also hold significant implications for its real-world applications. From wastewater treatment plants to biohydrogen production for clean energy generation, the insights gleaned in this chapter will inform and inspire a greener and more sustainable future.

2 Substrates in Microbial Electrolysis Cells

Biofuels and hydrogen are essential in modern society, offering versatile and sustainable energy solutions. Biofuels play a vital role in transportation and power generation, reducing emissions and air pollution. Hydrogen, as a clean and powerful energy carrier, boasts a high heating value and low density making it an efficient option. It serves as a raw material for various chemical processes and it is measured one of the least polluting fuels. Hydrogen fuel cell vehicles hold promise for the transportation sector. Moreover, hydrogen's energy density exceeds that of gasoline and coal, making it a promising candidate for a cleaner and more efficient energy future [2]. Hydrogen is receiving global recognition as an environmentally friendly energy carrier, offering the promise of replacing traditional fossil fuels. A state-of-the-art technology known as a Microbial Electrolysis Cell (MEC) has emerged as a viable method for sustainable and eco-friendly hydrogen production from a diverse array of renewable biomass sources. Various substrates have been explored as potential fuel sources in MECs, with a focus on materials such as wastewater, fermentable organic matter and non-fermentable organic materials [3]. Microbial Electrolysis Cells (MECs) have been investigated for a variety of purposes including the elimination of industrial pollutants, wastewater treatment, chemical synthesis and biosensing. However, MEC technology is still in its early stages and faces significant challenges when it comes to practical large-scale implementations. MECs are utilized for multiple functions, encompassing the generation of energy and the production of substances like hydrogen peroxide, methane, hydrogen or biohydrogen as well as the removal of pollutants. MECs are employed for energy generation and recycling, notably in the production of hydrogen, methane and hydrogen peroxide. They also used for the elimination of contaminants, particularly complex organic and inorganic pollutants. Additionally, MEC technology has been scrutinized in terms of novel concepts, optimizing configurations, understanding electron transfer pathways in biocathodes and integrating with other technologies to create value-added applications [4].

The Microbial Electrolysis Cell (MEC) presents an innovative and promising method for hydrogen generation from organic substances, encompassing wastewater and other sustainable resources. Hydrogen holds vast promise as a fuel and energy source. When hydrogen is burned, it doesn't contribute to greenhouse gas emissions, acid rain or ozone depletion as its oxidation by-product is simply water. Additionally, hydrogen is exceptionally efficient, boasting the highest energy content per unit weight among gaseous fuels, with 120 MJ/kg for H2 compared to 44 MJ/kg for gasoline. Its versatility is further enhanced by its potential to be derived from various energy sources, including organic waste materials [3]. This renders hydrogen production cost-effective, environmentally friendly, sustainable and renewable. It is worth seeing the majority of current hydrogen production relies on non-renewable fossil fuels like natural gas [5].

2.1 Types of Substrates for MECs

Substrates are a crucial component in MECs, as they serve as the source of organic matter for microbial degradation and the production of hydrogen or other valuable products. The choice of substrate can significantly influence. Numerous substrates are explored as potential fuel sources in Microbial Electrolysis Cells (MECs). These substrates encompass a wide range of materials, with a significant focus on various types of wastewaters and both fermentable and non-fermentable organic substances. In MECs, hydrogen production has been successfully demonstrated using a diverse array of substrates including acetic acid, butyric acid, lactic acid, glucose, cellulose and various forms of wastewater. This technology offers a promising alternative for the sustainable generation of hydrogen as a fuel in growing demand, while concurrently serving as an efficient method for treating and purifying waste effluents by harnessing their energy content [3].

2.2 The Characteristics of the Several Substrates Utilized in Microbial Electrolysis Cells (MECs)

Wastewater: Municipal and industrial wastewater, such as domestic sewage and agricultural overflow, can be used as a substrate in MECs. This approach can not only treat wastewater but also produce hydrogen or methane gases.

Municipal Wastewater is a common substrate for MECs and it contains a mixer of organic compounds such as organic acids, proteins and carbohydrates. MECs can effectively treat municipal wastewater while simultaneously producing hydrogen gas. The organic matter in sewage is readily biodegradable and makes suitable choice for MECs.

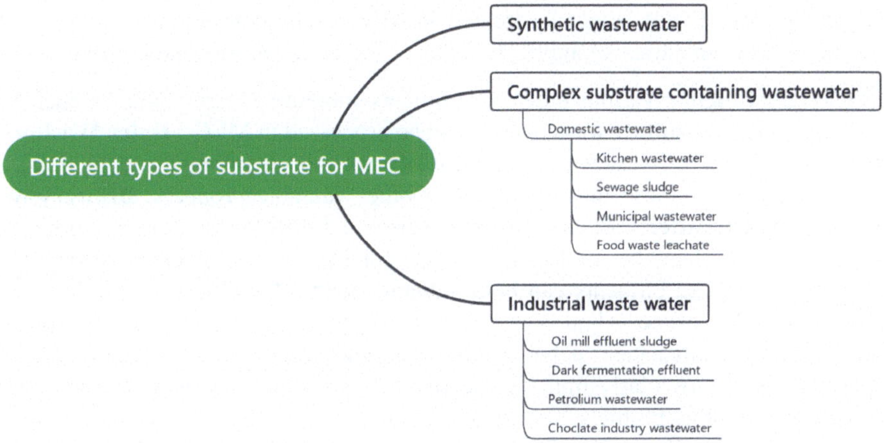

Fig. 1 Illustration of the various substrates used in MECs

Industrial Wastewater: MECs can also be applied to treat various industrial wastewaters such as those from food processing, pulp and paper mills and petrochemical plants. The type and complexity of organic compounds in industrial wastewater can vary widely and which may require specific adjustments in MEC operation and microbial consortia selection (Fig. 1).

Lignocellulosic Biomass: Lignocellulosic materials like agricultural residues, switchgrass and wood chips can be converted into biohydrogen or other valuable products through MECs. **Agricultural Residues**: Crop residues, straw and other agricultural by-products are rich in lignocellulose can be used as substrates in MECs. The lignocellulosic materials are more complex and less biodegradable than simple organic compounds. Pretreatment methods of acid or enzymatic hydrolysis are necessary to break down the complex structure into simple form to enhance microbial degradation. **Manure**: Animal manure such as swine or poultry manure can be used as a substrate for MECs. It is rich in organic matter, nutrients and making it a suitable feedstock for hydrogen or methane production using MECs. **Crop Residues** are corn stover or rice straw can be converted into bioenergy in MECs. The complexity of lignocellulosic materials in crop residues may require pre-processing techniques. **Wood Chips**: Wood chips and sawdust can serve as substrates for MECs. They contain cellulose, hemicellulose and lignin, which require specialized microbial communities for effective degradation. MECs can convert these materials into biohydrogen or other valuable products. **Food Waste**: Organic materials from food waste can be employed as a substrate in MECs, contributing to waste management and bioenergy production. **Organic Kitchen Waste**: Food waste from households and commercial kitchens is a readily available and biodegradable substrate for MECs. It helps reduce waste volume and can be an environmentally friendly way to generate bioenergy. **Restaurant Grease Trap Waste**: Grease trap waste, which contains fats, oils and grease (FOG), could be treated in MECs. The conversion of FOG into hydrogen gas is an attractive application due to the high-energy content of fats.

Industrial Effluents: Various industrial effluents containing organic matter, such as dairy, brewery or distillery waste waters can be processed in MECs. **Dairy Wastewater**: Dairy industry wastewater contains lactose and other organic compounds. MECs can convert these into hydrogen or other valuable products. **Brewery or Distillery Wastewater**: Effluents from breweries and distilleries are rich in organic matter, which perhaps be used as a substrate for MECs. The diversity of organic compounds in these effluents can be a valuable source for bioenergy production. **Algae and Microalgae**: Algal biomass as it may be used as a substrate in MECs and it offers the advantage of being renewable and sustainable. Algae are rich in organic matter, particularly lipids and carbohydrates which can be converted into biohydrogen or other biofuels. The choice of substrate for MECs depends on various factors, including substrate availability, biodegradability and the desired end products. Researchers are continuing to explore the new substrates and develop strategies to optimize MEC performance with different feedstock, leading to a broader range of applications in wastewater treatment and bioenergy production [6].

3 Suitable Substrates for Hydrogen Production in Microbial Electrolysis Cells

In the context of wastewater treatment and bioenergy production, the treatment of dark fermentation wastewater plays a crucial role. The performance of Microbial Electrolysis Cells (MECs) in the treatment of various substrates and a special focus on the treatment of volatile fatty acids (VFAs), which are common organic pollutants [7].

3.1 Volatile Fatty Acids (VFAs) as Organic Pollutants

Volatile fatty acids (VFAs) are organic compounds commonly found in wastewater streams and pose a significant environmental problem. They require effective treatment strategies to reduce their impact. In the following subsections, we look at studies investigating the potential of MECs in reducing VFAs.

3.1.1 Performance of MECs Using Different VFA Loads

Reference [8] looked into how MECs performed under various VFA loads. At an organic removal rate of eighty five percent and the maximum hydrogen production rate (HPR) was achieved, which was eighty-one mL/L/D. Additionally, the investigation demonstrated that, when compared to butyrate and propionate, acetate had the best performance for producing hydrogen.

3.2 Preference for Acetate as an MEC Substrate

According to studies by refs. [9] and [10], acetate is the suggested VFA for biodegradation in MEC. These results show that, in comparison with other VFAs, acetate improves MEC performance in addition to being readily biodegradable.

3.3 Comparing Sludge Treatment Methods for MEC Substrate Selection

A study comparing several sludge treatment techniques found that sludge produced from thermal hydrolysis (THP) was the optimum substrate for MEC treatment. Its higher acetate concentration and lower VFA content were major factors in its superior MEC performance.

3.4 Performance Enhancement with Unique Substrates

The use of specific substrates to combine dark fermentation with MECs was examined. These experiments yielded promising results. When coupled with dark fermentation and palm oil mill effluent. For example, displayed notable hydrogen performance at an external voltage of 0.7 V [11]. The mixture's HPR was 7.81 L/L/D. Dark fermentation and MEC linked systems were used with effluent from sugar beetroot juice and cassava starch processing in a similar manner with varying degrees of success. In the case of wastewater from the cassava starch processing industry, an external voltage of 0.6 V yielded a significant hydrogen output and overall hydrogen recovery rate [12, 13].

3.5 Optimizing MEC Performance for Bioenergy Production

The majority of these research emphasis the significance of substrate selection, namely the selection of VFA and treatment techniques in MEC performance for wastewater treatment and hydrogen production. Since acetate has a greater effect on MEC efficiency and biodegradability, it becomes a favored substrate. Researchers can produce bioenergy and clean up the environment more effectively and sustainably by gaining a knowledge of these elements.

4 Simulating Complex Industrial Wastewater

Researchers tested various substrates to recreate the complexity of industrial effluent [14]. Three different forms of industrial effluent were explored as potential MEC substrates in one significant study: Dairy industry effluent, potato industry wastewater (mostly starch) and biodiesel industry wastewater (high in glycerol). Furthermore, the study looked into using a mixture of these three wastewaters as a MEC substrate. The combined substrate demonstrated the best MEC performance, with higher current density, hydrogen production rate (HPR), cathodic gas recovery and hydrogen purity. Milk outperformed the other individual substrates, demonstrating the possibility of dairy industry effluent as a suitable MEC substrate. The mixed substrate's enhanced performance can be attributed to its higher biodegradability. Simultaneous scarification and fermentation were used to turn a combination of vegetable and rice straw waste into a MEC substrate. In terms of hydrogen production rate 2.46 ± 0.07 mmol/L/D and energy efficiency, which reached two hundred and fifteen percent and this blended substrate performed better than the separate substrates. These findings strongly imply that more hospitable conditions for microbial metabolism might be found in mixed substrates.

5 MEC Performances with Dark Fermentation Effluents from Various Industrial Wastewaters

According to [15], the efficacy of MECs in treating dark fermentation wastewater from six distinct industrial wastewater sources—cheese, fruit juice, paper, sugar, fruit processing and spirits—was examined. Fruit juice wastewater demonstrated the strongest MEC performance because to its great degradability. Because industrial wastewater may be utilized directly to produce energy, it has benefits over substrates with extract concentration. With this method, the greatest rate of removal of chemical oxygen demand was 78.5 ± 5.7% and a significant amount of hydrogen (1608 ± 266.2 ml) was produced for every gram of COD consumed. In a single-chamber MEC, the potential of cheese whey a dairy industry by-product as a MEC substrate was also examined [16]. Compared to wastewater treated by fermentation or raw cheese, whey wastewater treated by anaerobic digestion demonstrated superior performance. It attained an energy efficiency based on electrical input of 116.6% and a cathodic hydrogen recovery rate of 63%. However, compared to the cheese, whey that underwent dark fermentation around ninety three percent and the coulombic efficiency was far lower—just thirty two percent. Significant amounts of methane were also produced as a result of the latter process. In a different investigation, forty-nine-point eight percent of coulombic efficiency and an HPR of 0.8 L/L/D were obtained using fresh cheese whey as the MEC substrate [17].

6 Diverse Substrates for MEC-Based Hydrogen Production

Microbial Electrolysis Cells (MECs) use a variety of substrates, such as special feedstocks and industrial by-products, to produce hydrogen on a flexible platform. The vast array of substrates that have been examined for their ability to produce hydrogen using MECs is examined in this section.

6.1 Crude Glycerol

As a by-product of making biodiesel, crude glycerol is frequently seen as having little value and requires processing to make it more useful [18]. Crude glycerol has been demonstrated to be a feasible MEC substrate in trials utilizing an alkaline MEC with an external voltage of 0.8 to 1.2 V, yielding a hydrogen production rate (HPR) of 0.46 L/L/D. Its promise is demonstrated by the extraordinarily high cathodic hydrogen recovery rate of eighty five percent. The results of this study also revealed notable similarities in the MEC performance of acetate and crude glycerol.

6.2 Landfilled Leachate

To mitigate its negative environmental effects, landfill leachate, a hazardous waste needs to be thoroughly treated [19]. Experiments were carried out utilizing a membrane less single-chamber MEC at an applied voltage of 0.8 V, linking the treatment of landfill leachate with MECs. In these tests, a dairy industry effluent and landfill leachate were combined to create a MEC substrate. Although the performance declines at greater loads and it was discovered that landfill leachate may be treated by MECs and a stabilized HPR of 15 mL/L/D is reached. A different study that used landfill leachate as the MEC substrate was able to obtain an HPR of 0.148 L/L/D at 1.0 V external voltage. Additionally, a high hydrogen production and biofilm enrichment of an extraordinary concentration of pure hydrogen were recorded in this investigation.

6.3 Neutralized Bio-Oil Organic Phase (NBOOP)

Reference [20] investigated neutralized bio-oil organic phase as a slightly acidic product created after the neutralization of bio-oil acid by capacitive deionization as a MEC substrate. It had a high organic conversion rate and a maximal HPR of 4.3 L/L/D. It was also discovered that increasing the organic loading rates resulted in an increase in current density.

6.4 Pyrolysis-Based Substrates

Different pyrolysis by-products, including switchgrass, guayule, willow and pine sawdust, were studied for their ability to create hydrogen [21]. Switchgrass pyrolysis with an aqueous stream suited for MECs produced a maximum HPR of 4.4 L/L/D but with coulombic and energy efficiencies of fifty-four and forty eight percent, respectively. Pine sawdust had a high HPR of 5.8 0.18 L/L/D, with performance decreasing at organic loadings larger than 10 g/L/D.

6.5 Pig Manure Wastewater and Urban wastewater

Swine manure wastewater treated by hydrothermal liquefaction was shown to be a viable MEC substrate with an organic removal rate of more than ninety percent and an HPR of 168 ml/L/D [22]. Furthermore, the residue generated from the hydrothermal liquefaction of maize stalks served as a MEC substrate despite having a poor HPR of 3.92 mL/L/D when 1.0 V was applied. A preliminary test utilizing genuine city wastewater as the MEC substrate exhibited high hydrogen production with an HPR of 4.2 L/L/D and exceptional hydrogen purity of more than ninety five percent.

6.6 Other Substrates

The ability of a number of other substrates to produce hydrogen in MECs was examined. When compared to sodium acetate and xylose fermentation broth, maize cobs treated in a two-step procedure performed differently [23]. Reference [24] reported that Makgeolli wastewater fed with acetate produced hydrogen at a rate of 1.55 L/L/D. Reference [25] also looked into dewatered sludge, which performed somewhat worse. Furthermore, the research conducted by Cardena et al. [26] demonstrates that the food waste effluent obtained through dark fermentation has a remarkable cathodic hydrogen recovery rate of more than ninety three percent and a maximum HPR of 0.685 L/L/D. Furthermore, even when subjected to an external voltage of

3.0 V, the liquid fraction from compacted municipal solid waste was studied as a MEC substrate since it has a lower HPR [27].

7 Bacterial Community Composition and Dynamics in MECs

Based on the study of [28] bacterial community dynamics of MECs prominent phylotypes were observed in several bacterial classes. In the Deltaproteobacteria class, Geobacter dominated in both Acetate (A) and Propionate (P) reactor anodes, indicating electrochemical selection and a significant increase in abundance over time. Other members such as *Desulfovibrio*, *Pelobacteraceae* and *Desulfuromonadales* were also present. In the Clostridia class, seven phylotypes were consistently observed, with Copothermobacter being more abundant during the initial stages but decreasing later, alongside the presence of Sedimentibacter and Sohengenia. Two phylotypes from the Synergistia class, Dethiosulfovibrionaceae and Synergistaceae were consistently found with Dethiosulfovibrionaceae being dominant. Betaproteobacteria revealed four phylotypes with *Dechloromonas*are abundant, while Rhodocyclaceae exhibited relatively consistent abundance in AH-reactor suspensions. Within *Gammaproteobacteria*, *Pseudomonas* and *Pseudomonadaceae* were more prevalent. The Bacteroidia class featured four identified phylotypes with Porphyromonadaceae and Bacteroidales relatively more abundant. Parabacteroides was exclusive. Anaerolineae class a single phylotype and Anaerolinaceae displayed higher abundance. In terms of Archaeal Community Composition and Dynamics, Methanobacteriaceae specifically *Methanobacterium* and *Methanobrevibacter* dominated in both A- and P-reactors.

8 Electrochemical Performance and Substrate Choice in MECs

Microbial Electrolysis Cells (MECs) are promising bio-electrochemical devices that hold significant potential for sustainable hydrogen production and wastewater treatment. The electrochemical performance of MECs is a critical aspect that influences their efficiency in hydrogen generation and organic substrate degradation. This chapter explores the interplay between substrate choice and electrochemical performance, shedding light on the factors that impact the effectiveness of MECs.

8.1 Electrochemical Performance in Microbial Electrolysis Cells (MECs)

8.1.1 Substrate Choice and Its Impact on Electrochemical Performance

The choice of substrate in MECs plays a pivotal role in determining their electrochemical performance. Different organic substrates including volatile fatty acids (VFAs) industrial wastewater and agricultural residues employed as feedstocks. The selection of the substrate significantly influences key parameters such as the rate of hydrogen production, current density and overall energy efficiency. Furthermore, substrate choice shapes the composition and activity of the microbial community, which is closely intertwined with electrochemical reactions at the anode and cathode.

8.1.2 Factors Affecting Electrochemical Performance

Several factors affect the electrochemical performance of MECs, including:

Microbial Community: The composition and activity of the microbial community are crucial for efficient electron transfer and overall performance. Certain microorganisms, particularly exoelectrogens are key players in facilitating electron transfer to the anode.

Electrode Material: The choice of electrode materials significantly impacts the electrochemical reactions at the electrode–electrolyte interface. Conducive materials such as graphite, carbon cloth and conductive polymers are commonly employed. **Operating Conditions Parameters (OCP)** such as temperature, pH and applied voltage have a profound impact on electrochemical performance. Optimizing these conditions is essential for maximizing hydrogen production and substrate degradation rates.

8.2 Substrate Choice and Its Influence on Electrochemical Performance

The selection of a substrate has a direct influence on the electrochemical performance of MECs. Various substrates offer distinct advantages and challenges, including:

8.2.1 Hydrogen Production Rate

Different substrates result in varying rates of hydrogen production. Microbes leading to higher hydrogen production rates often metabolize substrates with simpler organic compounds such as acetate, more rapidly.

8.2.2 Current Density

The current density generated in MECs influenced by the substrate. Some substrates promote more efficient electron transfer, resulting in higher current densities.

8.2.3 Energy Efficiency

Substrate choice affects the overall energy efficiency of MECs. Some substrates converted into hydrogen with greater energy efficiency, while others may lead to energy losses due to incomplete substrate utilization.

8.2.4 Microbial Activity

The composition and biodegradability of the substrate influence microbial activity. Microbes metabolize substrates differently, influencing the rate of electron transfer and hydrogen production.

8.3 Optimizing Substrate Choice for Enhanced Electrochemical Performance

Researchers and engineers continually work to optimize substrate choice to enhance the electrochemical performance of MECs.

8.3.1 Biodegradability Assessment

Researchers assess the biodegradability of various substrates to understand how efficiently microorganisms can metabolize them. Substrates with higher biodegradability are often preferred.

8.3.2 Specific Applications

The choice of substrate should align with the specific application of the MEC. For instance, in wastewater treatment substrates that promote organic matter removal may be prioritized while in hydrogen production and high hydrogen yield is essential.

8.3.3 Microbial Community Engineering

Manipulating the microbial community within MECs used to optimize substrate utilization. Tailoring the microbial consortia to specific substrates can improve electrochemical performance.

9 Synergistic Effects of Substrate Mixtures

In MECs, the exploration of substrate mixtures has unveiled the fascinating world of synergistic effects. Combining different substrates in MECs can result in a remarkable enhancement of electrochemical performance and making it a promising avenue for both hydrogen production and wastewater treatment. This synergy arises from a delicate balance, where one substrate's strengths compensate for another weaknesses, leading to more efficient utilization of carbon sources, improved electron transfer and a balanced nutrient profile. Examples include the combination of acetate and butyrate for enhanced hydrogen production and the mixing of diverse industrial wastewaters to optimize organic matter removal. To harness these synergistic effects, researchers meticulously adjust substrate ratios, manage microbial communities and fine-tune composition, offering exciting prospects for sustainable hydrogen generation and environmental remediation in the world of bioenergy.

10 Challenges and Future Directions in Microbial Electrolysis Cells (MECs)

Microbial Electrolysis Cells (MECs) have made significant strides in the fields of bioenergy and wastewater treatment but are not without their challenges. To over the kinetic limitations associated with electrochemical reactions and ensuring efficient electron transfer processes remains a central challenge. The diversity of available substrates and varying biodegradability levels poses a concern. Managing and engineering microbial communities for desired outcomes, ensuring their dominance by electroactive bacteria and preventing the growth of undesirable microorganisms is an ongoing challenge. Additionally, transitioning MECs to larger practical applications raises engineering and logistical issues.

The cost of materials and energy input can hinder widespread adoption calling for improvements in cost-effectiveness and energy efficiency. Effective by-product management and the prevention of potential environmental harm are essential considerations. To address these challenges and guide future developments in MEC technology, several key directions have emerged. Research into advanced electrode materials and designs can significantly improve electron transfer kinetics increasing MEC efficiency. The integration of synthetic biology approaches and microbiome engineering can lead to tailor-made microbial communities optimized for MEC performance. Investigating the potential of MECs to handle a broader spectrum of substrates including waste streams and organic-rich feedstocks will expand the utility. Scaling up MEC technology for industrial and municipal applications and integrating with existing wastewater treatment plants and renewable energy systems offers promising avenues for future development. Reducing operational costs through the development of cost-effective materials and efficient processes is crucial. Sustainable MEC systems that generate energy while treating wastewater offer a compelling solution. Lastly, the development of rigorous monitoring and regulatory frameworks will ensure the safe and environmentally responsible operation of MECs. As MECs continue to advance addressing, these challenges and exploring new directions will contribute to their role in sustainable bioenergy production, wastewater treatment and environmental remediation, making them integral in addressing the energy and environmental challenges of the future.

11 Conclusion

Microbial Electrolysis Cells (MECs) have emerged as a dynamic and transformative technology that stands at the confluence of bioenergy production, wastewater treatment and environmental sustainability. This chapter has comprehensively explored the multifaceted world of MECs encompassing a wide spectrum of subjects, including the diverse substrates employed in MECs electrochemical performance, the intriguing synergistic effects resulting from substrate mixtures and the myriad challenges and promising future directions that illuminate the path ahead. MECs as highlighted and hold great promise as a pathway toward sustainable bioenergy production. Their versatility allows to efficiently harness an array of substrates, ranging from volatile fatty acids and industrial wastewaters to unconventional sources such as landfill leachate and pyrolysis by-products. The judicious selection and management of these substrates play a crucial role in optimizing MEC performance and ongoing research continues to unveil the potential of substrate mixtures and which can exhibit synergistic effects, strengthening both hydrogen production and wastewater treatment.

Comparing the performances of MECs from the existing literature can be challenging due to the variations in operational parameters, electrode types, substrate concentrations and applied voltages. MECs can be fine-tuned to maximize their electrochemical performance. The optimization process holds the key to unlocking the full potential of MECs in sustainable hydrogen production and effective wastewater treatment. To making them a valuable technology for a cleaner and more energy-efficient future. The predilection for acetate as an MEC substrate and given its superior biodegradability and impact on efficiency underscores the paramount importance of substrate choice. Looking to the future of MECs an array of challenges and opportunities stands before us. Conquering kinetic limitations effectively managing microbial communities and enhancing substrate flexibility are foundational to the advancement of MEC technology.

The imperative task of scaling up MECs for industrial and municipal applications was coupled with the imperative of enhancing cost-effectiveness and energy efficiency. It is pivotal for their widespread adoption. The establishment of regulatory frameworks and monitoring protocols is indispensable to ensure their environmentally responsible operation. MECs are primed to play a crucial role in addressing the energy and environmental challenges of the future. As unravel the particulars of substrates, electrode materials, microbial communities and scaling, MECs represent anideal of promise for sustainable bioenergy production, wastewater treatment and environmental remediation.

References

1. Jayaseelan A, Panneer Selvam S, Kirubanandam GP, Packiyadoss P, Sivaprasad S, Rangarajan G, Quynh Hoang L, Arivalagan P (2024) New insights into microbial electrolysis cells (MEC) and microbial fuel cells (MFC) for simultaneous wastewater treatment and green fuel (hydrogen) generation. 355:129530. https://doi.org/10.1016/j.fuel.2023.129530
2. Gautam R, Nayak JK, Ress NV, Steinberger-Wilckens R, Ghosh UK (2023) Bio-hydrogen production through microbial electrolysis cell: structural components and influencing factors, 455. https://doi.org/10.1016/j.cej.2022.140535
3. Kadier A, Simayi Y, Kalil M, Abdeshahian P, Hamid A (2014) A review of the substrates used in microbial electrolysis cells (MECs) for producing sustainable and clean hydrogen gas. Renewable Energy 71:466–472. https://doi.org/10.1016/j.renene.2014.05.052
4. Devi R, Shivakumar A, Kasai D, Koutavarapu R, Peera G (2022) Microbial electrolysis cell as a diverse technology: overview of prospective applications, advancements, and challenges. Energies 15:2611. https://doi.org/10.3390/en15072611
5. Logan BE, Call D, Cheng S, Hamelers HVM, Sleutels THJA, Jeremiasse AW (2008) Microbial electrolysis cells for high yield hydrogen gas production from organic matter. Environ Sci Technol 42:8630e40
6. Sonawane JM, Mahadevan R, Pandey A, Greener J (2022) Recent progress in microbial fuel cells using substrates from diverse sources. Heliyon 8(12):E12353. https://doi.org/10.1016/j.heliyon.2022.e12353
7. Tang JH (2021) A review of suitable substrates for hydrogen production in microbial electrolysis cells 621(1):012145. https://doi.org/10.1088/1755-1315/621/1/012145
8. Rivera I, Buitrón G, Bakonyi P, Nemestóthy N, Bélafi-Bakó K (2015) Hydrogen production in a microbial electrolysis cell fed with a dark fermentation effluent. J Appl Electrochem 45:1223–1229. https://doi.org/10.1007/s10800-015-0864-6
9. Yang N, Hafez H, Nakhla G (2015) Impact of volatile fatty acids on microbial electrolysis cell performance. Biores Technol 193:449–455. https://doi.org/10.1016/j.biortech.2015.06.124
10. Lin L, Zakaria B, Koupaie EH, Lakeh AAB, Hafez H, Elbeshbishy E, Dhar BR (2019) Evaluation of sludge liquors from acidogenic fermentation and thermal hydrolysis process as feedstock for microbial electrolysis cells. Int J Hydrogen Energy 44. https://doi.org/10.1016/j.ijhydene.2019.09.162
11. Khongkliang P, Jehlee A, Kongjan P, Reungsang A, Sompong O-T (2019) High efficient biohydrogen production from palm oil mill effluent by two-stage dark fermentation and microbial electrolysis under thermophilic condition. Int J Hydrogen Energy 44. https://doi.org/10.1016/j.ijhydene.2019.10.022

12. Dhar BR, Elbeshbishy E, Hafez H, Lee H-S (2015) Hydrogen production from sugar beet juice using an integrated biohydrogen process of dark fermentation and microbial electrolysis cell. Biores Technol 198:223–230. https://doi.org/10.1016/j.biortech.2015.08.048
13. Khongkliang P, Kongjan P, Utarapichat B, Reungsang A, Sompong O-T (2017) Continuous hydrogen production from cassava starch processing wastewater by two-stage thermophilic dark fermentation and microbial electrolysis. Int J Hydrogen Energy 42. https://doi.org/10.1016/j.ijhydene.2017.06.145
14. Montpart N, Rago L, Baeza JA, Guisasola A (2015) Hydrogen production in single chamber microbial electrolysis cells with different complex substrates. Water Res 68:601–615. https://doi.org/10.1016/j.watres.2014.10.026
15. Marone A, Ayala-Campos O, Trably E, Carmona A, Moscoviz R, Latrille E, Steyer J-P, Alcaraz-González V, Bernet N (2016) Coupling dark fermentation and microbial electrolysis to enhance bio-hydrogen production from agro-industrial wastewaters and by-products in a bio-refinery framework. Int J Hydrogen Energy 42. https://doi.org/10.1016/j.ijhydene.2016.09.166
16. Rivera I, Bakonyi P, Cuautle-Marín MA, Buitrón G (2017) Evaluation of various cheese whey treatment scenarios in single-chamber microbial electrolysis cells for improved biohydrogen production 174:253–259. https://doi.org/10.1016/j.chemosphere.2017.01.128
17. Rago L, Baeza J, Guisasola A (2016) Bioelectrochemical hydrogen production with cheese whey as sole substrate. J Chem Technol Biotechnol 92. https://doi.org/10.1002/jctb.4987
18. Badia-Fabregat M, Rago L, Baeza J, Guisasola A (2019) Hydrogen production from crude glycerol in an alkaline microbial electrolysis cell. Int J Hydrogen Energy. https://doi.org/10.1016/j.ijhydene.2019.03.193
19. Gini R, Zahid N, Banu JR, Yogalakshmi KN (2020) Batch fed single chambered microbial electrolysis cell for the treatment of landfill leachate 153:168–174. https://doi.org/10.1016/j.renene.2020.01.118
20. Park L, Satinover S, Yiacoumi S, Mayes R, Borole A, Tsouris C (2018) Electrosorption of organic acids from aqueous bio-oil and conversion into hydrogen via microbial electrolysis cells. Renew Energy 125. https://doi.org/10.1016/j.renene.2018.02.076
21. Lewis A, Ren S, Ye X, Kim P, Labbé N, Borole A (2015) Hydrogen production from switchgrass via an integrated pyrolysis–microbial electrolysis process. Bioresour Technol 195. https://doi.org/10.1016/j.biortech.2015.06.085
22. Shen R, Jiang Y, Ge Z, Lu J, Zhang Y, Liu Z, Ren Z (2018) Microbial electrolysis treatment of post-hydrothermal liquefaction wastewater with hydrogen generation. Appl Energy 212:509–515. https://doi.org/10.1016/j.apenergy.2017.12.065
23. Yan D, Yang X, Yuan W (2015) Electricity and H2 generation from hemicellulose by sequential fermentation and microbial fuel/electrolysis cell. J Power Sources 289. https://doi.org/10.1016/j.jpowsour.2015.04.164
24. Kim JH, Kim C, Jeon Y, Kim S (2019) Hydrogen production from Makgeolli wastewater using a single-chamber microbial electrolysis cell: hydrogen production from wastewater by MEC. Bull Korean Chem Soc 41. https://doi.org/10.1002/bkcs.11939
25. Hu K, Jia S-Q, Yang C, Sun X, Chen W, Wang W, Han F (2020) Combined freezing-thawing pretreatment and microbial electrolysis cell for enhancement of highly concentrated organics degradation from dewatered sludge. Bioengineered 11:301–310. https://doi.org/10.1080/21655979.2020.1736735
26. Cardeña R, Moreno-Andrade I, Buitrón G (2017) Improvement of the bioelectrochemical hydrogen production from food waste fermentation effluent by using a novel start-up strategy: bioelectrochemical hydrogen production from food waste. J Chem Technol Biotechnol. 93. https://doi.org/10.1002/jctb.5443
27. Zhen G, Kobayashi T, Lu X, Kumar G, Hu Y, Bakonyi P, Rózsenberszki T, Koók L, Nemestóthy N, Bélafi-Bakó K, Xu K (2016) Recovery of biohydrogen in a single-chamber microbial electrohydrogenesis cell using liquid fraction of pressed municipal solid waste (LPW) as substrate. Int J Hydrogen Energy 41:17896–17906. https://doi.org/10.1016/j.ijhydene.2016.07.112
28. Rao HA, Krishnaveni V, Katuri KP, Samik B, Saikaly PE (2017) temporal microbial community dynamics in microbial electrolysis cells—influence of acetate and propionate concentration 8. https://doi.org/10.3389/fmicb.2017.01371

Electrochemical and Theoretical Concepts of Microbial Electrolysis Cells

Humaira Rashid Khan, Abdul Latif Ahmad, Ahmed Shuja Syed, Imran Murtaza, Sadia Batool, and Asadullah Dawood

Abstract The novel combination of electrochemical and microbiological principles found in microbial electrolysis cells (MECs) provides a sustainable means of transforming energy and wastewater disposal. Microbes function as biocatalysts in these cell types, assisting in the anaerobic oxidation of organic matter by producing electrons that transit through an exterior circuit to the cathode generating hydrogen gas. Microbial electron transfers processes, electrode reactions, and microbial metabolism are all deeply ingrained in the theoretical and electrochemical roots of MECs. The fundamental mechanism of MECs is the utilization of extracellular electron transfer-capable bacteria or other electroactive microorganisms for their metabolic processes. Organic substrates are broken down and electrons are released during microbial oxidation, which takes place at the anode. To reach the cathode, where reduction events take place and hydrogen gas is produced, these electrons travel via an external circuit. In order to forecast and maximize MEC performance, theoretical models are

H. R. Khan (✉) · A. L. Ahmad
School of Chemical Engineering, Universiti Sains Malaysia, Engineering Campus, Seri Ampangan, 14300 Nibong Tebal, Pulau Pinang, Malaysia
e-mail: humairar_khan@usm.my

A. L. Ahmad
e-mail: chlatif@usm.my

H. R. Khan · A. Shuja Syed · I. Murtaza
Center for Advanced Electronics & Photovoltaic Engineering (CAEPE), International Islamic University, Islamabad 44000, Pakistan
e-mail: ahmed.shuja@iiu.edu.pk

I. Murtaza
e-mail: imran.murtaza@iiu.edu.pk

I. Murtaza
Flexible Electronics Laboratory (FEL), Department of Physics, International Islamic University, Islamabad 44000, Pakistan

S. Batool
Department of Chemistry, Rawalpindi Women University, Rawalpindi 46000, Pakistan

A. Dawood
Department of Physics, National Excellence Institute (University), Islamabad 04524, Pakistan

A. A. Yaqoob and A. Ahmad (eds.), *Microbial Electrolysis Cell Technology*, Sustainable Materials and Technology, https://doi.org/10.1007/978-981-97-3356-9_6

essential. These models take into account several aspects like mass transfer, electrochemical characteristics, and microbial kinetics. This chapter shows how MECs can use microbiology and electrochemistry to recover energy and remediate wastewater, with potential applications in renewable energy production and environmental remediation.

Keywords Sustainable energy sources · Hydrogen production · Volmer reaction · Extracellular electron transfer (EET)

1 Introduction

We can all agree in theory that we must shift away from fossil fuels and toward renewable energy sources that are not generating any greenhouse gas, but this is not clear how to make that transition. One attractive option is to generate electricity from renewable resources like the wind, sunlight, ocean currents, and floods. However, it is necessary to have effective storage mechanisms in order to mitigate the variations caused by these periodic resources. One promising avenue for the future of energy storage and conversion is the production of hydrogen using water electrolysis. This would allow us to convert electrical energy into chemical energy, which could then be utilized or converted back into electricity as needed [1, 2]. There are several social, economic, and environmental advantages to hydrogen, the most recent fuel in a long line of them. A lot of individuals think it could be a great way for the globe to generate renewable energy in future generations. Therefore, there can be substantial synergistic consequences with beneficial environmental implications from hydrogen production. Microbial generation of hydrogen has become significant for two primary reasons when compared to traditional methods:

(i) It can use organic waste materials as feedstock.
(ii) It can be run at the normal pressure and temperature.

Photosynthetic bacteria and algae may create hydrogen gas from water using sun energy. However, this method is presently inefficient, and most analysts believe this method may never be feasible because to its large surface area. Dark fermentation, in which fermentative bacteria convert carbohydrates to hydrogen [3], and photo-fermentation [4], in which photosynthetic bacteria convert organic acids to hydrogen, are two bioprocesses that have been the subject of substantial research into the generation of hydrogen from organic materials during the last 20 years [5].

An essential sustainable technology for harnessing energy from wastes is through the utilization of exoelectrogen microbes, also known as exoelectrogens. These microbes are distinguished by their ability to move electron beyond their cells, enabling them to donate or accept electrons [6]. According to Zhang et al. [7], this theory enables the treatment of sewage as well as the creation of energy from wastes. So, EETs, like organisms that participate in biological waste treatment, break down soluble organic or inorganic substances in streams to create alcohol, methane, or

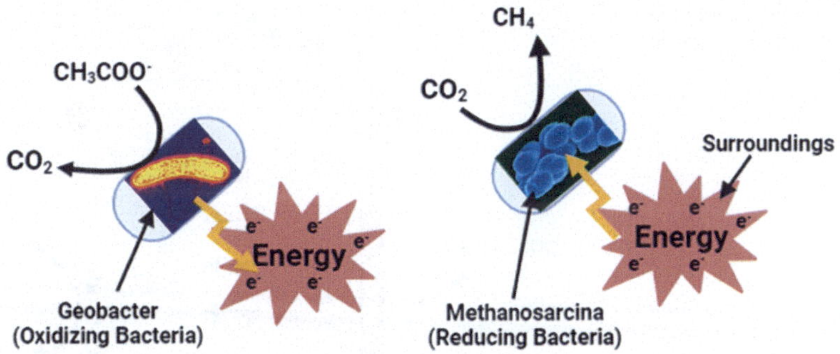

Fig. 1 Schematic diagram of electrons exchange in extracellular electron transfer (EET) at donor and acceptor sites (reproduce and redrawn in the text with permission from Ref. [9])

hydrogen, which are metabolic products [7]. The generation of these metabolites from trash is highly valuable in terms of environmental sustainability and energy preservation. Therefore, recognizing the significance of these electron transfer activities, electrochemists have employed certain natural bacterial processes to conserve energy in the anode and/or cathode chambers [8]. The donor–acceptor EET mechanism is simplified in Fig. 1.

Owing to its low cost and good microbial catalytic activity, cathodes used for developing acceptor extracellular electron transfer could replace platinum in several cathodic catalytic processes [10, 11]. Desulfovibrio vulgaris and other hydrogen-producing microbes are widespread and possess the enzymes necessary to catalyze the reversible reaction for the production of hydrogen [12, 13].

1.1 *Microbial Electrolysis Cell (MEC)*

In first decade of twenty-first century, the concept of microbial electrolysis cells (MECs) was presented [14]. The two main components of MEC are microbial anode and cathode, a site for hydrogen evolution. In order for microbes and their culture media to flourish, the electrolyte in the anodic section must have both. A biofilm of electroactive microbe's forms on the anode surface and serves as an electrocatalyst when some microbes colonize it randomly. Subsequently, allowing the anode to oxidize a wide range of carbon molecules [15, 16]. An advantage of MEC over non-biological water electrolysis is that it oxidizes the organic molecules at low redox potentials (Fig. 2).

The voltage of a MEC is significantly lower compared to the well-known 1.23 V required for electrolysis of water under the ordinary circumstances [18]. The voltage in the electrolysis cell is directly proportionate to the amount of power that cell

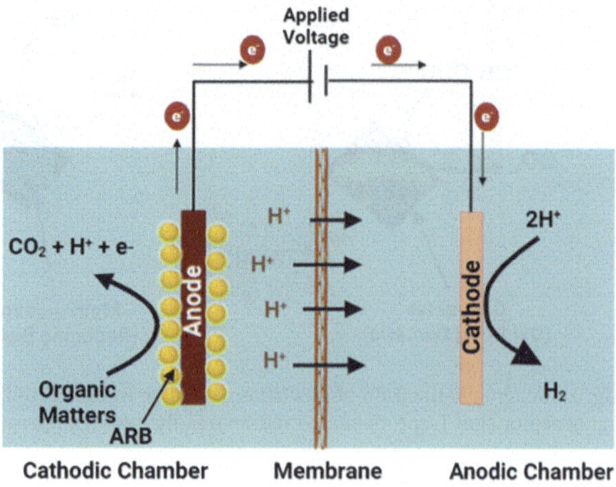

Fig. 2 Double chambered MEC for the production of H_2 [17]

consumes. Thermodynamic analysis has led some to believe that MECs may represent a game-changer in the quest for more cost-effective hydrogen production via electrolysis.

2 Working Concept of MEC

A voltage is given to MEC (bio-electrochemical systems (BESs), in order to carry out the bio-electrochemical reactions [19]. Microbial electrolysis cells (MECs) catalyze organic molecule oxidation in the anode chamber and hydrogen chemical evolution in the cathode chamber. Thus, the production of hydrogen gas closely relates with the current production in anaerobic conditions [20].

Exo-electrogenic bacteria emit electrons to speed hydrogen creation in a reactor like a microbial fuel cell. This phenomenon is known as electro-hydrogenesis, when bacteria release electrons that combine with protons to generate hydrogen gas, rather of producing electricity like in a microbial fuel cell (MFC). MEC produces H_2 gas by integrating microorganisms with electrochemically assisted elements in a reactor. The electro-chemical reactions occurring at the anodes and cathodes can be elucidated by the subsequent Eqs. (1)–(3) [11]:

$$\textbf{At anodic chamber}: CH_3COO^- + 4H_2O \rightarrow 2HCO_3{}^- + 9H^+ + 8e^- \quad (1)$$

$$\textbf{At cathodic chamber}: 8H^+ + 8e^- \rightarrow 4H_2 \quad (2)$$

$$\textbf{Overall reaction}: CH_3COO^- + 4H_2O \rightarrow 2HCO_3{}^- + H^+4H_2 \quad (3)$$

From the aforementioned equations, it is evident that the cell voltage is negative, indicating that non-spontaneous reaction. Therefore, extra voltage is necessary to ensure that the reactions can occur under thermodynamically favorable conditions.

3 The Electrochemical Conceptualization of Anodic Respiring Bacteria (ARBs)

The advancement of microbial electrochemical systems (MEC) depends on understanding electrochemical principles at the biofilm–electrode interface [21]. This requires examining interactions between microorganisms, substrates, and electrode surfaces during biofilm development stages: start-up [22], growth [23], and starvation [24]. Such expertise is essential for optimizing MEC design and efficiency. The connection between the current, ARB (anode-respiring bacteria), and the electrodes has been confirmed. This shows that the production of electrons results from the utilization of bio-catalytic substrates and the transmission of electrons through different electron exchange pathways to the electrode [25, 26]. However, when MECs integrate both living and non-living components, the flow of electrons within the cell will encounter impedance, leading to decreased electrical potential.

Ohmic losses in the electrode, biofilm, electrolyte, membrane, anode, and biofilm–electrode contact (cathode) cause cumulative voltage loss [27]. Hence, minimizing potential losses is crucial for optimizing the performance of an electrochemical system, and biofilm–electrode interface potential losses must be adjusted. There has been a scarcity of voltammetric studies conducted on the interface between the anodic electrode and the biofilm. The Butler–Volmer equation is commonly employed to describe the relationship between current and voltage in electrochemical systems. At the electrode–solution interface, the local current is generated by electron exchange [28]. On the other hand, when biofilm is created on an anode, it is vital to take into consideration both biotic and abiotic activities occurring on the surface of electrodes [29]. Therefore, the kinetic behavior of these anodes will differ from the abiotic system and, consequently, from the standard Butler–Volmer equation. This behavior is an outcome of the consumption of substrates through metabolic processes and the mechanism of electron transport. Various models have been used to study both the biotic reactions involving substrate utilization and the abiotic reactions involving EET. Among them, two significant types of the Monod model that have been modified are as follows:

(i) The Butler–Volmer–Monod model: It explains the connection between the substrate's over-potentials after metabolism [30].
(ii) Monod–Nernst model: It explains how the redox potential and species concentration are related to substrate metabolism rate [29].

A two-stage procedure was suggested by Hamelers et al. [30] for the development of the Butler–Volmer–Monod model:

The **first stage** is the intercellular biochemical utilization reaction, which is a catalyst for the production of both organic compounds and electrons. The rate of this reaction is governed by the kinetics of the enzymes, as defined by the Monod equation.

In the **second stage**, electrons leave the biofilm toward the electrode outside the cells. This process is outlined by using the standard electrochemical electron transfer kinetics equation, which was developed by Butler and Volmer. The inefficiency of the process is attributed to losses occurring at the bio-anode, resulting in conversion efficiencies below 100%. This can be assessed by examining the anode polarization curve, which illustrates the relationship between potential and current density.

The rates of microbial metabolism were characterized by Hamelers et al. [30] using Eq. (12). Butler–Volmer electrochemical kinetics and Monod enzyme kinetics are used to form its foundation. Results showed that the polarization model matched the observed curves of anode biofilm polarization. Furthermore, the Monod constant of an anode biofilm depends on the anode over-potentials [30].

$$j = j_{\text{lim}}\left(\frac{1 - e^{\frac{F}{RT}\eta_{\text{anode}}}}{K_1.e^{-(1-\alpha)\frac{F}{RT}\eta_{\text{anode}}} + K_2.e^{\frac{F}{RT}\eta_{\text{anode}}} + \left(\frac{K_M}{S} + 1\right)}\right)$$

The equation consists of F, R, T, η, j, j_{lim}, α, K_M, and S, describing the anode reaction, substrate affinity, and substrate concentration [30].

The Nernst–Monod model states that the biofilm anode regulates reaction rate and electric current by facilitating interfacial electron exchange. Consequently, as the electrochemical system becomes more efficient, the potential loss at the contact between the biofilm and electrolyte decreases. Anode potential losses (APL) arise at the interfaces between the biofilm and electrolyte, as well as between the electrode and biofilm. Therefore, it is crucial to assess APL from an electrochemical perspective in order to further the development of MEC systems. The literature does not fully describe APL, its relationship to anode biofilm performance, or the calculation method. The anode, also known as the anodic potential limit (APL), is linked to the electrocatalytic characteristics of bio-active sites. It is important for the APL to be modest in order to minimize the difference between the electrode potential and the redox potential of the substrate being oxidized. To make the process efficient, bacteria must lose little energy during substrate and EET use and electron transport to the anode. Torres et al. [31] provided a definition of APL as the potential losses resulting from the movement of electrons outside of cells and their transfer to the surface of an electrode. This definition suggests that a lower APL corresponds to a higher level of energy transfer, indicating efficient process. Hence, from an electrochemical perspective, the primary objective in optimizing the MEC process is to minimize potential dissipation and maximize the biologically active surface area to attain the highest possible current production. The primary objective of MECs is to develop a bio-electro catalytic system capable of utilizing certain organic substances (including waste materials) to generate a substantial current density while maintaining a low cell potential [31].

The biofilm cannot transmit electrons to the electrode when the anode voltage matches the electron donor voltage, preventing catalytic current shown by $j = 0$. Raising the anode potential to a higher positive value enhances the metabolic activity of the anode-respiring bacteria (ARB) and promotes the transfer of electrons between the biofilm and the electrode. This accelerates the process of substrate utilization, leading to the maximum current (known as the limiting current). The impedance at the interface can hinder the flow of electric current, leading to an increase in the average path length (APL). The value of APL (mV) is defined as the potential range between the electron donor and the onset potential, which is the point at which the current starts to increase.

The Nernst–Monod equation characterizes the ideal relationship between the current density and the applied potential in a bio-electrocatalytic system with a pure culture and electron transfer channels [32]. Anode biofilms formed in mixed culture conditions may not adhere to the Nernst–Monod paradigm. While the j–E voltammogram may exhibit a sigmoidal shape, it may not conform to the Nernst–Monod model, which makes it challenging to understand the kinetic data [33].

Experimental results may deviate from the Nernst–Monod model due to several factors, including grown on the electrode, and the existence of multiple electron transfer pathways that can take place at different electrode potentials.

4 Components of Microbial Electrolysis Cells (MEC)

The relationship between current, ARB (anode-respiring bacteria), and the electrode has been well established, indicating that electron production takes place due to the utilization of bio-catalytic substrates and the transmission of electrons to the electrode through multiple electron exchange pathways (EETs) [34, 35]. Electrons moving through MXCs that consist of both living and non-living components will experience resistance, leading to a decrease in voltage within the cell. The total voltage loss is the combined result of losses that occur at the interface between the biofilm and anode electrode, along with Ohmic losses that occur in the electrode, biofilm, electrolyte medium, membrane, and cathode over-potential (in the case of H_2 production or O_2 reduction) [27]. To achieve an efficient electrochemical system, it is crucial to minimize potential losses and optimize potential losses at the interface between the biofilm and electrode.

4.1 Anode

Carbon-based materials are certainly the most widely used and have frequently resulted in efficient microbial anodes [36, 37]. Unfortunately, because of the low electronic conductivity of these materials, metallic current collectors would be required if significant surface areas were to be adopted. As a result, metallic anodes are intriguing options [17]. Stainless steel has demonstrated a potential ability to generate microbial anodes, notably foam [38, 39]. Nonetheless, if a high salt concentration is utilized, chloride ions significantly increase the danger of pitting corrosion. If the anode's potential is not precisely managed, serious corrosion can ensue. Copper does not appear to be particularly suitable, despite its demonstrated ability to form microbial anodes, because, with a standard potential of dissolution of 0.35 V/SHE (0.10 V/ SCE), the anode potential would have to be precisely controlled, and any accidental drift toward high values could result in the release of Cu^{2+} ions into solution. These ions are extremely harmful to microbial cells. Copper–polymer hybrid compounds could be a viable alternative [40]. Titanium is recognized to be inappropriate for anode applications due to its weak conductive passivation layer. The difficulty was rectified by covering titanium with metal oxides including ruthenium and tantalum oxides [41] to create dimensionally stable anodes (DSA). A few research [9] have described DSA-supported microbial anodes, although the application of DSA to support microbial anodes is yet inadequately characterized.

The performance of MEC systems is influenced by anode materials and configuration [42, 43]. It can be challenging to contain enzymes, cellular organelles, and animal and plant cells without affecting their normal function. The process of restricting the movement of living organisms in the surrounding environment while maintaining their ability to catalyze reactions is called biological immobilization. Conductive carbonaceous materials such as carbon fabric, carbon fibers, carbon felt, carbon rods, carbon paper, carbonized cardboard, and granular activated carbon (GAC) are the perfect choice for immobilizing ARB on the anode. This is illustrated in Fig. 3. Carbon brush [44] and carbonized cardboard have superior microbial electrical efficiency due to their large surface area and low Ohmic resistance. In this regard, using a packed bed analyte compartment filled with GAC to increase surface area may increase the volume of the biofilm, and thus the number of ARB reactive sites. Some studies have suggested that using granular activated carbon (GAC) as a three-dimensional electrode could enhance the performance of microbial electrochemical cells (MECs) However, other studies have found that using sandwich-type electrodes (with only a membrane between the anode and cathode) in a similarly designed MEC reactor provided comparable results.

For instance, a MEC with a packed GAC bed between the electrodes (functioning as a 3-D electrode) had hydrogen yields of 3.55 mol H_2/mol acetate, hydrogen recovery efficiency of 90%, and COD removal rate of 80%, which were similar to those of a typical sandwich-type MEC [45]. Due to the utilization of a densely packed bed of granular activated carbon (GAC), the GAC particles are not adequately linked to the electrode (current collector) or have weak connections. This disconnection disrupts current flow from the newly established biofilm on the GAC surface.

Bacteria can rapidly stimulate biofilm growth on carbon electrode materials such as carbonized cardboard and brushes, connecting to the electrode body immediately and causing minimum harm compared to graphite anodes and acetate substrates [42, 46].

Extensive research has been conducted to improve the power generation of microbial fuel cells (MFCs) known as anaerobic baffled reactors (ABRs) by

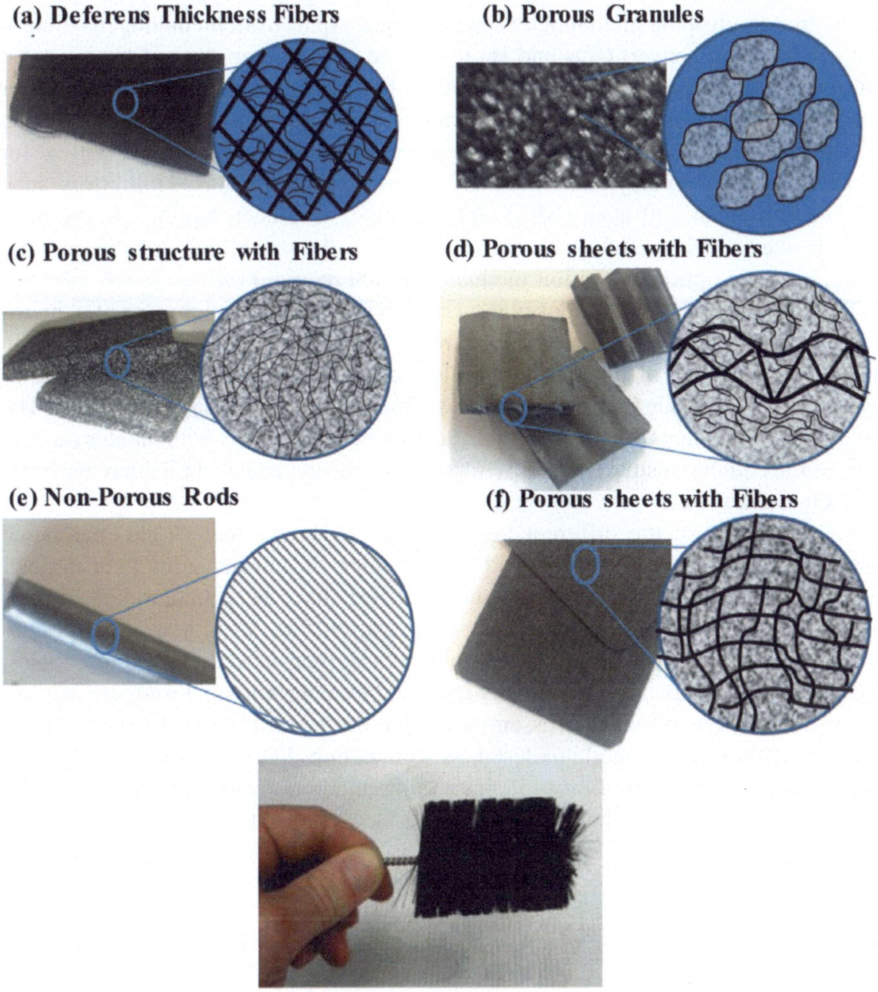

Fig. 3 Different carbonaceous materials used as ARB anode materials (reproduced with permission from Ref. [44])

using conductive polymers or nanocomposites. When a conductive poly(3,4-ethylenedioxythiophene) (PEDOT) film is applied to a carbon felt (CF) anode and introduced to palm oil mill waste, it enhances the bio-catalytic efficiency in facilitating acetate metabolism [47].

The microbial community that exists in the anode is also important in determining MEC hydrogen yields. The presence of methanogens causes the generation of methane along with hydrogen, which is a serious concern in MECs (mainly membrane-less MECs) [48]. This methane synthesis is largely related with the use of CO_2 and electrochemically produced H_2 rather than substrate utilization, implying that hydrogeno-trophic methanogens predominate. Apart from methanogens, homoacetogens, which convert CO_2 and H_2 to acetate, are another class of bacteria that reduces hydrogen yields [49]. These are more bothersome since they develop faster than electrogens and methanogens and can grow chemo-lithographically by using CO_2 and H_2. Thus, limiting the growth of non-electrogenic bacteria is critical for obtaining high hydrogen yields from MECs. Various chemical-based inhibitors, such as 2-bromomethanesulfonate (BES) and lumazine, have been reported in the literature. However, the use of such compounds raises the overall cost of the process; hence, low-cost options to inhibit methanogens are required [50].

The arrangement of electrodes plays a crucial role in the development of a microbial electrolysis cell (MEC) system and reducing the Ohmic voltage drop [51]. In their study, Liang et al. [52] employed two anodes placed on either one side or both sides of a cathode in a membrane-less microbial electrolysis cell (MEC). The researchers found that using a microbial electrolysis cell (MEC) with anodes on both sides of the cathode resulted in a 72% increase in current and a 118% increase in H_2 production rates [52].

A comparison of the different layouts (Fig. 4) reveals that single-chambered systems produce more hydrogen and have greater current densities than double-chambered systems. As a result, significant efforts have been undertaken to optimize this arrangement further for scale-up research [53]. MEC hydrogen production utilizes system architecture, which increases resistance and Ohmic losses. Conductive, biocompatible, chemically inert, and weatherproof materials are needed.

Pre-treatment of the anode is essential as it has the potential to modify the porosity and conductivity of the anode material, as well as introducing dopant particles that enhance the growth of biofilm [45]. Increasing anode porosity [55] increases biocatalytic sites and electrochemically active surface area [56].

4.2 *Cathode*

Most microbial electrolysis cells produce hydrogen at the cathode surface in the cathodic chamber. The hydrogen evolution reaction (HER), which involves mass transfer between the solid, liquid, and gas interfaces, is the rate-limiting stage in the MEC process. The HER mechanism consists of multistep reactions, the first of which is the electrochemical Volmer reaction, followed by a chemical Tafel reaction or an electrochemical Heyrovsky reaction (Fig. 5) [57]. Depending on the material, whether or not a catalyst is present, and the experimental conditions, the Volmer reaction can involve electron transfers from the anode to the cathode, leading to proton adsorption and absorbed hydrogen atoms on the surface of the electrode [58]. For the Heyrovsky process, the adsorbed H atom is recombined with another set of protons and electrons to generate hydrogen, or for the Tafel mechanism, two absorbed H atoms combine to produce hydrogen. During hydrogen synthesis in MECs, any of the previously described phases can be rate limiting. The accompanying over-potentials (activation and concentration over-potentials) primarily limit the HER,

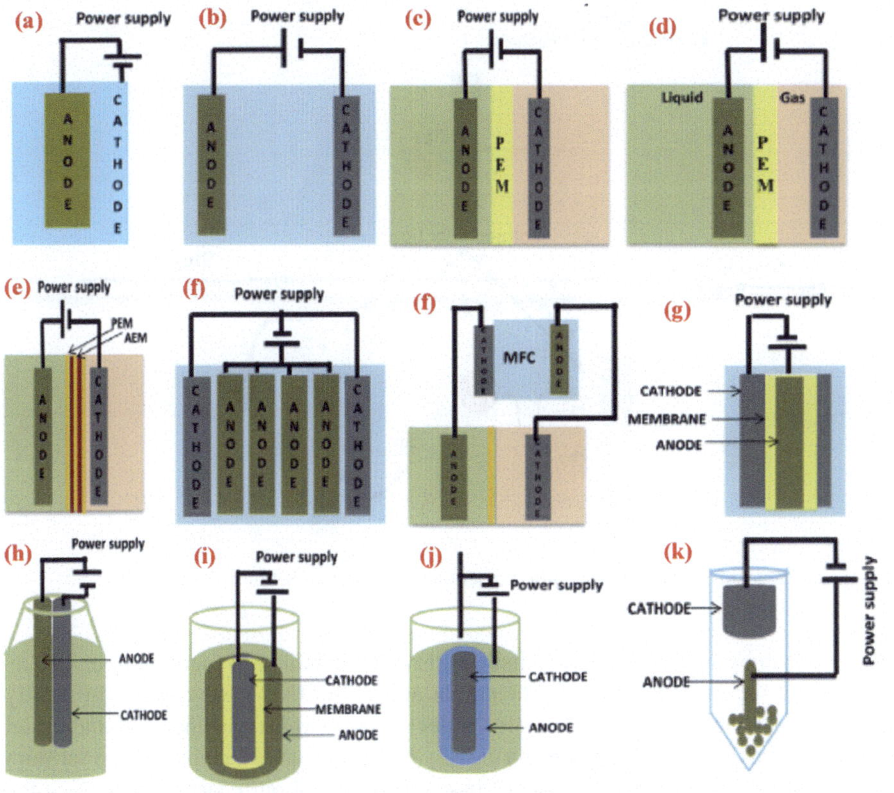

Fig. 4 Layouts of various configurations used in microbial electrolytic cells (MECS) for the production of H_2 (reproduced with permission from Ref. [54])

requiring an extra voltage to be applied beyond the thermodynamic limit to drive the reaction at a specific rate. These over-potentials can be enhanced by using specialized catalysts with active hydrogen evolution sites. Platinum has been identified as the most promising HER catalyst; however, the related costs make its application on a broadscale impracticable. Several alternatives to platinum have been described, including nickel-based alloys, gas diffusion electrodes, SS (stainless steel) woven mesh, and so on, albeit their efficiency is substantially lower when compared to platinum electrodes.

The hydrogen evolution potential of graphite electrodes is 420 mV compared to the standard hydrogen electrode (SHE) at pH 7 PBS. Additionally, the potential for dissolved O_2 reduction varies from 56 to 556 mV versus SHE. Considering these potential outcomes, it is feasible to achieve both O_2 reduction and H_2 evolution simultaneously. However, by controlling the environmental circumstances, it is possible to promote one process over the other [59].

Controlling the O_2 content in the cathodic chamber allows you to change the circumstances. Air saturation in the cathodic chamber favors O_2 reduction at lower

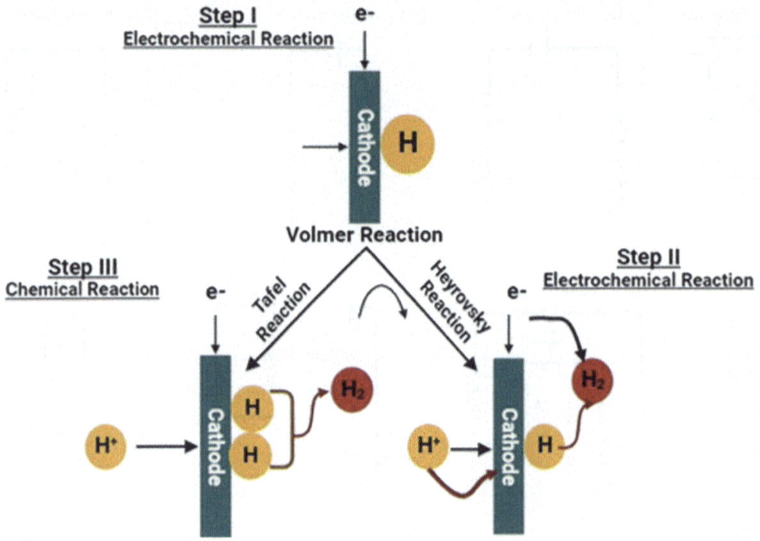

Fig. 5 Mechanism of hydrogen evolution reaction by different reactions

potential values; however, O_2 elimination allows H_2 synthesis if there is enough over-potential at the cathode surface. The potential loss in MXC systems is significantly attributed to voltage losses resulting from electrolyte Ohmic resistance and interfacial electron transfer at the cathode during O_2 reduction or H_2 generation. Platinum (Pt) is commonly employed as an electrocatalyst, doped onto carbon, to reduce the over-potential of the reaction. It is readily available in powder form and can be combined with a binder to create the electrode [60]. While Pt is highly effective as an electrocatalyst for hydrogen or oxygen reduction in MXCs, it is important to take into account the potential for deactivation caused by wastewater pollutants, such as sulfide. Researchers are exploring alternative options, such as different metals, alloys, doped carbon materials, and a reducing biofilm cathode, that could serve as substitutes for the expensive Pt [61].

Many investigations have also been conducted on the cathode material of MECs [62, 63]. Platinum is costly and experiences a decline in catalytic efficacy at pH values that are neither acidic nor basic. Possible contenders include stainless steel, nickel, and cobalt alloys. Porous materials used as cathodes in MECs enhance the available surface area [64, 65], although this advantage diminishes when subjected to high current densities [66]. A decrease in bubble coverage due to a high current density lowers the effective conductivity of the interstitial electrolyte and blocks the electrode. If gas development is to occur on spongy electrodes, it is best to drive current through them [67]. The use of weak acids allows for homogenous catalysis of hydrogen evolution [68]. Weak acids, such as phosphates [69], electrochemically deprotonate on some metallic surfaces, releasing hydrogen atoms:

$$H_2PO_4^- + e^- \rightarrow HPO_4^{2-} + {}^1\!/_2\, H_2 \quad (4)$$

Hydrogen and quick acid–base equilibrium result from this process:

$$HPO_4^{2-} + H_2O \rightarrow H_2PO_4^- + OH^- \quad (5)$$

Overall water reduction reaction:

$$H_2O + e^- \rightarrow {}^1\!/_2 H_2 + OH^- \quad (6)$$

Carbonates have demonstrated the same capacity [70]. In cathodic deprotonation [69], the weak acid molecule assures electron conduction with the electrode material, whereas the water molecule hardly interacts with the electrode. Hydrogen separation from acid molecules uses less energy than from water molecules. This catalytic route is more effective on the surfaces of mild steel [71] and stainless steel [72]. It reduces the over-potential required for hydrogen evolution by operating at neutral pH and use steel cathodes, which are less expensive than nickel alloys. In large-scale MECs, stainless steel has been commonly employed as a cathode material. However, selecting a steel grade is not an easy task [73].

Due to chloride ions, lowering the Ohmic drop in a very saline medium may oppose deterioration. MEC voltage protects the cathode against deterioration; therefore, low-grade stainless steel can be used. However, any unintentional interruption of the applied potential may result in rapid corrosion of low-grade stainless steel. The option could be an investment on the amount of trust that can be placed in the operational method.

Studies have shown that carbon cathodes with stainless steel, nickel alloys, and nanoparticles doped with MoS_2 or Pd are highly efficient for hydrogen evolution in a MEC system [74, 75]. In MFC systems, some carbon-supported electrocatalysts used for O_2 reduction include manganese oxide, composite compounds (such as metal macrocyclic compounds like iron phthalocyanines or cobalt tetra-methoxyphenyl-porphyrins), and nitrogen-doped carbon materials. In recent times, there has been a notable increase in the interest in developing cathodes that do not contain metal. This is primarily driven by the objective of lowering expenses and minimizing the ecological consequences associated with MXCs. There is a possibility that metal electro-catalyst materials could dissolve and contaminate the surrounding environment. Nevertheless, carbon compounds that have been infused with nitrogen, phosphate, sulfur, or a mixture of these elements have been created and utilized as cathode material for the reduction of O_2 and the creation of H_2 [76]. The introduction of doped moieties by doping foreign nonmetallic species on carbon substrates modifies the electron density on the substrate by either adding or removing electrons. Thus, heteroatoms in graphitic structures favor reactant and product adsorption/desorption, improving carbon's catalytic activity and some electrode interface redox reactions [77].

4.3 Separator and Membrane

Ion-exchange membranes separate the electrodes of MEC to preserve hydrogen gas purity and prevent anodic bacteria from absorbing electro-produced hydrogen [78]. Proton exchange membranes (PEM), anion exchange membranes (AEM), bipolar membranes, charge mosaic membranes, and other membranes are the most commonly utilized [79]. Aside from purity, another significant advantage of employing membrane is that it decreases fuel crossover and the chance of short circuit. They increase the internal resistance system, create a pH imbalance between the electrodes, and lower current densities at a given voltage. Moreover, the continued occurrence of contamination of the membrane might result in Ohmic losses and hinder the hydrogen evolution reaction (HER) at the cathode. If membranes are used in MEC designs, the cost of building rises as well. As a result, it is recommended that a membrane-less design be developed to help overcome these restrictions. The cations in CEM have the ability to freely exchange with other cations that are passing across the membrane, or they can move in response to an electric field that is provided. Therefore, the use of CEMs allows for the transportation of H^+ and other cations, including Na^+, K^+, NH^{+4}, Ca^{2+}, and Mg^{2+}, toward the cathodic chamber. On the other hand, in AEM, the exchange groups often consist of cations, which enables the transportation of anions such as OH, $PO_4{}^{-3}$, Cl^-, $SO_4{}^{2-}$, and anionic complexes toward the anodic chamber [80].

In a study, Sleutels et al. [55] shown that a two-compartment membrane electrode assembly (MEC) system, with an anion exchange membrane (AEM) between the compartments, was able to recover a greater amount of hydrogen (2.1 m^3 H_2 m^3/d) than a CEM (0.4 m^3 H_2 m^3/d) [55]. At 1 V, the study focused on acetate utilization losses and partial resistances. However, effluent and MEC inoculating media typically have a pH of pH 7. Consequently, the concentration of protons and hydroxide ions is approximately 10^7 mM, with a maximum of about 10^4 mM. This concentration is significantly lower than that of other cation species in wastewater, which typically have concentrations 4–5 orders of magnitude higher. Electro-dialysis takes place in MECs by the use of CEM, which allows for the movement of cation species from the anodic chamber and protons across the membrane. This process maintains the overall charge balance within the cell. Hydroxide (OH^-) anions from the cathode are transferred across the membrane into the anodic chamber to balance protons. Consequently, the introduction of an anion exchange membrane (AEM) will restrict the movement of metal cations to the cathodic chamber [61], effectively preventing any alteration in the composition of the surrounding medium in the anodic chamber. This stabilization of the environment promotes the growth and development of bacteria.

$$2H_2O + 2e^- \rightarrow 2OH^- + H_2$$

An electrochemical cell can create a pH gradient by inserting a membrane between the electrodes. The movement of charged particles toward the correct electrode generates a difference in pH between the cathodic and anodic chambers, leading to a decrease in the efficiency of microbial electrolysis cells (MECs) due to increased potential loss [61]. Rozendal et al. found that a difference in pH between the two sides of the membrane caused a significant amount of potential losses in a MEC system. They observed that the pH difference was larger in a system that included a CEM (pH = 6.4) compared to a system that included an AEM. The greater pH gradients observed in a CEM-enhanced system are associated with the potential for accelerated transport of cation species, including hydronium cations, to the cathode chamber. To minimize voltage losses, it is necessary to minimize pH variations across the membrane in practice. The Nernst equation states that a change of one unit in pH between the two chambers of an MEC leads to a decrease in the total cell voltage at pH = 2.303 RT/F.

Therefore, employing an AEM will create a steady environment that supports the growth of bacteria in the analyte chamber and reduces the adaptation of bacteria on the cathode, ensuring the production of pure H_2. However, a membrane separator increases internal resistance, mass transfer, contamination of the membrane, and pH imbalances, which complicates cell assembly and reduces voltage [61].

5 Operating Parameters

There are certain operating parameters that are responsible to greatly affect the electrochemical efficiency of MECs.

5.1 pH

The performance of MECs is primarily influenced by pH, as the pH level directly impacts the hydrogen evolution reaction (HER) occurring at the cathode. Any modification in the redox potentials at the cathode has the potential to result in elevated over-potentials. In addition to HER, microbial activity is additionally dependent upon pH. This is due to the sensitivity of microbes to the pH of their environment. Even a small change in pH can lead to changes in the metabolic processes of bacteria. Given that microorganisms are often most active in environments with a pH close to neutral, the majority of MEC research have been conducted at a pH of 7 [81]. Moreover, other additional parameters, such as solution conductance, transport of ions, oxidizing properties of the substrate, and so on, are in some way linked to the alteration in pH.

5.2 Temperature Impact

Similar to pH, temperature also exerts a substantial influence on the efficiency of microbial electrolysis cells (MECs). The fluctuating temperature during MEC operation results in alterations in the microbial community and activity, which in turn affects the efficiency of the reactor [82]. The majority of MEC experiments are carried out at standard temperatures, as it is anticipated that electrogenic microorganisms will exhibit greater levels of activity under the aforementioned conditions [83, 84]. The growth and mass transfer rate of the electrogenic bacteria are impacted by temperature changes [85]. Moreover, even a small pH shift can alter the electrode potentials and have an impact on the HER. As may be shown from the electrochemical investigations of water electrolysis, temperature has a significant influence on the kinetics of HER [86].

5.3 Impact of Potentials Applied

To overcome the thermodynamic threshold and make hydrogen production at the MEC cathode feasible, a voltage of at least 0.2 V is required. Unfortunately, the total effectiveness of the process is reduced since the cathode has huge over-potentials, which cause this figure to be elevated greatly. Hydrogen evolution increases with applied potential [87], although an appropriate value must be maintained for system sustainability. In addition to the HER, applied potential impacts electrogenic bacteria metabolism. Wang et al. found that applied voltage increased breakage of cells and decreased the metabolism of cells [88]. To comprehend the shift in microbial activity and community composition, it is necessary to conduct additional research into the impact of applied potential. To ascertain the process's economic performance, the methods of the voltage supply and utilization of energy are critical. Research on a small scale often employs a direct current (DC) power source. In any case, these are not going to cut it for industrial use. To achieve the goal of industrialization, voltage must be monitored and controlled in real time. A variety of energy sources with minimal electrical consumption have been created over the past decade by multiple research organizations.

5.4 Substrates Effect

The electrogenic bacteria use the substrate as their only source of electrons, which they then use to oxidize the substrate and transmit to the cathode via the anode, creating hydrogen gas. Therefore, the ultimate destiny of hydrogen yields produced in MECs is dependent on selecting the appropriate substrate. From basic carbs to complicated biodegradable materials like biomass and pollutants, MECs may use nearly any organic material [79]. In this particular cell, the theoretical amount of hydrogen produced per mole of glucose is almost 12 mol. On the other hand, the overall production of hydrogen is roughly from 2.55 to 3.81 mol/mol has been obtained so far [89]. This is because the MEC system is subject to various overpotentials. However, the kind of substrate used affects the conversion performance. Acetate was found to have greater hydrogen production and efficiency of conversion when used as a substrate in comparison with glucose [90]. In addition to substrate form, the load rate is an important factor in illustrating MEC performance. The saturation point may result from more substantial loading, while the overall current density of the system may be reduced by lower rate of loading.

5.5 MEC Design and Control

As an early observation, it is worth noting that exploring the potential of implementing high pressure MEC is an intriguing avenue that has not yet been explored, to the best of our knowledge. A novel system design that separates the bio-anode from the hydrogen evolution cathode could be highly intriguing. Each electrode is placed in a distinct electrochemical reactor in this scenario. The hydraulic loop connecting the two reactors utilizes a redox mediator, which undergoes a reduction in the reactor equipped with the bio-anode and oxidation in the reactor responsible for the evolution of hydrogen. The decoupled approach has been effectively utilized in water electrolysis [91], providing the key benefit of segregating hydrogen and oxygen producing reactors. For MECs, this method enables the utilization of distinct electrolytes for the bio-anode and the hydrogen evolving cathode. This solution has the potential to address the issue of cathode deposit and significantly contribute to the regulation of pH levels in the two distinct reactors. However, it comes at the expense of a slight decrease in thermodynamic efficiency. This system is currently being implemented in MECs and should receive significant attention in the future. One important aspect of MEC management is the use of Faradaic yields to detect malfunctions like hydrogen cycling, substantial hydrogen utilization, and fermentation-based hydrogen production. In order to effectively control MEC operation, it is necessary to construct theoretical tools that are straightforward and easy to use. To do this, it would be beneficial to enhance the model by incorporating calculations of local pH and integrating it with experimental measurements of local pH [92]. As a result, it will be possible to identify with accuracy the particular operational parameters of every electrode and provide preliminary suggestions for controlling MEC performance.

An effective control approach can address the significant constraint caused by acidification of the microbial anodes. Intermittently changing the polarity of the system, causing the anode to function as a cathode, can be an effective method for using the protons that have built up inside and surrounding the biofilm [93]. The assessment of polarity inversion's length, frequency, and their impact on MEC production is crucial in this new area of research. The latest developments in reversible electrodes, or more accurately bidirectional electrodes, could serve as a valuable resource for novel approaches in controlling microbial electrolysis cell (MEC) [94]. These approaches would enable the mitigation of microbial acidification by repeated reversal of polarity.

An MEC can be seen as a component of a more intricate production system. MECs can be utilized to produce methane and other complex chemicals through synthesis. Hydrogen serves as an intermediary reactant in this scenario [95]. To maximize efficiency, an electrolyzer designed for massive production must run at a high current density. Because of this, hydrogen must undergo rapid evolution, leading to the formation of large bubbles with brief residence periods. On the other hand, gas–liquid mass transfer limits necessitate the adoption of particular gas–liquid contactors, which contributes to the low reaction rates observed in state-of-the-art microbially catalyzed hydrogenation reactions [96]. Consequently, for best results, the MEC and synthesis reactor must be separated. The microbially catalyzed synthesis would take place in a gas–liquid contactor, with the gas outlet of the MEC serving as the input. The MEC would create hydrogen, either pure or combined with CO_2 and maybe CH_4. This two-step approach optimizes contactor gas–liquid transmission and MEC hydrogen evolution. The need to maximize the reaction rate of MEC synthesis would be eliminated. We remove this limitation from MEC optimization. If the downstream synthesis reactor hydrogenates CO_2, MEC CO_2 production is no longer an issue. With this two-step technique, CO_2 can be introduced into the cathodic compartment for pH regulation without polluting the resultant stream.

6 Research Challenges and Instant Prospects

The MEC approach possesses highly appealing theoretical advantages, such as a conventional cell equilibrium potential of 0.123 V at pH 7.0 and energy and thermal energy outputs. There is still a significant amount of research that needs to be conducted to apply such theoretical benefits to potential technological advances. Primarily, an inhibitory feedback chain occurs in the microbial electrolysis cell (MEC) as the current density rises. This results in a fall in pH at the anode and a boost in pH at the cathode, with mobility unable to counterbalance the pH gradients. Minor fluctuations in pH result in obvious adverse physicochemical impacts. Furthermore, the acidity of the anode hinders or entirely eradicates its electrocatalytic characteristics, whereas the alkalinization of the cathode results in the clogging of the cathode. It is important to consider this adverse response to higher current density while conducting research focused on developing large-scale microbial electrolysis cells (MECs) for industrial use.

6.1 Anodic Materials

Researchers have achieved substantial progress in comprehending the basic processes behind the improvement of bio-anode kinetics, which is a popular area of study. Assuming the techniques may be used reasonably for electrodes with a wide surface area, MEC development will undoubtedly reap the benefits of these advancements. In particular, to create microbial anodes that can withstand acidity, it is important to seek out acidophilic electrogenic microorganisms [97, 98]. When comparing the field of abiotic electrochemical technology, it appears that there are a few areas that have not been thoroughly explored and could be suggested for further investigation. The morphological aspects of the electrode are a crucial factor in enhancing abiotic electrodes. There has been a significant amount of research focused on studying the micro-roughness of electrodes. The field of microbial anodes is still struggling to make use of this fundamental understanding, but it could facilitate quick progress in surface topography optimization. Since the influence of the electrode topography is not immediately apparent in the case of microbial anodes, the scenario is not entirely clear [99]. This effect may be eliminated for completely grown biofilms, according to certain research.

Multiple research investigations argue that this phenomenon loses its significance in well-established biofilms. One can hypothesize that the influence of surface topography, which is evident in the first stage of developing a biofilm, may later be overshadowed by the emergence of constraining factors in fully developed biofilms [100]. This topic warrants additional investigation to arrive at conclusive findings. Although the extensive use of 3D permeable anodes in many research projects, it remains challenging to determine the optimum range of hole diameters that prevent biofilm-induced pore blockage and maximize the benefits of permeability.

The theoretical techniques created in electrochemical technology, specifically for simulating reactive diffusion within pores, can be valuable in providing directions for reflections [101]. This has already been initiated to enhance the optimization of surface micro-structuring. A corresponding theoretical strategy for 3-dimensional porous electrodes could assist with the design of ideal porous frameworks for microbial anodes, especially by considering global acidic conditions. Ultimately, there is a potential for increased and in-depth investigation into metallic electrodes to perhaps eliminate the necessity of a two-dimensional current collector, which is now necessary when using carbon-based electrodes. For example, employing DSA could be an intriguing option for investigation. DSA offers a significant benefit as they are specifically intended for application in massive electrochemical reactors, making them highly suitable for this application. They merit additional investigation in the domain of MECs. Maintaining microbial anodes could be a promising application of a titanium peeling process that has recently demonstrated outstanding performance in PEM water electrolysis [102].

6.2 Cathodic Materials

It appears that working on the bio-anode kinetics is just as important as improving the cathode kinetic mechanism. The typical metallic catalysts that function well in acidic or alkaline environments are rendered ineffective at the almost neutral pH that the bio-anodes require. It is still necessary to look for acidophilic or alkalophilic microbial anodes, but there are not much effective approaches available at this time [97]. As a result, it seems that the cathode kinetics of hydrogen development around a neutral pH require additional exploration. The catalytic evolution of hydrogen by microbes is an interesting field of research. For catalysis of hydrogen evolution at high current densities, it is not an appropriate option, although it can be helpful in anaerobic bio-corrosion. The biofilm might be completely removed from the electrode surface by the high current density necessary for an electrolysis process to be economical [103]. Since there is still need for improvement in the catalysts used for traditional water electrolysis, the quest for catalysts efficient at neutral pH will undoubtedly be a continual effort [104]. The homogeneous catalysis method shows promise for MECs because it is suitable for the pH range. Another important benefit is that it is eco-friendly.

The presence of cations like Ca^{2+} and Mg^{2+} in the cathode area poses a significant challenge due to the alkalinization process. These cations, that are necessary for the maintenance of the bio-anode, accumulate on the surface of the cathode [105, 106]. There is a possibility that the presence of hydrogen bubbles could reduce the accumulation of impurities on the cathode [107]. This positive effect is expected to be much more pronounced when operating at high current density.

Hydrogen bubble production on electrodes has been a research area in electrochemical engineering for decades [108] and currently continues [109, 110], particularly as it affects cathode over-potential. This theoretical data was not previously utilized in the realm of MECs and could be valuable for developing appropriate cathode topologies to reduce deposition. To prevent the formation of hydroxide deposits, it is advisable to restrict the alkalinization of the cathode. In pursuit of this goal, carbon dioxide (CO_2) has been effectively employed to lower the overall pH, but its impact on the cathode over-potential is a somewhat restricted [111]. It would be beneficial to start research using this approach by concentrating on the local effect on the cathode surface [112]. In this case, the goal of employing gas diffusion cathodes might be to disperse CO_2 as near to the cathode surface as feasible rather than to minimize it. Furthermore, the possibility of bicarbonate acting as a homogenous catalyst for the evolution of hydrogen makes this option even more fascinating. Bicarbonate would also get to the anode and lessen anode acidity in the process. Due to all of these factors, it makes sense to do research on the CO_2 delivery to the cathode chamber [113]. However, it pollutes the hydrogen created with CO_2; therefore, it would only be entirely suitable in situations where the purity of the hydrogen is not a crucial factor, such as in the synthesis of methane or other chemicals.

7 Electrolytes to Reduce pH Gradient

The main challenge in the development of large-scale, operational MECs was the inadequate conductivity of ions of the solution. The utilization of highly concentrated saline electrolytes has been contemplated as a potential resolution [114–116]. However, it is important to note that the presence of excessive levels of salts might reduce the ability of the electrolyte to maintain a stable pH, and can specifically worsen the local acidification of the biofilm on the anode. Exploring the adaptation of microbial anodes to high buffer concentrations, rather than NaCl, is a crucial research pathway that can be pursued to enhance MEC performance [117].

In addition, the materials to be used will be more expensive due to the increased chance of corrosion caused by using a hypersaline electrolyte, especially one that contains chlorides. During operation, the applied voltage protects the electrodes from corrosion [118, 119]. However, during maintenance or if polarization is accidentally disturbed, they are still vulnerable. The materials that can be used are severely restricted by the presence of an electrolyte that is extremely saline. Because of these reasons, the electrolyte composition is the main emphasis [120]. By increasing buffering species and tailoring the bio-anode to this solution, the electrolyte composition should be improved.

8 Conclusions

When developing a MEC, we are attempting to achieve dual objectives. In general, a MEC needs to establish the integration of two hostile techniques. Massive electrolyzers are most suited for implementation with pristine electrolytes that have a simple chemical composition, strong ionic conductivity, and elevated pH values. Conversely, microbial processes necessitate culture media that are abundant in chemicals and have a neutral pH. Moreover, they typically do not tolerate solutions with high conductivity. There is a difference between the two types of processes in terms of the spectrum of ideal circumstances. Once the current densities are low enough to sustain metabolic processes, this becomes irrelevant. The disparity between the ideal and actual circumstances causes the system to fail when the engineer attempts to push the process toward industrial production using high current densities. The MEC becomes less efficient as the current density rises due to a negative feedback loop. Can MECs eventually become a technology used in manufacturing? This is still an intriguing problem with a challenging answer. In an effort to find a solution to this mystery, this critical evaluation suggests potential avenues of investigation.

References

1. Götz M, Lefebvre J, Mörs F, Koch AM, Graf F, Bajohr S, Reimert R, Kolb T (2016) Renewable power-to-gas: a technological and economic review. Renew Energy 85:1371–1390
2. Khan HR, Akram R, Aamir M, Malik MA, Tahir AA, Choudhary MA, Akhtar J (2023) Investigations of photoelectrochemical performance of polycrystalline Bi-doped ZnO thin films. J Phys Chem Solids 181:111529
3. Sarangi PK, Nanda S (2020) Biohydrogen production through dark fermentation. Chem Eng Technol 43:601–612
4. Zhang T, Jiang D, Zhang H, Jing Y, Tahir N, Zhang Y, Zhang Q (2020) Comparative study on bio-hydrogen production from corn stover: photo-fermentation, dark-fermentation and dark-photo co-fermentation. Int J Hydrogen Energy 45:3807–3814
5. Osman AI, Deka TJ, Baruah DC, Rooney D (2020) Critical challenges in biohydrogen production processes from the organic feedstocks. Biomass Conv Bioref, 1–19
6. Yadav M, Sehrawat N, Singh M, Kumar V, Sharma AK, Kumar S (2021) Thermophilic microbes-based fuel cells: an eco-friendly approach for sustainable energy production. In: Bioremediation for environmental sustainability. Elsevier, pp 235–246
7. Zhang Y, Liu M, Zhou M, Yang H, Liang L, Gu TJR (2019) Microbial fuel cell hybrid systems for wastewater treatment and bioenergy production: synergistic effects, mechanisms and challenges. Renew Sustain Energy Rev 103:13–29
8. Nawaz A, ul Haq I, Qaisar K, Gunes B, Raja SI, Mohyuddin K, Amin H (2022) Microbial fuel cells: insight into simultaneous wastewater treatment and bioelectricity generation. Process Saf Environ Prot 161:357–373
9. Nercessian O, Parot S, Délia M-L, Bergel A, Achouak WJPO (2012) Harvesting electricity with Geobacter bremensis isolated from compost. PLoS ONE 7:e34216
10. Jeremiasse AW, Hamelers HV, Buisman C (2010) Microbial electrolysis cell with a microbial biocathode. Bioelectrochemistry 78:39–43
11. Lim SS, Fontmorin J-M, Izadi P, Daud WRW, Scott K, Yu E (2020) Impact of applied cell voltage on the performance of a microbial electrolysis cell fully catalysed by microorganisms. Int J Hydrogen Energy 45:2557–2568
12. Guiral-Brugna M, Giudici-Orticoni M-T, Bruschi M, Bianco P (2001) Electrocatalysis of the hydrogen production by [Fe] hydrogenase from Desulfovibrio vulgaris Hildenborough. J Electroanal Chem 510:136–143
13. Yaqoob AA, Al-Zaqri N, Yaakop AS, Umar K (2022) Potato waste as an effective source of electron generation and bioremediation of pollutant through benthic microbial fuel cell. Sustain Energy Technol Assess 53:102560
14. Yaqoob AA, Bakar MABA, Kim HC, Ahmad A, Alshammari MB, Yaakop AS (2022) Oxidation of food waste as an organic substrate in a single chamber microbial fuel cell to remove the pollutant with energy generation. Sustain Energy Technol Assess 52:102282
15. Idris MO, Noh NAM, Ibrahim MNM, Yaqoob AA (2023) Sustainable microbial fuel cell functionalized with a bio-waste: a feasible route to formaldehyde bioremediation along with bioelectricity generation. Chem Eng J 455:140781
16. Yaqoob AA, Fadzli FS, Ibrahim MNM, Yaakop AS (2023) Benthic microbial fuel cells: a sustainable approach for metal remediation and electricity generation from sapodilla waste. Int J Environ Sci Technol 20(4):3927–3940
17. Idris MO, Kim HC, Yaqoob AA, Ibrahim MNM (2022) Exploring the effectiveness of microbial fuel cell for the degradation of organic pollutants coupled with bio-energy generation. Sustain Energy Technol Assess 52:102183
18. Rousseau R, Etcheverry L, Roubaud E, Basséguy R, Délia M-L, Bergel A (2020) Microbial electrolysis cell (MEC): strengths, weaknesses and research needs from electrochemical engineering standpoint. Appl Energy 257:113938
19. Liu H, Hu H, Chignell J, Fan Y (2010) Microbial electrolysis: Novel technology for hydrogen production from biomass. Biofuels 1:129–142

20. Rozendal RA, Hamelers HV, Euverink GJ, Metz SJ, Buisman C (2006) Principle and perspectives of hydrogen production through biocatalyzed electrolysis. Int J Hydrogen Energy 31:1632–1640
21. Strycharz SM, Malanoski AP, Snider RM, Yi H, Lovley DR, Tender L (2011) Application of cyclic voltammetry to investigate enhanced catalytic current generation by biofilm-modified anodes of Geobacter sulfurreducens strain DL1 vs. variant strain KN400. Energy Environ Sci 4:896–913
22. Paitier A, Godain A, Lyon D, Haddour N, Vogel TM, Monier J (2017) Microbial fuel cell anodic microbial population dynamics during MFC start-up. Biosens Bioelectron 92:357–363
23. Choi S, Kim JR, Cha J, Kim Y, Premier GC, Kim C (2013) Enhanced power production of a membrane electrode assembly microbial fuel cell (MFC) using a cost effective poly [2, 5-benzimidazole](ABPBI) impregnated non-woven fabric filter. Bioresour Technol 128:14–21
24. Zhu X, Yates MD, Hatzell MC, Ananda Rao H, Saikaly PE, Logan B (2014) Microbial community composition is unaffected by anode potential. Envoron Sci Technol 48:1352–1358
25. Aiyer K (2020) How does electron transfer occur in microbial fuel cells? World J Microbiol Biotechnol 36:19
26. Chabert N, Ali OA, Achouak W (2015) All ecosystems potentially host electrogenic bacteria. Bioelectochemistry 106:88–96
27. G. Zhen, X. Lu, G. Kumar, P. Bakonyi, K. Xu, Y. Zhao (2017) Microbial electrolysis cell platform for simultaneous waste biorefinery and clean electrofuels generation: current situation, challenges and future perspectives. Progr Energy Combustion Sci 63:119–145
28. Torres CI, Marcus AK, Lee H-S, Parameswaran P, Krajmalnik-Brown R, Rittmann B (2010) A kinetic perspective on extracellular electron transfer by anode-respiring bacteria. FEMS Microbial Rev 34:3–17
29. Ibrahim MNM, Rodríguez-Couto S, Ahmad A (2021) Preparation, characterization, and application of modified carbonized lignin as an anode for sustainable microbial fuel cell. Process Saf Environ Prot 155:49–60
30. Hamelers HV, Ter Heijne A, Stein N, Rozendal RA, Buisman C (2011) Butler–Volmer–Monod model for describing bio-anode polarization curves. Bioresour Technol 102:381–387
31. Grattieri M, Minteer S (2018) Microbial fuel cells in saline and hypersaline environments: advancements, challenges and future perspectives. Bioelectrochemistry 120:127–137
32. Lee H (2018) Electrokinetic analyses in biofilm anodes: Ohmic conduction of extracellular electron transfer. Bioresour Technol 256:509–514
33. Yasri NG, Gunasekaran S (2017) Electrochemical technologies for environmental remediation, pp 5–73
34. Yaakop AS, Ahmad A (2021) Application of microbial fuel cells energized by oil palm trunk sap (OPTS) to remove the toxic metal from synthetic wastewater with generation of electricity. Appl Nanosci 11(6):1949–1961
35. Reguera G, McCarthy KD, Mehta T, Nicoll JS, Tuominen MT, Lovley D (2005) Extracellular electron transfer via microbial nanowires. Nature 435:1098–1101
36. Zhou M, Chi M, Luo J, He H, Jin T (2011) An overview of electrode materials in microbial fuel cells. J Power Sources 196:4427–4435
37. Santoro C, Arbizzani C, Erable B, Ieropoulos I (2017) Microbial fuel cells: From fundamentals to applications. A review. J Power Sources 356:225–244
38. Pocaznoi D, Calmet A, Etcheverry L, Erable B, Bergel A (2012) Stainless steel is a promising electrode material for anodes of microbial fuel cells. Energy Environ Sci 5:9645–9652
39. Erable B, Byrne N, Etcheverry L, Achouak W, Bergel A (2017) Single medium microbial fuel cell: stainless steel and graphite electrode materials select bacterial communities resulting in opposite electrocatalytic activities. Int J Hydrogen Energy 42:26059–26067
40. Baudler A, Langner M, Rohr C, Greiner A, Schröder U (2017) Metal–polymer hybrid architectures as novel anode platform for microbial electrochemical technologies. ChemSusChem 10:253–257
41. Trasatti A (2000) Electrocatalysis: understanding the success of DSA®. Electrochimica Acta 45:2377–2385

42. Baudler A, Schmidt I, Langner M, Greiner A, Schröder U (2015) Does it have to be carbon? Metal anodes in microbial fuel cells and related bioelectrochemical systems. Energy Environ Sci 8:2048–2055
43. Wagner RC, Porter-Gill S, Logan B (2012) Immobilization of anode-attached microbes in a microbial fuel cell. AMB Express 2:1–6
44. Logan B, Cheng S, Watson V, Estadt G (2007) Graphite fiber brush anodes for increased power production in air-cathode microbial fuel cells. Environ Sci Technol 41:3341–3346
45. Yasri NG, Nakhla G (2016) Electrochemical behavior of anode-respiring bacteria on doped carbon electrodes. ACS Appl Mater Interfaces 8:35150–35162
46. Chen S, He G, Liu Q, Harnisch F, Zhou Y, Chen Y, Hanif M, Wang S, Peng X, Hou H (2012) Layered corrugated electrode macrostructures boost microbial bioelectrocatalysis. Energy Environ Sci 5:9769–9772
47. Rafatullah M, Chua YS, Ahmad A, Umar K (2020) Recent advances in anodes for microbial fuel cells: an overview. Materials 13(9):2078
48. Chae K-J, Choi M-J, Kim K-Y, Ajayi F, Chang I-S, Kim I (2010) Selective inhibition of methanogens for the improvement of biohydrogen production in microbial electrolysis cells. Int J Hydrogen Energy 35:13379–13386
49. Lu L, Xing D, Liu B, Ren N (2012) Enhanced hydrogen production from waste activated sludge by cascade utilization of organic matter in microbial electrolysis cells. Water Res 46:1015–1026
50. Fu Q, Wang D, Li X, Yang Q, Xu Q, Ni B-J, Wang Q, Liu X (2021) Towards hydrogen production from waste activated sludge: principles, challenges and perspectives. Renew Sustain Energy Rev 135:110283
51. Li X-H, Liang D-W, Bai Y-X, Fan Y-T, Hou H (2014) Enhanced H_2 production from corn stalk by integrating dark fermentation and single chamber microbial electrolysis cells with double anode arrangement. Int J Hydrogen Energy 39:8977–8982
52. Liang D-W, Peng S-K, Lu S-F, Liu Y, Lan F, Xiang Y (2011) Enhancement of hydrogen production in a single chamber microbial electrolysis cell through anode arrangement optimization. Bioresour Technol 102:10881–10885
53. Wang H, Ren Z (2013) A comprehensive review of microbial electrochemical systems as a platform technology. Biotecnol Adv 31:1796–1807
54. Varanasi JL, Veerubhotla R, Pandit S, Das D (2019) Biohydrogen production using microbial electrolysis cell: recent advances and future prospects, pp 843–869
55. Sleutels TH, Hamelers HV, Buisman C (2011) Effect of mass and charge transport speed and direction in porous anodes on microbial electrolysis cell performance. Bioresour Technol 102:399–403
56. Dhand V, Rao MV, Prasad J, Mittal G, Rhee KY, Kim HJ, Jung D (2014) Carbon nanospheres synthesized via solution combustion method: their application as an anode material and catalyst for hydrogen production. Carbon Lett 15:198–202
57. Jeremiasse AW (2011) Cathode innovations for enhanced H_2 production through microbial electrolysis. Wageningen University and Research
58. Desmarais M, Kraljić D (2021) Clean biological hydrogen production in microbial electrolysis cell. In: Waste biorefinery. Elsevier, pp 491–505
59. Nissim R, Compton R (2013) Superoxide generation from the reduction of oxygen at the carbon–oil–water triple phase boundary. Phys Chem Chem Phys 15:11918–11925
60. Cheng S, Liu H, Logan B (2006) Power densities using different cathode catalysts (Pt and CoTMPP) and polymer binders (Nafion and PTFE) in single chamber microbial fuel cells. Environ Sci Technol 40:364–369
61. Rozendal RA, Hamelers HV, Molenkamp RJ, Buisman C (2007) Performance of single chamber biocatalyzed electrolysis with different types of ion exchange membranes. Water Res 41:1984–1994
62. Ribot-Llobet E, Nam J-Y, Tokash JC, Guisasola A, Logan B (2013) Assessment of four different cathode materials at different initial pHs using unbuffered catholytes in microbial electrolysis cells. Int J Hydrogen Energy 38:2951–2956

63. Li F, Liu W, Sun Y, Ding W, Cheng S (2017) Enhancing hydrogen production with Ni–P coated nickel foam as cathode catalyst in single chamber microbial electrolysis cells. Int J Hydrogen Energy 42:3641–3646
64. Selembo PA, Merrill MD, Logan B (2009) The use of stainless steel and nickel alloys as low-cost cathodes in microbial electrolysis cells. J Power Sources 190:271–278
65. Selembo PA, Merrill MD, Logan B (2010) Hydrogen production with nickel powder cathode catalysts in microbial electrolysis cells. Int J Hydrogen Energy 35:428–437
66. Vogt H, Balzer R (2005) The bubble coverage of gas-evolving electrodes in stagnant electrolytes. Electrochim Acta 50:2073–2079
67. Bumroongsakulsawat P, Kelsall G (2015) Tinned graphite felt cathodes for scale-up of electrochemical reduction of aqueous CO_2. Electrochimica Acta 159:242–251
68. Hurlen T, Gunvaldsen S, Blaker F (1984) Effects of buffers on hydrogen evolution at iron electrodes. Electrochim Acta 29:1163–1164
69. Da Silva S, Basséguy R, Bergel A (2004) Electrochemical deprotonation of phosphate on stainless steel. Electrochim Acta 49:4553–4561
70. Roubaud E, Lacroix R, Da Silva S, Bergel A, Basséguy R, Erable B (2018) Catalysis of the hydrogen evolution reaction by hydrogen carbonate to decrease the voltage of microbial electrolysis cell fed with domestic wastewater. Electrochim Acta 275:32–39
71. Munoz LDS, Bergel A, Basséguy R (2007) Role of the reversible electrochemical deprotonation of phosphate species in anaerobic biocorrosion of steels. Corrosion Sci 49:3988–4004
72. Rousseau R, Ketep SF, Etcheverry L, Délia M-L, Bergel A (2020) Microbial electrolysis cell (MEC): a step ahead towards hydrogen-evolving cathode operated at high current density. Bioresour Technol Rep 9:100399
73. Raveendran A, Chandran M, Dhanusuraman R (2023) A comprehensive review on the electrochemical parameters and recent material development of electrochemical water splitting electrocatalysts. RSC Adv 13:3843–3876
74. Hou Y, Zhang B, Wen Z, Cui S, Guo X, He Z, Chen J (2014) A 3D hybrid of layered MoS_2/nitrogen-doped graphene nanosheet aerogels: an effective catalyst for hydrogen evolution in microbial electrolysis cells. J Matr Chem A 2:13795–13800
75. Tenca A, Cusick RD, Schievano A, Oberti R, Logan B (2013) Evaluation of low cost cathode materials for treatment of industrial and food processing wastewater using microbial electrolysis cells. Int J Hydrogen Energy 38:1859–1865
76. Ghasemi M, Daud WRW, Hassan SH, Oh S, Ismail M, Rahimnejad M, Jahim J (2013) Nanostructured carbon as electrode material in microbial fuel cells: a comprehensive review. J Alloys Compd 580:245–255
77. Abrol I, Yadav J, Massoud F (1988) Salt-affected soils and their management. Food & Agriculture Org
78. Zhang Y, Angelidaki I (2012) Energy recovery from waste streams with microbial fuel cell (MFC)-based technologies
79. Kadier A, Simayi Y, Abdeshahian P, Azman NF, Chandrasekhar K, Kalil M (2016) A comprehensive review of microbial electrolysis cells (MEC) reactor designs and configurations for sustainable hydrogen gas production. Alexandria Eng J 55:427–443
80. Chaudhary A, Donaldson J, Grimes S, Yasri N (2000) Separation of nickel from cobalt using electrodialysis in the presence of EDTA. J Appl Electrochem 30:439–445
81. Merrill MD, Logan B (2009) Electrolyte effects on hydrogen evolution and solution resistance in microbial electrolysis cells. J Power Sources 191:203–208
82. Omidi H, Sathasivan A (2013) Optimal temperature for microbes in an acetate fed microbial electrolysis cell (MEC). Int Biodeterioration Biodegradation 85:688–692
83. Dange P, Pandit S, Jadhav D, Shanmugam P, Gupta PK, Kumar S, Kumar M, Yang Y-H, Bhatia S (2021) Recent developments in microbial electrolysis cell-based biohydrogen production utilizing wastewater as a feedstock. Sustainability 13:8796
84. Jensen LS, Kaul C, Juncker NB, Thomsen MH, Chaturvedi T (2022) Biohydrogen production in microbial electrolysis cells utilizing organic residue feedstock: a review. Energies 15:8396

85. Zhu G, Wu T, Jha AK, Zou R, Liu L, Huang X, Liu C, Treatment W (2014) Review of bio-hydrogen production and new application in the pollution control via microbial electrolysis cell. Desal Water Treatment 52:5413–5421
86. Madram AR, Fathollahi Zonouz A, Pouretedal H (2015) Effect of temperature on kinetics of the hydrogen evolution reaction on Ni-PC cathodes in alkaline solution. Iran J Hydrogen Fuel Cell 1:215–222
87. Call D, Logan B (2008) Hydrogen production in a single chamber microbial electrolysis cell lacking a membrane. Environ Sci Technol 42:3401–3406
88. Wang K, Sheng Y, Cao H, Yan K, Zhang Y (2017) Impact of applied current on sulfate-rich wastewater treatment and microbial biodiversity in the cathode chamber of microbial electrolysis cell (MEC) reactor. Chem Eng J 307:150–158
89. Escapa A, Mateos R, Martínez E, Blanes J, Reviews SE (2016) Microbial electrolysis cells: an emerging technology for wastewater treatment and energy recovery. From laboratory to pilot plant and beyond. Renew Sustain Energy Rev 55:942–956
90. Kim IS, Yang E, Choi M-J, Chae K (2015) Bioelectrochemical production of hydrogen from organic waste, pp 249–281
91. Luo Y, Zhang Z, Yang F, Li J, Liu Z, Ren W, Zhang S, Liu B, Science E (2021) Stabilized hydroxide-mediated nickel-based electrocatalysts for high-current-density hydrogen evolution in alkaline media. Energy Environ Sci 14:4610–4619
92. Savla N, Khilari S, Pandit S, Jung S (2020) Effective cathode catalysts for oxygen reduction reactions in microbial fuel cell, pp 189–210
93. Jiang Y, Liang P, Zhang C, Bian Y, Sun X, Zhang H, Yang X, Zhao F, Huang X (2016) Periodic polarity reversal for stabilizing the pH in two-chamber microbial electrolysis cells. Appl Energy 165:670–675
94. Wang Y-X, Hou N, Liu X-L, Mu Y (2022) Advances in interfacial engineering for enhanced microbial extracellular electron transfer. Bioresour Technol 345:126562
95. Blanchet E, Duquenne F, Rafrafi Y, Etcheverry L, Erable B, Bergel A (2015) Importance of the hydrogen route in up-scaling electrosynthesis for microbial CO_2 reduction. Energy & Environ Sci 8:3731–3744
96. Zabranska J, Pokorna D (2018) Bioconversion of carbon dioxide to methane using hydrogen and hydrogenotrophic methanogens. Biotechnol Adv 36:707–720
97. Dopson M, Ni G, Sleutels T (2016) Possibilities for extremophilic microorganisms in microbial electrochemical systems. FEMS Microbiol Rev 40:164–181
98. Shrestha N, Chilkoor G, Vemuri B, Rathinam N, Sani RK, Gadhamshetty V (2018) Extremophiles for microbial-electrochemistry applications: a critical review. Bioresour Technol 255:318–330
99. Champigneux P, Delia M-L, Bergel A (2018) Impact of electrode micro-and nano-scale topography on the formation and performance of microbial electrodes. Biosens Bioelectron 118:231–246
100. Champigneux P, Renault-Sentenac C, Bourrier D, Rossi C, Delia M-L, Bergel A (2019) Effect of surface roughness, porosity and roughened micro-pillar structures on the early formation of microbial anodes. Bioelectrochemistry 128:17–29
101. Ryan EM, Mukherjee P (20196) Mesoscale modeling in electrochemical devices—a critical perspective. Progr Energy Combustion Sci 71:118–142
102. Bystron T, Vesely M, Paidar M, Papakonstantinou G, Sundmacher K, Bensmann B, Hanke-Rauschenbach R, Bouzek K (2018) Enhancing PEM water electrolysis efficiency by reducing the extent of Ti gas diffusion layer passivation. J Appl Electrochem 48:713–723
103. Roy M, Yadav R, Chiranjeevi P, Patil S (2021) Direct utilization of industrial carbon dioxide with low impurities for acetate production via microbial electrosynthesis. Bioresour Technol 320:124289
104. Kim J, Jung H, Jung S-M, Hwang J, Kim DY, Lee N, Kim K-S, Kwon H, Kim Y-T, Han J (2020) Tailoring binding abilities by incorporating oxophilic transition metals on 3D nanostructured Ni arrays for accelerated alkaline hydrogen evolution reaction. J Am Chem Soc 143:1399–1408

105. An J, Li N, Wan L, Zhou L, Du Q, Li T, Wang X (2017) Electric field induced salt precipitation into activated carbon air-cathode causes power decay in microbial fuel cells. Water Res 123:369–377
106. Sugnaux M, Happe M, Cachelin CP, Gasperini A, Blatter M, Fischer F (2017) Cathode deposits favor methane generation in microbial electrolysis cell. Chem Eng J 324:228–236
107. Katuri KP, Werner CM, Jimenez-Sandoval RJ, Chen W, Jeon S, Logan BE, Lai Z, Amy GL, Saikaly P (2014) A novel anaerobic electrochemical membrane bioreactor (AnEMBR) with conductive hollow-fiber membrane for treatment of low-organic strength solutions. Environ Sci Technol 48:12833–12841
108. Hreiz R, Abdelouahed L, Fünfschilling D, Lapicque F (2015) Electrogenerated bubbles induced convection in narrow vertical cells: a review. Chem Eng Res Des 100:268–281
109. Chandran P, Bakshi S, Chatterjee D (2015) Study on the characteristics of hydrogen bubble formation and its transport during electrolysis of water. Chem Eng Sci 138:99–109
110. Khalighi F, Deen NG, Tang Y, Vreman A (2023) Hydrogen bubble growth in alkaline water electrolysis: an immersed boundary simulation study. Chem Eng Sci 267:118280
111. Rossi R, Baek G, Logan B (2021) Vapor-fed cathode microbial electrolysis cells with closely spaced electrodes enables greatly improved performance. Environ Sci Technol 56:1211–1220
112. Diaz-Sainz D, Alvarez-Guerra M, Irabien A (2022) Continuous electroreduction of CO_2 towards formate in gas-phase operation at high current densities with an anion exchange membrane. J CO_2 Util 56:101822
113. Korth B, Kuchenbuch A, Harnisch F (2020) Availability of hydrogen shapes the microbial abundance in biofilm anodes based on geobacter enrichment. ChemElectroChem 7:3720–3724
114. Rodríguez-Couto S (2020) Development and modification of materials to build cost-effective anodes for microbial fuel cells (MFCs): an overview. Biochem Eng J 164:107779
115. Logan BE, Call D, Cheng S, Hamelers HV, Sleutels TH, Jeremiasse AW, Rozendal R (2008) Microbial electrolysis cells for high yield hydrogen gas production from organic matter. Environ Sci Technol 42:8630–8640
116. Logan BE (2008) Microbial fuel cells. Wiley.
117. Lusk BG, Peraza I, Albal G, Marcus AK, Popat SC, Torres C (2018) pH dependency in anode biofilms of Thermincola ferriacetica suggests a proton-dependent electrochemical response. J Am Chem Soc 140:5527–5534
118. Antolini E (2015) Composite materials for polymer electrolyte membrane microbial fuel cells. Biosens Bioelectron 69:54–70
119. Yang G, Erbay C, Yi S-I, de Figueiredo P, Sadr R, Han A, Yu C (2016) Bifunctional nano-sponges serving as non-precious metal catalysts and self-standing cathodes for high performance fuel cell applications. Nano Energy 22:607–614
120. Nam J-Y, Logan B (2012) Optimization of catholyte concentration and anolyte pHs in two chamber microbial electrolysis cells. Int J Hydrogen Energy 37:18622–18628

Microbiology in Focus: Unveiling the Dynamics of Microbial Electrolysis Cells

Thamby Rajah Mahendran, Ooi Kah Hong, and Amira Suriaty Yaakop

Abstract In the intricate realm of microbial electrolysis cells (MECs), the synergy between microbiology and electrochemistry unfolds a fascinating narrative. This chapter delves into the pivotal role of microbiology in the study of MECs, unraveling the essential interplay between microorganisms and electrochemical processes. In this chapter, we will explore the microbial communities thriving within MECs, their metabolic activities, and their profound impact on the efficiency and performance of this groundbreaking technique. Join us in unlocking the secrets of microbial intricacies that underpin the success of microbial electrolysis, as we bridge the gap between microbiology and electrochemistry in this vital exploration.

Keywords Microbial electrolysis cells · Electrogenic bacteria · Electrosynthesis cell · 16S rRNA sequencing

1 Introduction

1.1 Understanding Microbial Community Structures and Functions in Microbial Electrosynthesis Cell

Microbial communities involved in microbial electrosynthesis are diverse and frequently include electroactive bacteria capable of extracellular electron transfer (EET). Electroactive bacteria play a key role by generating electron through metabolic process and establishing direct or indirect electron transfer with the electrode which facilitates the electricity generation [101]. Concurrently, other microbial community members which do not directly involve in the EET process may contribute in subsequent metabolic pathways leading to the production of current, specific chemicals or fuels [58, 61, 75]. In summary, the microbial community's metabolic pathways dictate the type and yield of the final products generated from the

T. R. Mahendran · O. K. Hong · A. S. Yaakop (✉)
School of Biological Sciences, Universiti Sains Malaysia, 11800 Minden, Penang, Malaysia
e-mail: amirasuriaty@usm.my

A. A. Yaqoob and A. Ahmad (eds.), *Microbial Electrolysis Cell Technology*, Sustainable Materials and Technology, https://doi.org/10.1007/978-981-97-3356-9_7

electrochemical reactions. Consequently, it is imperative to comprehend the roles of diverse microbes within the community to optimize the overall process and enhance the yields of the products. Various techniques are employed to study the diversity, structure, and role of various microbes involved in a MEC.

1.1.1 16S rRNA Sequencing

16S rRNA gene sequencing stands out as a widely employed technique for examining and characterizing microbial communities within a microbial electrolysis cell (MEC) configuration [21, 51, 77]. This gene, responsible for encoding the 16S ribosomal RNA (rRNA) molecule in bacteria and archaea, is integral to the small subunit of the prokaryotic ribosome which is responsible for protein synthesis [38]. The 16S rRNA gene is ubiquitous in the genomes of all bacteria and archaea and features both conserved regions (unchanged part of sequence vital for its ribosomal function) and multiple hypervariable regions (part of sequence which differ across species). The distinctive variation in specific regions is what makes the 16S rRNA gene a valuable molecular marker, enabling the distinguishing of diverse microbial species [16]. In a study, the 16S rRNA gene amplification along with Illumina sequencing was employed to examine the structure of the microbial community on the anode surface within four dual-chamber bioelectrochemical systems. The key findings of this study suggest that the microbial community structures in BES are intricate, with MFCs exhibiting a more diverse community compared to MECs. Despite differences in the community structure between MFCs and MECs, both shared the same dominant species. Furthermore, the analysis of microbial community structure revealed that functional bacteria from genera such as *Desulfovibrio*, *Butyricicoccus*, *Petrimonas*, and *Propionivibrio* were predominant in the anodic biofilm of MFCs. In contrast, *Desulfovibrio*, *Acidaminococcus*, *Desulfocapsa*, and *Acetobacterium* were predominant in the anodic biofilm of MECs [4].

1.1.2 DGGE Profiling

Denaturing gradient gel electrophoresis (DGGE) is a method employed to segregate a blend of DNA fragments based on their respective melting points, enabling the analysis of culturable and unculturable microbial communities [88]. In DGGE profiling for microbial identification, universal genes such as the 16S rRNA gene are usually targeted. Despite this gene being universally present and having similar base pair size across most microbes, variations in GC content and composition among different microbial rRNAs lead to distinct denaturation properties in these DNA molecules [100]. Therefore, the resulting pattern derived from the migration of DNA fragments in DGGE profiling varies and can serve as a means to discern the diversity and composition of microbial communities found in diverse samples, including environments or experimental reactors like MEC. DGGE profiling has been used in various studies to determine the diversity and abundance of bacteria at the MEC anode particularly in research investigating the impact of specific treatments or modifications on the bacterial community. [20, 21, 94].

1.1.3 Metagenomic Analysis

Metagenomic analysis enables researchers to examine the genetic material (DNA) of all microorganisms present in a sample, offering insights into their diversity and richness of microbial communities and their potential functional roles and interactions in diverse ecosystems [15, 29, 53, 93]. The hypervariable sections of the 16S rRNA gene, sometimes referred to as variable regions or V regions, are a common target of metagenomic investigation. These sections provide the necessary sequence diversity needed to distinguish between different kinds of microorganisms. For amplification and sequencing, researchers frequently choose particular hypervariable areas (e.g., V3–V4) to assess the richness and diversity of microbial communities in environmental samples [49, 66]. In an investigation examining the impact of altering anode potential on the microbial communities on biofilms within a dual-chambered MEC, the primer set consisting of 341F and 805R was applied to amplify the V3–V4 regions of the 16S rRNA gene. Additionally, qPCR analysis was employed to quantify the total microbial count. The analysis indicated that shifting the anode potential toward more positive values had a significant influence on microbial communities. This effect was characterized by the suppression of anode-respiring bacteria, *Geobacter*, as well as the hydrolytic *Bacteroides*. Meanwhile, there was a selective enrichment of fermentative bacteria such as the *Paludibacter* species in the biofilms [91].

1.1.4 Metatranscriptomics

Metatranscriptomics is a molecular biology technique used to investigate the collective RNA transcripts within a microbial community. This offers insights into the functional profile of its members [2]. In the context of MEC, metatranscriptomics is utilized to decipher gene expression patterns of microorganisms participating in electrochemical processes, unveiling their roles in electron transfer, substrate utilization, and other pivotal activities. This knowledge is essential for fine-tuning MEC performance, pinpointing critical contributors to electrochemical processes and formulating strategies to optimize bioenergy generation or the enhanced production of targeted compounds [35, 36]. In a MEC study to enhance biogas upgrading by direct electromethanogenesis, transcriptomics analysis revealed that *Methanothrix* species present on the cathode and in the bulk sludge used different pathways, namely CO_2 reduction pathway and acetate decarboxylation pathway, respectively, for production of methane. The application of transcriptomics analysis in this study offered a glimpse into the variations in *Methanothrix* metabolism between growth on the cathode and planktonic growth in the surrounding sludge which may contribute to enhancing the direct electromethanogenesis process [60].

1.1.5 Metaproteomics

Metaproteomics analysis focuses on detailed analysis of proteins present in target microbial communities. In MEC, metaproteomics analysis provides means for exploring the functional aspects of microbial communities by identifying and characterizing proteins synthesized which could provide insight on the impact of various experimental settings or conditions and their implication on the MEC process [17, 32, 56]. In a study, proteomics analysis was employed to elucidate proteins that facilitate methane production in the presence of ammonia inhibition within MEC system coupled with anaerobic digestion. The findings indicated an up-regulation of proteins associated with direct electron transfer (DIET), including cytochrome C, and proteins related to ammonium detoxification. Conversely, proteins involved in ammonia transfer were found to be down-regulated [98].

1.1.6 T-RFLP

Terminal restriction fragment length polymorphism (T-RFLP) is a molecular technique employed for the analysis and characterization of microbial communities in various environments based on the variation in the gene of interest, often the16S rRNA gene in the case of bacterial sample.

In T-RFLP, the DNA is amplified with fluorescently labeled primer during the PCR step. The use of fluorescently labeled primers facilitates the visualization and examination of the resulting fragments. Subsequently, the amplified DNA is subjected to treatment with restriction enzymes and terminal restriction fragments (TRFs) which cleave DNA at specific recognition sites, thereby creating fragments of diverse lengths based on specific sequences. Following this enzymatic digestion, the DNA fragments are separated by size through gel electrophoresis, producing a distinctive pattern of bands that represents the various DNA fragments resulting from the restriction enzyme activity. The ensuing pattern, often referred to as a "fingerprint," is analyzed to ascertain the lengths of terminal restriction fragments (TRFs), each of which corresponds to a distinct microbial population within the original sample. The analysis involves comparing the observed TRF lengths to a database containing information on known microbial species for identification [23, 72].

In MEC studies, T-RFLP is utilized to evaluate the variability and proportional presence of microorganisms communities. Microbial communities engaged in electrochemical processes, aiding researchers in comprehending the composition and dynamics of microbial populations within the anodic biofilm of MECs. Monitoring changes in T-RFLP profiles over time or under diverse experimental conditions offers insights into how microbial communities adapt to variations in operational parameters or substrates in MECs. In a research endeavor exploring the influence of methanogens on biofilms predominantly composed of *Geobacter* spp. in microbial electrosynthesis cells (MECs), T-RFLP was utilized to track alterations in the composition of electroactive biofilms triggered by anaerobic digestion effluents. The outcomes, coupled with the visual examination of biofilm electrodes at the conclusion of the experiment, confirm that the inhibitory impact of anaerobic digestion effluent is more noticeable in terms of both integrity and community structure in young biofilms compared to their mature counterparts [69].

1.1.7 Microscopy Techniques

Different microscopy techniques are applied in microbial electrolysis cell (MEC) configurations, primarily aimed at examining microbial communities, processes within the cells, biofilm structure, and more. Typically, the main aim is to acquire a better understanding of microbial interactions, either among themselves or with the electrodes. The most commonly used microscopy techniques in MEC studies are scanning electron microscopy (SEM), confocal laser scanning microscopy (CLSM), and fluorescence in situ hybridization (FISH). SEM utilizes a focused electron beam for scanning a specimen's surface, creating high-resolution images through the detection of electron–atom interactions. This technique is commonly employed to examine the morphology and spatial distribution of microbial biofilms on electrodes in MECs, yielding detailed insights into microbial community structures and the efficiency of the formed biofilm [86]. In similar fashion, CLSM employs laser light to illuminate fluorescently labeled specimens, generating three-dimensional images which could then be analyzed to understand the microbial community dynamics or evaluate the properties of the biofilm formed at the anode [70]. On the other hand, FISH utilizes specific fluorescent probes to selectively hybridize specific DNA sequences within the sample. Subsequently, the sample undergoes examination using a fluorescence microscope. If the sample contains the target sequences, they emit fluorescent signals upon excitation by specific wavelengths of light. These fluorescence signals are then captured using fluorescence microscopy and subjected to analysis to ascertain the presence, abundance, and spatial distribution of the targeted sequences, thereby revealing the composition of microbial communities in a MEC reactor [72, 82].

In a research investigation, the scanning electron microscope (SEM) was employed to assess the configuration and deposition of biofilm on the anodic surface of a biofilm-based microbial electrosynthesis cell (MEC) that operated continuously for over 100 days. The analysis revealed that the majority of carbon felt fibers were encased in amorphous layers, signifying the adhesion of a microbial biofilm to the electrode material. Furthermore, the biofilm was predominantly colonized by rod-shaped structures resembling Deltaproteobacteria, aligning with the results of the microbial analysis conducted [13]. In a separate study investigating the breakdown of 2,4-dichloronitrobenzene (DClNB) in a single-chamber microbial electrosynthesis cell (MEC), the utilization of a fluorescence staining method and confocal laser scanning microscopy allowed for the examination of live and deceased cell distributions. The results indicated that the adaptation of the bioanode substantially increased the percentage of live cells within the biofilm, contributing to the stabilization of the anode's performance when exposed to a DClNB shock [62]. In a separate study, FISH analysis alongside 16S rRNA gene clone libraries was employed to explore the microbial communities on anodes during the transition from microbial fuel cell to microbial electrolysis cell operation using two different wastewaters, namely potato and dairy. The examination indicated that the anode biofilm in the microbial electrolysis cell, supplied with potato wastewater, displayed a significant concentration of *Geobacter sulfurreducens* cells. In contrast, the system fed with dairy manure wastewater demonstrated a lower abundance of *G. sulfurreducens*. This underscores the pivotal role of anodic *Geobacteraceae* populations in both the generation of current in MFCs and, notably, in gas production in MECs [51]. In another study, all three microscopy techniques, namely CLSM, SEM, and FISH analysis, were conducted to study the evolution of *D. acetexigens* and *G. sulfurreducens* within the anodic biofilm of MECs supplied with acetate or domestic wastewater. The results of this investigation indicate that in MECs supplied with domestic wastewater, *D. acetexigens* and *G. sulfurreducens* coexist despite the latter dominance. Moreover, augmenting MECs with both *D. acetexigens* and *G. sulfurreducens* could be a promising approach to bolster functional redundancy within the anodic biofilm, thereby enhancing the functional stability of MECs [80].

1.1.8 DNA Stable Isotope Probing (DNA-SIP)

DNA stable isotope probing (DNA-SIP) is a molecular biology technique employed to identify and investigate active microbial populations participating in specific metabolic processes within diverse microbial communities without the need to cultivate them. This method involves introducing stable-isotope-labeled carbon (13C) or nitrogen (15N) to the target microorganisms. As these microorganisms consume the labeled substrate, the stable isotopes become integrated into cellular components. Subsequent separation and molecular analysis of labeled nucleic acids (DNA or RNA) offer valuable insights into the phylogenetic and functional traits of the microorganisms engaged in metabolizing a particular substrate [68].

In a work, a dual-working electrode bioelectrochemical system (BES) was constructed and employed for the complete degradation of pentachlorophenol (PCP). In this investigation, DNA-SIP in conjunction with high-throughput sequencing analysis was employed to analyze the biofilms adhered to both the anodic and cathodic electrodes in the BES to explore the microorganisms engaged in the processes of dechlorination and mineralization. It was revealed that the cathodic working electrode hosted potential dechlorinators, namely *Comamonas*, *Pseudomonas*, *Methylobacillus,* and *Dechlorosoma*, while the anodic working electrode enriched potential intermediate mineralizing bacteria specifically *Comamonas*, *Stenotrophomonas,* and *Geobacter*. This suggests that the complete degradation of PCP was achieved through the synergistic action of these functional microorganisms [11].

2 Microbial Electrosynthesis of By-Products: Insight into General Mechanism

Microbial electrolysis cell (MEC) is a bioelectrochemical system which integrates microbial metabolism with electrochemical reactions for the generation of electrical energy or other valuable products. MECs are comprised of an anode and a cathode, occasionally separated by a membrane. Exoelectrogenic bacteria inhabit the anode, where they engage in metabolic processes, commonly involving the oxidation of organic matter or other compounds. In the course of this microbial metabolism, electrons are generated as by-products. These electrons have the capacity to traverse an external circuit and ultimately reach the cathode, where they are accepted, often participating in the reduction of a specific compound or molecule, yielding valuable products such as hydrogen and methane (Fig. 1).

2.1 *Electrohydrogenesis*

Hydrogen gas is currently recognized as the most promising green fuel for the future, primarily because it is a clean energy source. In an electrochemical fuel cell system, the oxidation of hydrogen gas results in the production of pure water as a by-product, with no emission of carbon dioxide [24]. In addition, it holds distinction as the most abundant element in the universe and stands out as the lightest fuel, boasting a high energy density per unit mass. Unlike electricity, hydrogen can be conveniently stored, adding to its appeal as a versatile and efficient energy carrier [37]. The utilization of hydrogen is increasingly extending to various sectors such as transportation, power

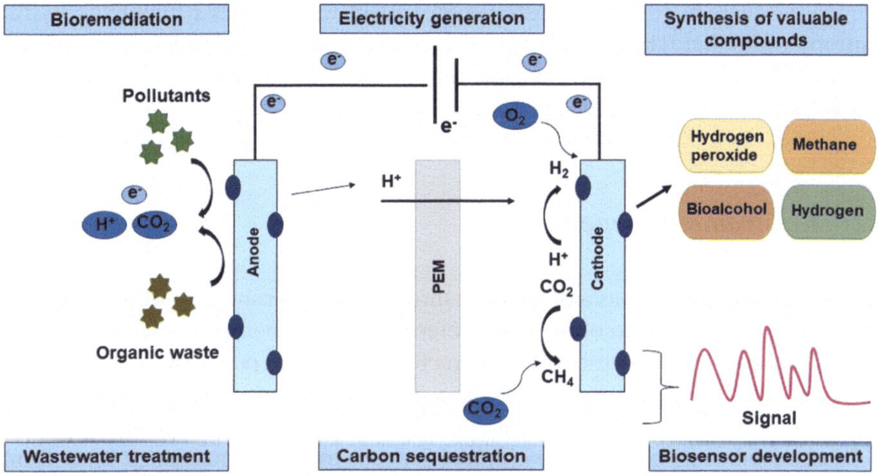

Fig. 1 Various applications of MEC

generation, and military equipment, owing to its notable benefits of high efficiency and minimal emissions [24].

Electrohydrogenesis refers to the ability of electrogenic bacteria in a MEC setup to oxidize organic substrates and convert them into biohydrogen. Electrohydrogenesis establishes itself as an economically viable and energy-efficient process for biohydrogen production, outperforming other biological methods in terms of yield, rate, and energy efficiency [74].

The electrohydrogenesis process typically involves a MEC setup with an external voltage applied across the MEC reactor. Electrogenic bacteria play a crucial role in this process, as their ability to generate and release electrons extracellularly facilitates the oxidation of various substrates. Hydrogen and electron are example of by-products which can be produced during this oxidation process. The generated electrons, influenced by the external voltage, traverse from the anodic electrode to the cathode. Concurrently, hydrogen can either permeate through the semipermeable membrane (dual-chamber MEC) or diffuse through the electrolyte (single-chamber MEC) and reach the cathode. At the cathode, these electrons catalyze the reduction of hydrogen ions, leading to the formation of hydrogen gas (H_2) [74].

Several factors influence the ultimate yield of hydrogen, with one of the most crucial considerations being the choice of substrate. The anodic microbial community is recognized for its flexibility, versatility, and adaptability, enabling it to utilize various substrates for hydrogen production, including organic matter and wastewater. Notably, using wastewater or industrial effluents as a substrate serves a dual purpose of wastewater treatment and simultaneous hydrogen production [33]. Wastewaters, being predominantly industrial effluents with no inherent economic value, play a significant role in shaping the economic feasibility of the hydrogen production process. However, the term "wastewater" lacks specificity, as it does not account for the types of organic matter present, their complexity, and degradability [57]. Consequently, the efficiency of hydrogen production is impacted, given that microbes may struggle to degrade complex substrates present in wastewaters without successful syntropy with other fermenters [81, 99]. To successfully scale up hydrogen production in MEC, it is imperative to consider the nature of the substrate and the metabolic pathways involved in microbial processes.

2.2 Electromethanogenesis

Methane can be obtained as a by-product through the anaerobic digestion of organic matter in MEC. This form of methane, referred to as biomethane, is highly favored as a biofuel due to its clean-burning characteristics and its potential as a renewable energy source. Biomethane can serve as a direct alternative to natural gas, finding applications in residential heating, industrial processes, and power generation [48].

Electromethanogenesis is a biological process in which methanogens, a form of methane-producing microbes, are utilized to convert CO_2 generated from the oxidation of substrates at the anode into methane (CH_4) at the cathode in a MEC

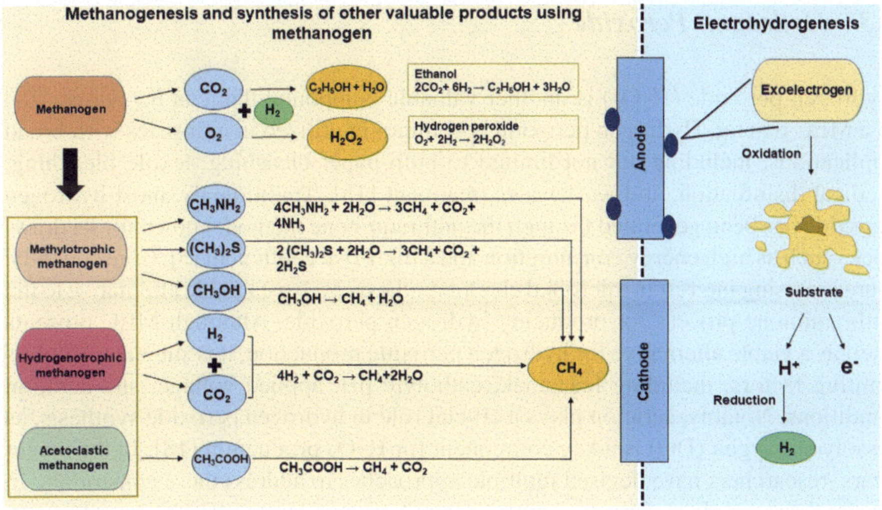

Fig. 2 Synthesis of various valuable by-products via MEC

setup [90]. Similar to electrohydrogenesis, the overall reaction is thermodynamically not favorable as it needs electrical energy to drive the reaction [10]. There are three distinct pathways for methanogenesis, namely hydrogenotrophic methanogenesis, acetoclastic methanogenesis, and methylotrophic methanogenesis (Fig. 2).

In the process of hydrogenotrophic methanogenesis, methanogens utilize hydrogen gas (H_2) and carbon dioxide (CO_2) to generate methane according to: $4\,H_2 + CO_2 \rightarrow CH_4 + 2\,H_2O$. Acetoclastic methanogenesis, on the contrary, entails the transformation of acetate into methane following: $CH_3COOH \rightarrow CO_2 + CH_4$ [25].

On the other hand, methylotrophic methanogens have the capacity to employ one-carbon compounds like methanol (CH_3OH), methylamines, or methylated sulfur compounds as substrates for the production of methane. The chemical formula for the reaction may differ based on the substrate, but the final outcomes typically encompass the expected methane with some other by-products such as carbon dioxide, water, ammonia, or hydrogen sulfide [47]. Much like electrohydrogenesis, various constraints impact the efficiency of CO_2 electroreduction. Factors like the poised cathodic potential, characteristics of the cathode material, configurations of MECs, and operational modes must be improved to enable the widespread adoption of CO_2 electromethanogenesis [95].

2.3 *Hydrogen Peroxide*

Hydrogen peroxide (H_2O_2) is another valuable compound that can be synthesized in a MEC reactor. Hydrogen peroxide is commonly utilized in industries with broad applications, including but not limited to pulp paper bleaching, textile bleaching, medical disinfection, and wastewater treatment [18]. Traditionally, most hydrogen peroxide has been generated through the anthraquinone method, known for its drawbacks such as high energy consumption and a high-risk coefficient [5]. Consequently, there is a rising interest in microbial electrosynthesis as a sustainable alternative to the anthraquinone process for producing hydrogen peroxide. Although MEC presents itself as a viable alternative for hydrogen peroxide production, it is subject to various limiting factors, including feedstock, catholyte pH, applied voltage, and aeration conditions. Notably, aeration plays a crucial role in hydrogen peroxide synthesis, as dissolved oxygen (DO) is a key component for H_2O_2 production [18]. In the recent years, researchers have devised multiple approaches to address these constraints. In one study, researchers addressed poor oxygen transmission in MEC by introducing a superhydrophobic natural air diffusion floating electrode (NADFE). This innovation led to a significantly increased H_2O_2 production rate, reaching 2.2 times higher than that of a gas diffusion electrode [89]. In another study, an MES with multiple graphite electrodes effectively upscaled H_2O_2 production. The research also underscores the significance of electrode material in the scale-up of H_2O_2 production, emphasizing that the selection of electrode material plays a pivotal role in determining current efficiency (CE), selectivity in the two-electron oxygen reduction reaction (ORR), and energy consumption during the in situ electrosynthesis of H_2O_2 at the cathode [101].

3 Synthesis of Other Valuable Products

The MEC system can be employed to synthesize a range of valuable compounds, including but not limited to alcohols, organic acids, and even biopolymers [6, 9, 28]. Table 1 outlines several enhancements made in the recent years to increase the yield of valuable products through MEC.

The ultimate determination of end products in a MEC system is heavily influenced by factors such as the specific microorganism engaged and the experimental conditions. This can be seen in the study where a MEC leveraging the synergistic metabolic processes of the *exoelectrogen Geobacter sulfurreducens* and the bacterium *Clostridium cellobioparum* efficiently transformed glycerol into ethanol, achieving high yields (90%) while simultaneously generating fermentative by-products that acted as electron donors for *G. sulfurreducens* [84]. In a different study, the production of alcohols from volatile fatty acids (VFAs) was examined within single-chambered microbial electrosynthesis systems (MES). The study employed

Table 1 Recent advances in synthesis of value-added products via MEC

Value-added products obtained	Type of MEC setup	Type of inoculum/ substrate (s)	Modification/ optimization done	Outcome	Reference
Hydrogen	Double chamber	• Cane sugar wastewater • Raw sugar wastewater	• Evaluation of sugar and raw sugar as substrates • Evaluation of different cathode materials, namely nickel plate, nickel foam, stainless steel mesh	At an applied voltage of 1.0 V, Ni-foam showed the highest hydrogen production, reaching 1.59 mmol/L/D in cane sugar effluent and 1.43 mmol/L/D in raw sugar effluent which were approximately twice as much as SS mesh and 1.2 times that of Ni plate	[40]
Hydrogen	Double chamber	Synthetic fermentation effluent	Optimization of cathodes using nickel powder (pNi) was blended with activated carbon (AC)	• AC-pNi4.8 (with a nickel loading of 4.8 mg/ cm^2) yielded higher rates of hydrogen production (0.38 ± 0.04 L-H_2/L-d) compared to pNi77 (0.28 ± 0.02 L-H_2/L-d), despite having 16 times less nickel loading • Additionally, cathodic hydrogen recovery using AC-pNi4.8 (98 ± 5%) surpassed that of pNi77 (82 ± 4%), suggesting an improvement in catalytic activities through the blending of activated carbon (AC)	[52]

(continued)

Table 1 (continued)

Value-added products obtained	Type of MEC setup	Type of inoculum/ substrate (s)	Modification/ optimization done	Outcome	Reference
• Methane • Ammonium	Triple chamber	• Activated sludge • Synthetic organic substrates	Development of MEC with dual-sided cathode configuration, connecting two identical cathodic chambers in parallel through a titanium wire separated from an intermediate anodic compartment by an anion exchange membrane (AEM) and a cation exchange membrane (CEM)	• The new MEC configuration improved cathodic operations, specifically methane production and CO_2removal, compared to a previous setup featuring a three-chamber MEC with an accumulation chamber • The dual-sided configuration allowed for the migration and subsequent accumulation of ammonium nitrogen exclusively in the CEM cathode. This isolation helps prevent potential inhibition of methanogens in this compartment	[92]
• Hydrogen • Methane • Organic acids, namely acetic, butyric, propionic, and lactic acid	Single	• Palm oil mill effluent (POME) wastewater • Anaerobic seed sludge	Mesophilic MEC setup combined with a thermophilic continuous stirred tank reactor (CSTR) in an integrated approach	• The CSTR reactor run resulted in a hydrogen yield of 205 mL H_2 $gCOD^{-1}$at optimum conditions • The continuous, single-chambered MECs fed with dark fermentation effluents generated 290 mL CH_4 $gCOD^{-1}$ and 2700 ml CH_4 L^{-1} at optimum conditions	[54]

(continued)

Table 1 (continued)

Value-added products obtained	Type of MEC setup	Type of inoculum/ substrate (s)	Modification/ optimization done	Outcome	Reference
• Hydrogen • Methane	Double chamber	Anaerobic sludge	Evaluation of MEC activity using the coenzyme M (CoM) and 2-bromoethanesulfonate (2-BES)	• The introduction of CoM into MECs resulted in a substantial improvement, increasing their methane production rate, purity, and yield by 4.5 times, 14.5%, and 76.1%, respectively, in comparison with the control • Microbial community analysis revealed that *Methanosaeta*, the primary acetoclastic methanogen, maintained dominance in the microbial community but experienced a gradual decline in relative abundance following the CoM injection. Conversely, notable increases were observed in hydrogenotrophic methanogens, including *Methanoculleus*and *Methanolinea*, along with the potential presence of syntrophic acetate-oxidizing bacteria • Conversely, the trials injected with 2-BES exhibited a substantial inhibition of CH_4 production by 79.5%, leading to a concurrent increase of 145.5% in H_2 production compared to the control	[73]
Hydrogen	Single chamber	• Palm oil mill effluent (POME)• Anaerobic microbial sludge	MEC under thermophilic condition with two-stage dark fermentation	The continuous two-stage dark fermentation and microbial electrolysis system achieved a hydrogen yield of 236 mL-H_2·gCOD^{-1}which was a threefold increase in hydrogen yield compared to dark fermentation alone	[50]

(continued)

Table 1 (continued)

Value-added products obtained	Type of MEC setup	Type of inoculum/ substrate (s)	Modification/ optimization done	Outcome	Reference
Hydrogen	Single chamber	• Different organic substrates including glucose, sucrose, and cellobiose, along with volatile fatty acids such as sodium acetate, sodium butyrate, sodium propionate, and sodium lactate • Water hyacinth	• Evaluation of water hyacinth as sole substrate • Development of two-step dark fermentation-MEC system	The two-stage process demonstrated enhanced performance compared to the single-stage MEC process, achieving an overall hydrogen yield of 67.69 L H_2/kg COD $_{consumed}$, a COD removal rate of 70.33%, and an energy recovery of 46%	[85]
Methane	MEC-AD coupled reactors	Waste activated sludge (WAS)	MEC and anaerobic digestion (AD) system combined with ultrasound and alkali pretreatment	The coupled reactors of MEC-AD attained a methane yield of 808 ± 8 mL, representing a 97.0% ± 1.85% increase compared to the control AD with a yield of 410 m	[7]
Hydrogen	Single chamber	Anaerobic digested sludge	Sequential dark fermentation coupled with MEC	With anaerobic digested sludge, the rates of biogas production were 0.83 L/L d, and the proportion of hydrogen was 92.58%, surpassing those observed with anaerobic granular sludge (0.55 L/L d and 86.21%)	[43]

(continued)

Table 1 (continued)

Value-added products obtained	Type of MEC setup	Type of inoculum/ substrate (s)	Modification/ optimization done	Outcome	Reference
Hydrogen	Single chamber	Food waste	Integrated reactor that merges anaerobic digestion with microbial electrolysis cell (AD-MEC) with anaerobic digestion (AD) for the efficient recovery of hydrogen, utilizing food waste (FW) as the substrate	The continuous operation of AD-MEC resulted in higher hydrogen production (511.02 mL H_2g^{-1} VS) compared to the production achieved through AD alone (49.39 mL H_2 g^{-1} VS)	[34]
Hydrogen	Double chamber	Growth medium	Cathode modified using nanophosphating cobalt coated nickel foam (CoP-NF)	Hydrogen evolution performance of the CoP-NF cathode resulted in a significantly improved hydrogen production rate of 222 ± 20.3 mL H_2L^{-1} d^{-1} at an applied voltage of 0.7 V in fully assembled MECs. This performance was three times superior to the bare NF and even surpassed that of Pt/C cathode	[59]
Hydrogen	Double chamber	Sugar wastewater	Cathode modification using nickel foam and synthesized catalyst of $NiCo_2O_4$graphene nanocomposites	Enhanced cathodic hydrogen recovery at 27.9%, with an overall hydrogen recovery of 18.4%	[39]
Hydrogen	Modified double chamber	Distillery wastewater	Construction of a compact retrofit design of MEC within an anaerobic digester	The MEC attained a maximum hydrogen production of 30.2 ± 1.2 mL at a current density of 811.33 ± 20 mA/m^{2}which was slightly lower than the conventional double chamber MEC, but the adapted MEC is space-efficient, requiring less area compared to the traditional design and serves as a retrofit for an existing anaerobic digester system	[67]

(continued)

Table 1 (continued)

Value-added products obtained	Type of MEC setup	Type of inoculum/ substrate (s)	Modification/ optimization done	Outcome	Reference
Hydrogen	Double chamber	Anaerobic granular sludge	Separating membrane made of poly (vinyl) alcohol/chitosan (PVA/CS)	• The results obtained suggested that there were no significant differences in MEC performance for hydrogen production between PVA/CS and Nafion membranes. MEC with PVA/CS demonstrated a hydrogen production rate and yield of 1277 ± 46 mL $H_2 L_{cat}^{-1} d^{-1}$ and 974 ± 116 mL H_2 $g_{acetate}^{-1}$, respectively • The use of the PVA/CS membrane facilitated acetate removal, showing a 7% improvement compared to Nafion	[30]
Hydrogen	Double chamber	Sugar wastewater	Cathode modification using $NiMoO_4$/Ni-Foam	Under favorable conditions, a higher hydrogen production rate of 0.12 ± 0.01 L L^{-1} D^{-1} was attained with nickel molybdate ($NiMoO_4$) coated NF compared to the control	[41]
Methane	Double chamber	Anaerobic sludge	Development of novel anaerobic osmotic membrane bioreactor-microbial electrolysis cell (AnOMBR-MEC)	In contrast to AnOMBR, the methane production increased from 55 to 90% by the conclusion of the operation, and the methane yield in the AnOMBR-MEC witnessed a 1.6-fold increase	[27]

(continued)

Table 1 (continued)

Value-added products obtained	Type of MEC setup	Type of inoculum/ substrate (s)	Modification/ optimization done	Outcome	Reference
Methane	Single chamber	• Secondary sludge • Bovine serum albumin (BSA)	Testing various applied voltage and BSA concentrations to optimize methane production during MEC-AD of proteins	• Applied voltage increased the methane production rate from 23.8% to 45.6% under low and medium organic loading conditions (BSA concentration of 500 mg/L and 4 g/L) and resulted in a remarkable enhancement of methanogenesis efficiency by 225.4% at a high BSA concentration (20 g/L) when a voltage of 0.6 V was applied, compared to the performance observed under open circuit conditions • Analysis of the microbial community revealed that the application of voltage led to a 46.7% increase in the abundance of fermentative bacteria at the anode. Simultaneously, at the cathode, the abundance of *Methanobacterium*surged from 10.4% to 84.3%, signifying a shift in the methanogenesis pathway from acetoclastic to hydrogenotrophic	[101]
Hydrogen	Single chamber	• Palm oil mill effluent (POME) anaerobic seed sludge • POME wastewater	Optimization of experimental conditions using empirical model generated with response surface methodology (RSM) and central composite design (CCD)	Following optimal conditions determined by experimental model, a maximum HPR of 1.1747 $m^3 H_2/m^3$ d was produced	[45]

continuous pH control to inhibit methanogenesis. Further analysis revealed the presence of fermented bio-alcohols, including ethanol, methanol, and propanol, with ethanol as the predominant product [28].

In a nutshell, MEC-based systems possess the potential to synthesize a variety of products depending on the conditions and the type of microbes employed. Continuous research efforts are dedicated to exploring and enhancing MEC systems, aiming to produce valuable compounds with increased yield and reduced production costs.

4 Microbial Community and Their Processes that Drive Microbial Electrolysis Cells (MECs)

4.1 Microbial Community in MEC

Microorganisms, specifically hydrogen-producing bacteria, and methanogens are among the integral components of microbial electrolysis cells (MECs) that establish their core operating principles. The bacteria consume and metabolize the organic substrates, typically into hydrogen gas (H_2) or methane gas (C_2H_6) which were achieved with an external supply of electrical voltage within a MEC under anaerobic conditions while simultaneously releases electrons to a terminal electron acceptor (electrode) to generate an electrical output [42]. The vast biodiversity of the hydrogen-producing bacteria and methanogens further opens up limitless research opportunities for the optimization of MEC.

An interesting aspect to explore in the context of MEC would be the physiological differences in the microbial community within a MEC. Some common bacterial genera that were found to reside dominantly in wastewater or sludge samples used in MEC studies include the obligate anaerobes *Desulfovibrio* sp. and *Geobacter* sp. MEC studied by [4] using activated sludge found that it attained a maximum hydrogen production rate of 0.28 m^3-H_2/m^3-d at 1.1 V and ~ 90% chemical oxygen demand (COD) removal efficiency with *Desulfovibrio* sp. occupying 13.2% of the total microbial diversity in the sample. Similarly, another study focusing on biocathode in MEC resulted in the production of 1.1 A m^{-2} current density and 0.63 m^3 H_2 m^{-3} cathode liquid volume per day at 0.7 V. *Desulfovibrio vulgaris* and *Geobacter sulfurreducens* were main bacterial occupants in the MEC [21].

Comparative studies between single species and mixed consortium of microbial community in MEC optimization were another key point to address in order to maximize hydrogen output. A comparative study regarding the hydrogen production rates and hydrogen recoveries of pure and mixed cultures in MEC was done using the hydrogen utilizing exoelectrogenic bacterium, *Geobacter sulfurreducens* against a mixed consortium. Results indicated that at an applied voltage of 0.7 V, both *G. sulfurreducens* and the mixed culture generated similar current densities (~ 160 A/m^3), resulting in hydrogen production rates of ~ 1.9 m^3 H_2/m^3/day [12].

5 Factors Affecting Microbial Community and Microbial Electrolysis Cells (MECs) System

The primary purpose of a microbial electrolysis cell (MEC) would be to liberate hydrogen gas (H_2) or methane gas (C_2H_6) from the degradation of organic substrates via the redox reactions performed physiologically by electrogenic bacteria (EB). As such, factors concerning the viability and enrichment of the microbial community are crucial variables that determine the efficiency of MEC.

5.1 Anaerobic Condition

First and foremost, the bacteria utilized in MEC must undoubtedly be grown in strictly anaerobic conditions. This is to ensure the optimal growth of the anaerobic bacteria involved while also preventing the formation of water instead of hydrogen gas which would otherwise occur in an aerobic cathode [19, 46].

5.2 Types and Concentration of Substrates

Optimization of a MEC with regard to its microbial community also takes into consideration the type of substrates they may utilize efficiently for optimal electrical energy output. Generally, simpler, easily digestible substrates were favored over more complex and less digestible ones. In a study done by [83] using different substrates in MEC, they concluded that hydrogen production correlates with current generation. Another conclusion was that sodium salt were more outstanding than sugars since the hydrolysis of sugar raw materials does not promote the growth and reproduction of electroactive microorganisms. Sodium acetate was crowned the best current and hydrogen production, achieving a maximum output of 23.4 mA and a maximum hydrogen content of 25.85%, respectively. The subsequent trends with regard to decreasing performance continue with sodium butyrate, sodium propionate, glucose, and finally starch. Glucose and starch bottom the ranking as the system environment is difficult to be controlled and they were easily acidified. This coincides with another study which had concluded acetate-fed reactors exhibited higher performance in terms of maximum current density (I: 4.25 ± 0.23 A/m^2), coulombic efficiency (CE: 95 ± 8%), and substrate degradation rate (98.8 ± 1.2%) than propionate-fed reactors (I: 2.7 ± 0.28 A/m^2; CE: 68 ± 9.5%; substrate degradation rate: 84 ± 13%) [31]. Concentration of substrates also influenced the performance of MEC [79] carried out a study and determined that concentrations of acetic acid and propionic acid at 1500 mg/L and 250 mg/L, respectively, resulted in the best hydrogen production rate (265 mL/d/Lreactor). The outcome of the study was also facilitated with the presence of *Proteobacteria* and *Firmicutes* in the inoculum that are well-known fermenters and syntrophic bacterial partners that can ferment volatile fatty acids, such as acetate, propionate, and butyrate [26, 51].

5.3 PH

Furthermore, pH is a key element that impacts the microbial community. Because of the concentration of protons in the anode chamber, the pH surrounding the anode would fall owing to reduction–oxidation (redox) processes. Solution pH has a large impact on the bacterial community and bacterial development; thus, any changes in the ambient pH may alter the microbial population. Low pH decreases microbial activity and biofilm development, as well as lowering system performance. Moreover, changes in solution pH might also modify the distribution of bacteria on the electrode, changing the microbial structure [14].

5.4 Temperature

Temperature is regarded as one of the key parameters crucial in sustaining biological life by influencing the regulation of physiological activities. Biological organisms possess different viable temperature ranges which also include their optimal activity temperature. The operational temperature for reactors based on mesophilic bacteria was intended to be between 30 and 38 °C, whereas for electrochemically active microorganisms in a continuous reactor, a lower temperature range of 30–31 °C was discovered to be the most efficient temperature settings [55, 65]. Likewise, a temperature of 31 °C in MEC was observed to induce the highest rate of chemical oxygen demand (COD) removal and the highest decline in biomass concentration within wastewater sample [71].

5.5 Applied Voltage

Another key player in the operations of the MEC is the external power supply required for it to maintain an operational state [22] agreed to a threshold of 0.8 V of provided input, which increasing it, would be detrimental to the microbial community in the MEC. This optimal value promotes growth of the microbial community and contaminant degrading capabilities, hence an increase in COD removal and methane yield. Adverse effects such as cell apoptosis, lower growth rate, and reduced metabolism ensue when the supplied voltage went beyond 0.8 V. These were deduced to be the aftermath of the disruption of electrical charges and ionic balance in the bacterial cell membrane. Similar conclusions were obtained by [44] in their work using waste activated sludge for methanogenesis.

5.6 *Presence of Methanogens*

Presence of methanogens in MEC intended for biohydrogen production were mostly inevitable due to the favorable anaerobic conditions. Such a situation often led to the contamination of the reactor and decreased yield of hydrogen. Thus, multiple strategies were developed to cease methanogenesis in MEC such as temperature shock, low pH, carbonate limitation, methanogenic inhibitors, and UV–IR radiation [64]. A study by [8] used a methanogenic inhibitor, sodium 2-bromoethanesulfonate, to chemically prevent methanogenic growth and activity, which proved to be successful.

6 Sustainability with Microbial Electrolysis Cells (MECs)

6.1 *Source of Green Fuel*

Demands for green energy and technology are on the rise due to the increasing concerns of global climate change. Productions of energy are required to be sustainable, as in producing minimal greenhouse gases and hazardous wastes. MECs bear the capability to generate hydrogen gas, which is a highly sought after green energy source, with a relatively high energy output value, approximately 120 MJ or 33.33 kWh per 1 kg of hydrogen in comparison with conventional fossil fuels (Table 2) [1]. The method of harvesting hydrogen via MEC benefits through its ability as a mass reproducible renewable energy source with a more environmental friendly and cost-effective production method where waste can be transformed into value-added products [3, 63]. To elaborate, hydrogen only produces water vapor after combustion as well, as opposed to fossil fuels that produce carbon dioxide which is a greenhouse gas that contribute to global warming. The aforementioned properties collectively promote the potential usage of hydrogen in the power and automobile industry (hydrogen fuel cells), effectively relieving the dependence on conventional fossil fuels which are currently facing depletion crisis.

$$\text{Combustion of hydrogen: } 2H_2 + O_2 \rightarrow 2H_2O$$

6.2 *Bioremediation in Wastewater Treatment*

MEC system triumphs in the aspect that it was not selective in the types of organic matter that it could utilize. This trait poses a significant advantage where it could be used to treat wastewater from various domestic and industrial settings. The hazardous composition of wastewater containing solid organic waste, toxic chemicals, heavy

Table 2 Comparison of energy content among several fuel types

Fuel	Energy content (MJ/kg)	
	Lower heating value	Higher heating value
Gaseous hydrogen	119.96	141.88
Liquid hydrogen	120.04	141.77
Natural gas	47.13	52.21
Liquefied petroleum gas (LPG)	46.60	50.14
Conventional gasoline	43.44	46.52
Conventional diesel	42.78	45.76

metals, etc. may be degraded via the use of MEC, specifically through the metabolic actions of the microorganisms within the MEC.

In a study involving the removal of nickel ions (Ni^{2+}) from contaminated water with an initial concentration of 50 mg/L, 99% Ni^{2+} removal efficiency was attained using MEC with carbon felt anode and stainless steel cathode for an applied external voltage of 0.9 V [76]. MEC has also been used to treat highly concentrated and complicated wastewater streams, including landfill leachate and dairy plant effluent. The team used MEC with a carbon electrode to treat mixed leachate and dairy wastewater with an initial COD of 8400 mg L^{-1} and achieved a COD removal efficiency of 73% at an externally applied cell voltage of 0.8 V and a hydrogen generation rate of 15 ml L^{-1} day^{-1} [78]. In another investigation with respect low strength wastewater, after applying an anodic potential of + 0.2 V and continually feeding the setup with acetate as substrate for 8.35 days, the cathodic ammonium concentration reached 318 mg L^{-1} compared to 44.5 mg L^{-1}. As a result, in steady state, this dual-chamber MEC achieved a maximum 91% Coulombic efficiency and 79% cathode capture efficiency as methane, as well as 91.2 and 50.9 meq L^{-1} d^{-1} of methane and biohydrogen generation rates, respectively [87].

7 Conclusion

In concluding our exploration of the microbiology behind microbial electrolysis cells (MECs), it becomes evident that understanding the intricate relationships between microorganisms and electrochemical processes is paramount to advancing this innovative technique. The microbial communities thriving within MECs serve as architects of energy conversion, influencing the efficiency and sustainability of the overall system. By unraveling the mysteries of microbial intricacies, we pave the way for enhanced performance, improved yields, and broader applications of MECs. As we embrace the symbiotic dance between microbiology and electrochemistry, the future of microbial electrolysis holds the promise of cleaner energy production and environmental sustainability. This chapter underscores the indispensable nature of microbiological studies within the realm of MECs, shaping the landscape of a technology poised to revolutionize the energy paradigm.

References

1. Abe JO, Popoola A, Ajenifuja E, Popoola OM (2019) Hydrogen energy, economy and storage: review and recommendation. Int J Hydrog Energy 44:15072–15086
2. Aguiar-Pulido V, Huang W, Suarez-Ulloa V, Cickovski T, Mathee K, Narasimhan G (2016) Metagenomics, metatranscriptomics, and metabolomics approaches for microbiome analysis. Evol Bioinform. https://doi.org/10.4137/EBO.S36436
3. Ahluwalia RK, Wang X, Rousseau A, Kumar R (2004) Fuel economy of hydrogen fuel cell vehicles. J Power Sour 130:192–201
4. Almatouq A, Babatunde AO, Khajah M, Webster G, Alfodari M (2020) Microbial community structure of anode electrodes in microbial fuel cells and microbial electrolysis cells. J Water Process Eng. https://doi.org/10.1016/j.jwpe.2020.101140
5. Anantharaj S, Pitchaimuthu S, Noda S (2021) A review on recent developments in electrochemical hydrogen peroxide synthesis with a critical assessment of perspectives and strategies. Adv Colloid Interface Sci. https://doi.org/10.1016/j.cis.2020.102331
6. Banu JR, Kumar MD, Gunasekaran M, Kumar G (2019) Biopolymer production in bio electrochemical system: literature survey. Bioresour Technol Rep. https://doi.org/10.1016/j.biteb.2019.100283
7. Bao H, Yang H, Zhang H, Liu Y, Su H, Shen M (2020) Improving methane productivity of waste activated sludge by ultrasound and alkali pretreatment in microbial electrolysis cell and anaerobic digestion coupled system. Environ Res. https://doi.org/10.1016/j.envres.2019.108863
8. Baeza JA, Martínez-Miró À, Guerrero J, Ruiz Y, Guisasola A (2017) Bioelectrochemical hydrogen production from urban wastewater on a pilot scale. J Power Sour 356:500–509
9. Bian C, Chen X, Wang J, Xiao B, Liu R, Li L, Liu J (2023) Simultaneous biohythane and volatile fatty acids production from food waste in microbial electrolysis cell-assisted acidogenic reactor. J Clean Prod. https://doi.org/10.1016/j.jclepro.2023.138370
10. Blasco-Gómez R, Batlle-Vilanova P, Villano M, Balaguer MD, Colprim J, Puig S (2017) On the edge of research and technological application: a critical review of electromethanogenesis. Int J Mol Sci. https://doi.org/10.3390/ijms18040874
11. Cai X, Li J, Guan F, Luo X, Yu Z, Yuan Y (2023) Complete pentachlorophenol biodegradation in a dual-working electrode bioelectrochemical system: performance and functional microorganism identification. Water Res. https://doi.org/10.1016/j.watres.2022.119529
12. Call DF, Wagner RC, Logan BE (2009) Hydrogen production by Geobacter species and a mixed consortium in a microbial electrolysis cell. Appl Environ Microbiol 75:7579–7587
13. Carmona-Martínez AA, Trably E, Milferstedt K, Lacroix R, Etcheverry L, Bernet N (2015) Long-term continuous production of H2 in a microbial electrolysis cell (MEC) treating saline wastewater. Water Res. https://doi.org/10.1016/j.watres.2015.05.041
14. Chae K-J, Choi M-J, Lee J-W, Kim K-Y, Kim IS (2009) Effect of different substrates on the performance, bacterial diversity, and bacterial viability in microbial fuel cells. Bioresour Technol 100:3518–3525
15. Chakraborty A, DasGupta CK, Bhadury P (2014) Application of molecular techniques for the assessment of microbial communities in contaminated sites. In: Microbial biodegradation and bioremediation. https://doi.org/10.1016/B978-0-12-800021-2.00004-2

16. Chakravorty S, Helb D, Burday M, Connell N, Alland D (2007) A detailed analysis of 16S ribosomal RNA gene segments for the diagnosis of pathogenic bacteria. J Microbiol Methods. https://doi.org/10.1016/j.mimet.2007.02.005
17. Chignell JF, De Long SK, Reardon KF (2018) Meta-proteomic analysis of protein expression distinctive to electricity-generating biofilm communities in air-cathode microbial fuel cells. Biotechnol Biofuels. https://doi.org/10.1186/s13068-018-1111-2
18. Chung TH, Meshref MNA, Hai FI, Al-Mamun A, Dhar BR (2020) Microbial electrochemical systems for hydrogen peroxide synthesis: critical review of process optimization, prospective environmental applications, and challenges. Bioresour Technol. https://doi.org/10.1016/j.biortech.2020.123727
19. Cotterill S, Heidrich E, Curtis T (2016) Microbial electrolysis cells for hydrogen production. Microbial electrochemical and fuel cells, pp 287–319
20. Croese E, Keesman KJ, Geelhoed JS, Plugge CM, Sleutels TH, Stams AJ, Euverink GJW (2013) Relating MEC population dynamics to anode performance from DGGE and electrical data. Syst Appl Microbiol 36(6):408–416. https://doi.org/10.1016/j.syapm.2013.05.004
21. Croese E, Pereira MA, Euverink GJW, Stams AJM, Geelhoed JS (2011) Analysis of the microbial community of the biocathode of a hydrogen-producing microbial electrolysis cell. Appl Microbiol Biotechnol. https://doi.org/10.1007/s00253-011-3583-x
22. Ding A, Yang Y, Sun G, Wu D (2016) Impact of applied voltage on methane generation and microbial activities in an anaerobic microbial electrolysis cell (MEC). Chem Eng J 283:260–265
23. Ercolini D, Cocolin L (2014) Identification methods: culture-independent techniques. In: Encyclopedia of food microbiology, 2nd edn. https://doi.org/10.1016/B978-0-12-384730-0.00438-9
24. Fan L, Tu Z, Chan SH (2021) Recent development of hydrogen and fuel cell technologies: a review. Energy Rep. https://doi.org/10.1016/j.egyr.2021.08.003
25. Fenchel T, King GM, Blackburn TH (2012) Bacterial biogeochemistry: the ecophysiology of mineral cycling, bacterial biogeochemistry: the ecophysiology of mineral cycling. https://doi.org/10.1016/B978-0-12-415836-8
26. Freguia S, Teh EH, Boon N, Leung KM, Keller J, Rabaey K (2010) Microbial fuel cells operating on mixed fatty acids. Bioresour Technol 101:1233–1238
27. Gao T, Zhang H, Xu X, Teng J (2021) Integrating microbial electrolysis cell based on electrochemical carbon dioxide reduction into anaerobic osmosis membrane reactor for biogas upgrading. Water Res. https://doi.org/10.1016/j.watres.2020.116679
28. Gavilanes J, Noori MT, Min B (2019) Enhancing bio-alcohol production from volatile fatty acids by suppressing methanogenic activity in single chamber microbial electrosynthesis cells (SCMECs). Bioresour Technol Rep. https://doi.org/10.1016/j.biteb.2019.100292
29. Ghosh S, Das AP (2018) Metagenomic insights into the microbial diversity in manganese-contaminated mine tailings and their role in biogeochemical cycling of manganese. Sci Rep. https://doi.org/10.1038/s41598-018-26311-w
30. González-Pabón MJ, Cardeña R, Cortón E, Buitrón G (2021) Hydrogen production in two-chamber MEC using a low-cost and biodegradable poly(vinyl) alcohol/chitosan membrane. Bioresour Technol. https://doi.org/10.1016/j.biortech.2020.124168
31. Hari AR, Venkidusamy K, Katuri KP, Bagchi S, Saikaly PE (2017) Temporal microbial community dynamics in microbial electrolysis cells–influence of acetate and propionate concentration. Front Microbiol 8:1371
32. Herschend J, Damholt ZBV, Marquard AM, Svensson B, Sørensen SJ, Hägglund P, Burmølle M (2017) A meta-proteomics approach to study the interspecies interactions affecting microbial biofilm development in a model community. Sci Rep. https://doi.org/10.1038/s41598-017-16633-6
33. Hua T, Li S, Li F, Zhou Q, Ondon BS (2019) Microbial electrolysis cell as an emerging versatile technology: a review on its potential application, advance and challenge. J Chem Technol Biotechnol. https://doi.org/10.1002/jctb.5898

34. Huang J, Feng H, Huang L, Ying X, Shen D, Chen T, Shen X, Zhou Y, Xu Y (2020) Continuous hydrogen production from food waste by anaerobic digestion (AD) coupled single-chamber microbial electrolysis cell (MEC) under negative pressure. Waste Manag. https://doi.org/10.1016/j.wasman.2019.12.015
35. Ishii S, Suzuki S, Norden-Krichmar TM, Tenney A, Chain PSG, Scholz MB, Nealson KH, Bretschger O (2013) A novel metatranscriptomic approach to identify gene expression dynamics during extracellular electron transfer. Nat Commun. https://doi.org/10.1038/ncomms2615
36. Ishii S, Suzuki S, Tenney A, Nealson KH, Bretschger O (2018) Comparative metatranscriptomics reveals extracellular electron transfer pathways conferring microbial adaptivity to surface redox potential changes. ISME J. https://doi.org/10.1038/s41396-018-0238-2
37. Jain IP (2009) Hydrogen the fuel for 21st century. Int J Hydrog Energy. https://doi.org/10.1016/j.ijhydene.2009.05.093
38. Jay ZJ, Inskeep WP (2015) The distribution, diversity, and importance of 16S rRNA gene introns in the order thermoproteales. Biol Direct. https://doi.org/10.1186/s13062-015-0065-6
39. Jayabalan T, Manickam M, Naina Mohamed S (2020) $NiCo_2O_4$-graphene nanocomposites in sugar industry wastewater fed microbial electrolysis cell for enhanced biohydrogen production. Renew Energy. https://doi.org/10.1016/j.renene.2020.03.071
40. Jayabalan T, Matheswaran M, Naina Mohammed S (2019) Biohydrogen production from sugar industry effluents using nickel based electrode materials in microbial electrolysis cell. Int J Hydrog Energy. https://doi.org/10.1016/j.ijhydene.2018.09.219
41. Jayabalan T, Matheswaran M, Radhakrishnan TK, Naina Mohamed S (2021) Influence of nickel molybdate nanocatalyst for enhancing biohydrogen production in microbial electrolysis cell utilizing sugar industrial effluent. Bioresour Technol. https://doi.org/10.1016/j.biortech.2020.124284
42. Jensen LS, Kaul C, Juncker NB, Thomsen MH, Chaturvedi T (2022) Biohydrogen production in microbial electrolysis cells utilizing organic residue feedstock: a Review. Energies 15:8396
43. Jia X, Li M, Wang Y, Wu Y, Zhu L, Wang X, Zhao Y (2020) Enhancement of hydrogen production and energy recovery through electro-fermentation from the dark fermentation effluent of food waste. Environ Sci Ecotechnol. https://doi.org/10.1016/j.ese.2019.100006
44. Joicy A, Seo H, Lee M-E, Kim D-H, Cho SK, Ahn Y (2022) Enhanced methane production using pretreated sludge in MEC-AD system: performance, microbial activity, and implications at different applied voltages. Int J Hydrog Energy 47:40731–40741
45. Kadier A, Wang J, Chandrasekhar K, Abdeshahian P, Islam MA, Ghanbari F, Bajpai M, Katoch SS, Bhagawati PB, Li H, Kalil MS, Hamid AA, Abu Hasan H, Ma PC (2022) Performance optimization of microbial electrolysis cell (MEC) for palm oil mill effluent (POME) wastewater treatment and sustainable bio-H2 production using response surface methodology (RSM). Int J Hydrog Energy. https://doi.org/10.1016/j.ijhydene.2021.09.259
46. Kadier A, Simayi Y, Abdeshahian P, Azman NF, Chandrasekhar K, Kalil MS (2016) A comprehensive review of microbial electrolysis cells (MEC) reactor designs and configurations for sustainable hydrogen gas production. Alex Eng J 55:427–443
47. Keltjens JT, Vogels GD (1993) Conversion of methanol and methylamines to methane and carbon dioxide. In: Methanogenesis. https://doi.org/10.1007/978-1-4615-2391-8_6
48. Khan MU, Lee JTE, Bashir MA, Dissanayake PD, Ok YS, Tong YW, Shariati MA, Wu S, Ahring BK (2021) Current status of biogas upgrading for direct biomethane use: a review. Renew Sustain Energy Rev. https://doi.org/10.1016/j.rser.2021.111343
49. Khati P, Sharma A, Chaudhary P, Singh AK, Gangola S, Kumar R (2019) High-throughput sequencing approach to access the impact of nanozeolite treatment on species richness and evenness of soil metagenome. Biocatal Agric Biotechnol. https://doi.org/10.1016/j.bcab.2019.101249
50. Khongkliang P, Jehlee A, Kongjan P, Reungsang A, O-Thong S (2019) High efficient biohydrogen production from palm oil mill effluent by two-stage dark fermentation and microbial electrolysis under thermophilic condition. Int J Hydrog Energy. https://doi.org/10.1016/j.ijhydene.2019.10.022

51. Kiely PD, Cusick R, Call DF, Selembo PA, Regan JM, Logan BE (2011) Anode microbial communities produced by changing from microbial fuel cell to microbial electrolysis cell operation using two different wastewaters. Bioresour Technol. https://doi.org/10.1016/j.biortech.2010.05.019
52. Kim KY, Logan BE (2019) Nickel powder blended activated carbon cathodes for hydrogen production in microbial electrolysis cells. Int J Hydrog Energy. https://doi.org/10.1016/j.ijhydene.2019.04.041
53. Kouzuma A, Kasai T, Nakagawa G, Yamamuro A, Abe T, Watanabe K (2013) Comparative metagenomics of anode-associated microbiomes developed in rice paddy-field microbial fuel cells. PLoS ONE. https://doi.org/10.1371/journal.pone.0077443
54. Krishnan S, Md Din MF, Taib SM, Nasrullah M, Sakinah M, Wahid ZA, Kamyab H, Chelliapan S, Rezania S, Singh L (2019) Accelerated two-stage bioprocess for hydrogen and methane production from palm oil mill effluent using continuous stirred tank reactor and microbial electrolysis cell. J Clean Prod. https://doi.org/10.1016/j.jclepro.2019.04.365
55. Kyazze G, Popov A, Dinsdale R, Esteves S, Hawkes F, Premier G, Guwy A (2010) Influence of catholyte pH and temperature on hydrogen production from acetate using a two chamber concentric tubular microbial electrolysis cell. Int J Hydrog Energy 35:7716–7722
56. Leary DH, Hervey WJ, Malanoski AP, Wang Z, Eddie BJ, Tender GS, Vora GJ, Tender LM, Lin B, Strycharz-Glaven SM (2015) Metaproteomic evidence of changes in protein expression following a change in electrode potential in a robust biocathode microbiome. Proteomics. https://doi.org/10.1002/pmic.201400585
57. Leicester D, Amezaga J, Heidrich E (2020) Is bioelectrochemical energy production from wastewater a reality? Identifying and standardising the progress made in scaling up microbial electrolysis cells. Renew Sustain Energy Rev. https://doi.org/10.1016/j.rser.2020.110279
58. Li Z, Fang A, Cui H, Ding J, Liu B, Xie G, Ren N, Xing D (2021) Synthetic bacterial consortium enhances hydrogen production in microbial electrolysis cells and anaerobic fermentation. Chem Eng J. https://doi.org/10.1016/j.cej.2020.127986
59. Liang D, Zhang L, He W, Li C, Liu J, Liu S, Lee HS, Feng Y (2020) Efficient hydrogen recovery with CoP-NF as cathode in microbial electrolysis cells. Appl Energy. https://doi.org/10.1016/j.apenergy.2020.114700
60. Liu C, Sun D, Zhao Z, Dang Y, Holmes DE (2019) Methanothrix enhances biogas upgrading in microbial electrolysis cell via direct electron transfer. Bioresour Technol. https://doi.org/10.1016/j.biortech.2019.121877
61. Liu Q, Ren ZJ, Huang C, Liu B, Ren N, Xing D (2016) Multiple syntrophic interactions drive biohythane production from waste sludge in microbial electrolysis cells. Biotechnol Biofuels. https://doi.org/10.1186/s13068-016-0579-x
62. Liu Y, Wang C, Zhang K, Zhou Y, Xu Y, Xu X, Zhu L (2020) Rapid degradation of 2,4-dichloronitrobenzene in single-chamber microbial electrolysis cell with pre-acclimated bioanode: a comprehensive assessment. Sci Total Environ. https://doi.org/10.1016/j.scitotenv.2020.138053
63. Logan BE, Call D, Cheng S, Hamelers HV, Sleutels TH, Jeremiasse AW, Rozendal RA (2008) Microbial electrolysis cells for high yield hydrogen gas production from organic matter. Environ Sci Technol 42:8630–8640
64. Lu L, Ren ZJ (2016) Microbial electrolysis cells for waste biorefinery: a state of the art review. Biores Technol 215:254–264
65. Metcalf L, Eddy HP & Tchobanoglous G (1991) Wastewater engineering: treatment, disposal, and reuse, McGraw-Hill New York
66. Mohd Shaufi MA, Sieo CC, Chong CW, Gan HM, Ho YW (2015) Deciphering chicken gut microbial dynamics based on high-throughput 16S rRNA metagenomics analyses. Gut Pathog. https://doi.org/10.1186/s13099-015-0051-7
67. Samsudeen N, Spurgeon J, Matheswaran M, Satyavolu J (2020) Simultaneous biohydrogen production with distillery wastewater treatment using modified microbial electrolysis cell. Int J Hydrog Energy 45:18266–18274. https://doi.org/10.1016/j.ijhydene.2019.06.131

68. Neufeld JD, Vohra J, Dumont MG, Lueders T, Manefield M, Friedrich MW, Murrell CJ (2007) DNA stable-isotope probing. Nat Protoc. https://doi.org/10.1038/nprot.2007.109
69. Ngoumelah DD, Harnisch F, Kretzschmar J (2021) Benefits of age-improved resistance of mature electroactive biofilm anodes in anaerobic digestion. Environ Sci Technol. https://doi.org/10.1021/acs.est.0c07320
70. Ogunseitan O (2007) Microbial diversity: form and function in prokaryotes, microbial diversity: form and function in prokaryotes. https://doi.org/10.1002/9780470750490
71. Omidi H, Sathasivan A (2013) Optimal temperature for microbes in an acetate fed microbial electrolysis cell (MEC). Int Biodeterior Biodegrad 85:688–692
72. Panigrahi S, Velraj P, Subba Rao T (2019) Functional microbial diversity in contaminated environment and application in bioremediation. In: Microbial diversity in the genomic era. https://doi.org/10.1016/B978-0-12-814849-5.00021-6
73. Park SG, Rhee C, Shin SG, Shin J, Mohamed HO, Choi YJ, Chae KJ (2019) Methanogenesis stimulation and inhibition for the production of different target electrobiofuels in microbial electrolysis cells through an on-demand control strategy using the coenzyme M and 2-bromoethanesulfonate. Environ Int. https://doi.org/10.1016/j.envint.2019.105006
74. Parkhey P, Gupta P (2017) Improvisations in structural features of microbial electrolytic cell and process parameters of electrohydrogenesis for efficient biohydrogen production: a review. Renew Sustain Energy Rev. https://doi.org/10.1016/j.rser.2016.09.101
75. Pillot G, Davidson S, Auria R, Combet-Blanc Y, Godfroy A, Liebgott PP (2020) Production of current by syntrophy between exoelectrogenic and fermentative hyperthermophilic microorganisms in heterotrophic biofilm from a deep-sea hydrothermal chimney. Microb Ecol. https://doi.org/10.1007/s00248-019-01381-z
76. Qin B, Luo H, Liu G, Zhang R, Chen S, Hou Y, Luo Y (2012) Nickel ion removal from wastewater using the microbial electrolysis cell. Biores Technol 121:458–461
77. Rago L, Ruiz Y, Baeza JA, Guisasola A, Cortés P (2015) Microbial community analysis in a long-term membrane-less microbial electrolysis cell with hydrogen and methane production. Bioelectrochemistry. https://doi.org/10.1016/j.bioelechem.2015.06.003
78. Rani G, Nabi Z, Banu JR, Yogalakshmi K (2020) Batch fed single chambered microbial electrolysis cell for the treatment of landfill leachate. Renew Energy 153:168–174
79. Ruiz V, Ilhan ZE, Kang D-W, Krajmalnik-Brown R, Buitrón G (2014) The source of inoculum plays a defining role in the development of MEC microbial consortia fed with acetic and propionic acid mixtures. J Biotechnol 182:11–18
80. Sapireddy V, Katuri KP, Muhammad A, Saikaly PE (2021) Competition of two highly specialized and efficient acetoclastic electroactive bacteria for acetate in biofilm anode of microbial electrolysis cell. NPJ Biofilms Microbiomes. https://doi.org/10.1038/s41522-021-00218-3
81. Saravanan A, Karishma S, Kumar PS, Yaashikaa PR, Jeevanantham S, Gayathri B (2023) Microbial electrolysis cells and microbial fuel cells for biohydrogen production: current advances and emerging challenges. Biomass Convers Biorefin. https://doi.org/10.1007/s13399-020-00973-x
82. Shakoori AR (2017) Fluorescence in situ hybridization (FISH) and its applications. In: chromosome structure and aberrations. https://doi.org/10.1007/978-81-322-3673-3_16
83. Shao Q, Li J, Yang S, Sun H (2019) Effects of different substrates on microbial electrolysis cell (MEC) anodic membrane: biodiversity and hydrogen production performance. Water Sci Technol 79:1123–1133
84. Speers AM, Young JM, Reguera G (2014) Fermentation of glycerol into ethanol in a microbial electrolysis cell driven by a customized consortium. Environ Sci Technol. https://doi.org/10.1021/es500690a
85. Varanasi JL, Das D (2020) Maximizing biohydrogen production from water hyacinth by coupling dark fermentation and electrohydrogenesis. Int J Hydrog Energy. https://doi.org/10.1016/j.ijhydene.2019.06.030
86. Verea L, Savadogo O, Verde A, Campos J, Ginez F, Sebastian PJ (2014) Performance of a microbial electrolysis cell (MEC) for hydrogen production with a new process for the biofilm formation. Int J Hydrog Energy. https://doi.org/10.1016/j.ijhydene.2014.03.203

87. Villano M, Scardala S, Aulenta F, Majone M (2013) Carbon and nitrogen removal and enhanced methane production in a microbial electrolysis cell. Biores Technol 130:366–371
88. Walker LM, Hoppe T, Silliker ME (2021) Molecular techniques and current research approaches. In: Myxomycetes: biology, systematics, biogeography and ecology. https://doi.org/10.1016/B978-0-12-824281-0.00006-3
89. Wang G, Yao Y, Tang K, Wang G, Zhang W, Zhang Y, Rasmus Andersen H (2021) Cost-efficient microbial electrosynthesis of hydrogen peroxide on a facile-prepared floating electrode by entrapping oxygen. Bioresour Technol. https://doi.org/10.1016/j.biortech.2021.125995
90. Zakaria BS, Dhar BR (2019) Progress towards catalyzing electro-methanogenesis in anaerobic digestion process: fundamentals, process optimization, design and scale-up considerations. Bioresour Technol. https://doi.org/10.1016/j.biortech.2019.121738
91. Zakaria BS, Lin L, Dhar BR (2019) Shift of biofilm and suspended bacterial communities with changes in anode potential in a microbial electrolysis cell treating primary sludge. Sci Total Environ. https://doi.org/10.1016/j.scitotenv.2019.06.519
92. Zeppilli M, Simoni M, Paiano P, Majone M (2019) Two-side cathode microbial electrolysis cell for nutrients recovery and biogas upgrading. Chem Eng J. https://doi.org/10.1016/j.cej.2019.03.119
93. Zhang H, Chen X, Braithwaite D, He Z (2014) Phylogenetic and metagenomic analyses of substrate-dependent bacterial temporal dynamics in microbial fuel cells. PLoS ONE. https://doi.org/10.1371/journal.pone.0107460
94. Zhang J, Zhang Y, Quan X, Chen S (2013) Effects of ferric iron on the anaerobic treatment and microbial biodiversity in a coupled microbial electrolysis cell (MEC)—anaerobic reactor. Water Res. https://doi.org/10.1016/j.watres.2013.06.056
95. Zhang Z, Song Y, Zheng S, Zhen G, Lu X, Takuro K, Xu K, Bakonyi P (2019) Electro-conversion of carbon dioxide (CO_2) to low-carbon methane by bioelectromethanogenesis process in microbial electrolysis cells: the current status and future perspective. Bioresour Technol. https://doi.org/10.1016/j.biortech.2019.01.145
96. Zhao J, Li F, Cao Y, Zhang X, Chen T, Song H, Wang Z (2021) Microbial extracellular electron transfer and strategies for engineering electroactive microorganisms. Biotechnol Adv. https://doi.org/10.1016/j.biotechadv.2020.107682
97. Zhao L, Wang XT, Chen KY, Wang ZH, Xu XJ, Zhou X, Xing DF, Ren NQ, Lee DJ, Chen C (2021) The underlying mechanism of enhanced methane production using microbial electrolysis cell assisted anaerobic digestion (MEC-AD) of proteins. Water Res. https://doi.org/10.1016/j.watres.2021.117325
98. Zhao ZJ, Wang YR, Wang YX, Zhang W, Li ZH, Mu Y (2022) Electrical stimulation enhancing anaerobic digestion under ammonia inhibition: a comprehensive investigation including proteomic analysis. Environ Res. https://doi.org/10.1016/j.envres.2022.113006
99. Zhen G, Lu X, Kumar G, Bakonyi P, Xu K, Zhao Y (2017) Microbial electrolysis cell platform for simultaneous waste biorefinery and clean electrofuels generation: current situation, challenges and future perspectives. Prog Energy Combust Sci. https://doi.org/10.1016/j.pecs.2017.07.003
100. Zijnge V, Welling GW, Degener JE, Van Winkelhoff AJ, Abbas F, Harmsen HJM (2006) Denaturing gradient gel electrophoresis as a diagnostic tool in periodontal microbiology. J Clin Microbiol. https://doi.org/10.1128/JCM.00122-06
101. Zou R, Hasanzadeh A, Khataee A, Yang X, Xu M, Angelidaki I, Zhang Y (2021) Scaling-up of microbial electrosynthesis with multiple electrodes for in situ production of hydrogen peroxide. iScience. https://doi.org/10.1016/j.isci.2021.102094

Microbial Electrolysis Cell Integration with Other Bioelectrochemical Fuel Cells for Sustainable Applications

Babak Jaleh, Ensiye Shabnlou, Mahtab Eslamipanah, and Ghazaleh Ashrafi

Abstract In the past two decades, microbial fuel cell (MFC) and microbial electrolysis cell (MEC) as emerging bioelectrochemical methods have been intensively investigated. These two methods utilizing microorganisms as biocatalysts are basically developed for wastewater treatment and generation of electric energy or hydrogen (H_2), respectively. An attractive and varied approach to improve the possibility and efficiency of these systems is the MEC integrated with MFCs which usually the MFC acts as the power supply of MEC. These systems are designed to overcome the specific challenges of stand-alone biochemical setups effectively. Some of the remarkable advantages of this MEC-MFC system are more sustainable voltage production, better overall treatment performance, energy-saving potential, and cost-effective wastewater treatment operation. However, these systems are still in the initial stages of development. In this chapter, the sustainable application of MEC-MFC systems, their structural elements, and factors affecting their efficiency are investigated and discussed.

Keywords Bioelectrochemical systems (BESs) · Microbial electrolysis cell (MEC) · Microbial fuel cell (MFC) · Wastewater treatment · Hydrogen (H_2) production

1 Introduction

Over the past few decades, there has been a notable and substantial rise in global energy consumption. In response to the diminishing fossil fuel sources namely oil, coal, and natural gas, there is an expanding requirement to develop superseding energy sources. This imperative is further accentuated by significant cost variations and the resulting environmental effects [1]. Nowadays, 80% of the energy needed worldwide is supplied by fossil fuels. Although the depletion of fossil fuel

B. Jaleh (✉) · E. Shabnlou · M. Eslamipanah · G. Ashrafi
Department of Physics, Faculty of Science, Bu-Ali Sina University, 65174 Hamedan, Iran
e-mail: jaleh@basu.ac.ir

A. A. Yaqoob and A. Ahmad (eds.), *Microbial Electrolysis Cell Technology*, Sustainable Materials and Technology, https://doi.org/10.1007/978-981-97-3356-9_8

resources is occurring rapidly, the growing energy demand is amplifying the pressure on daily available resources. Energy production from carbon-containing fossil fuel sources harms the environment such as greenhouse gases (GHGs) releasing and global warming [2]. Considering this, green fuels stand out as a promising and supportable energy source, thanks to their high efficiency, uncomplicated production methods, and environmentally friendly operation [1]. Hydrogen (H_2) is universally accepted as a vital environmentally friendly renewable energy source as well as an alluring selection for fossil fuels' substitute. This is due to its nature as a carbon-free fuel, boasting the highest energy content of any fuel on a global scale [2]. Up to now, the H_2 has been often extracted from fossil fuel sources using steam reforming, gasification, and pyrolysis methods that cause emitting CO_2 [3, 4]. Moreover, wastewater has been introduced as an essential source of energy, nutrients (phosphorus and nitrogen), and materials (fibers, alginate, metals, cellulose, and bioplastics) [1]. While current wastewater treatment methods are reliable in their treatment capabilities, they are not without drawbacks, as they assist the release of greenhouse gases and emerging water quality concerns [1]. To face these challenges, it has been presented environmentally friendly processes. Recently, the production of H_2 from water by electrolysis and biomass (organic food wastes, algae, energy crops, and agricultural residue) in biological processes was introduced as a sustainable technology [5]. Wastewater treatment approaches can be generally categorized into anaerobic and aerobic procedures, and the choice between them is determined by the characteristics of the wastewater sources. A preliminary examination reveals that both well-established techniques and emerging process configurations, such as fermentation and dark fermentation (DF) methods, anaerobic digestion (AD), and bioelectrochemical systems (BESs), have been employed as highly efficient technologies. These technologies play a significant part in the advancement of chemically, energy, and cost-effective sustainable wastewater treatment systems [6]. These methods are much cleaner processes due to energy production and waste minimization at the same time [7]. Two common BESs are microbial electrolysis cells (MEC) and microbial fuel cells (MFCs) [8, 9]. BESs can be employed as a practical technology in the waste remediation, green electricity generation, energy recovery, and production of valuable chemicals under various MEC and MFC processes. MFC is divided into galvanic cells in which exoelectronic electrochemical reactions (electron generator) are carried out in the anode compartment and it flows by a conductive circuit and receives in the cathode compartment with an electron acceptor. The theoretical cell voltage or electromotive force (emf) of the overall reaction, which is the difference between the anode and cathode potentials, the determining factor for whether the system can generate the electricity, is mentioned below [6].

$$\Delta E^{\circ}_{Cell} = \Delta E^{\circ}_{Cathode} - \Delta E^{\circ}_{Anode}$$

A negative free reaction energy results in a positive standard cell voltage. This difference is crucial in differentiating between the electrolytic and galvanic cells. The former operates spontaneously, generating positive cell voltage, while the latter necessitates external electrical input, accompanied by positive free reaction energy and, consequently, a negative cell voltage. Indeed, MFCs are bioelectrochemical devices that leverage microorganisms as catalysts. Their primary function is to convert the chemical energy stored in inorganic and organic substrates and generate an electric energy ultimately [1, 2, 6]. In other words, MFCs have the capacity to convert chemical energy from industrial and wastewaters as well as generate electrical energy using specific electroactive bacteria (EAB). The EAB acts on the oxidant reduction and substrate oxidation reactions leading to the production of energy and minimizing the costs of wastewater treatment [10]. Figure 1a indicates the working principle of a typical MFC. Besides, MEC uses the EAB at the anode for oxidizing organic materials using an electrode as an acceptor of electrons. Electrons traverse the external power source and subsequently unite with protons and oxidants at the cathode to facilitate the electrosynthesis of chemicals [11]. The H_2 production from MEC after wastewater treatment is shown in Fig. 1b. Although each of these methods alone has special advantages, the integration of the MEC and MFC processes can be beneficial in terms of resource recovery (water and other chemicals), energy recovery, and ultimate goals based on improving water quality [6]. For example, the MFC with a high voltage of about 0.80 V can supply the extra energy requirement of the MEC, and H_2 can entirely be collected from the substrate within the microbial cells [12]. Also, one or two more MFCs can be used in the MEC-MFC coupled systems (MEC-MFC CSs) for enhancing the voltage supply [12, 13]. Apart from the generation of H_2, MEC-MFC systems can be also applied to recover heavy metal particles [14], ions [15], or mixtures of different metals [16] from wastewater at the same time using degradation of contaminant [12, 17]. MEC-MFC CSs have been employed for the improvement of decolorization of azo dyes from wastewater [18]. The removal of nutrients from wastewater was also carried out by MEC-MFC systems [12]. MFCs can also be used to power CO_2 reduction in the MECs' cathode [19]. The schematic of the MEC-MFC coupled system's configuration is shown in Fig. 1c. This chapter will discuss some of the sustainable applications of H_2 production and resource-efficient wastewater treatment using integrated MEC-MFC CSs as promising technologies. Also, the factors that affect the efficiency of this system will be investigated.

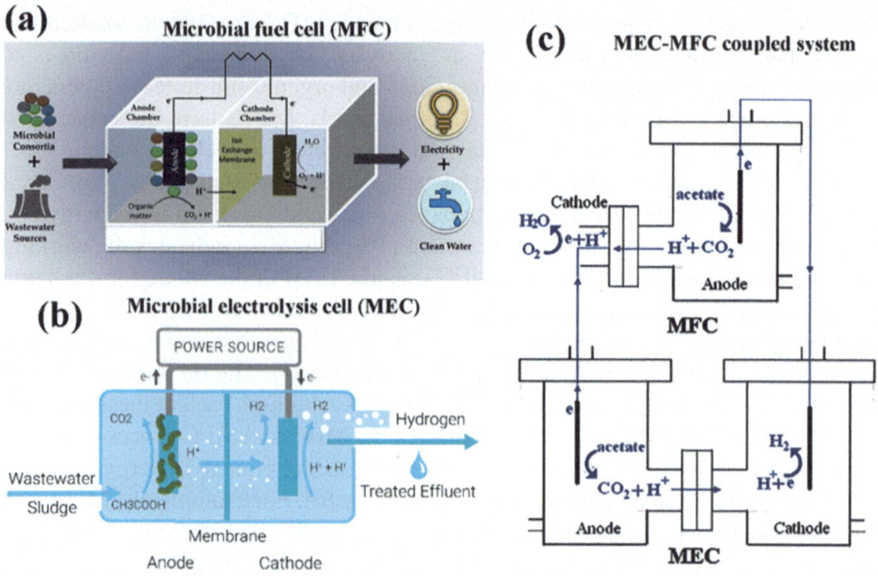

Fig. 1 **a** Configuration of typical MFC. Reproduced with permission from Ref. [10]. Copyright 2023, MDPI. b MEC for treating wastewater to produce H_2. Reproduced with permission from Ref. [20]. Copyright 2021, MDPI. c MEC-MFC coupled system. Reproduced with permission from Ref. [21]. Copyright 2022, Elsevier

2 Hydrogen (H_2) Production by MEC-MFC Systems

H_2 gas emerges as a prospective environmentally friendly energy carrier and a sustainable alternative energy source, widely employed as a primary resource in substantial quantities across the chemical, food, and petroleum sectors. In contrast to traditional energy sources (natural gas, coal, and crude oil), H_2 stores a higher amount of energy per unit volume. The demand for production of H_2 energy has increased in recent years, globally. The majority of H_2 production techniques like gasification, steam reforming, and pyrolysis processes necessitate thermal or electric input from energy sources [2]. As such techniques produce carbon emissions alongside the consumption of fossil fuels, biological H_2 production is increasingly receiving attention [22]. There are different biohydrogen producing methods, namely biophotolysis, photofermentation, dark fermentation, indirect photolysis, and microbial electrolysis [23]. In the microbial electrolysis method, the electrical energy is added through an electrohydrogenesis process, in MECs. The organic matter is usually oxidized through exoelectrogenic bacteria, and electrons are produced in the so-called MFC, simultaneously [24]. In an integrated MEC-MFC-coupled system, hydrogen is generated in an MEC from wastes, and the necessary electricity is prepared in the MFC [13].

2.1 Hydrogen (H_2) Generation Within an MEC-MFC CS

In order to omit the external power supply for producing H_2 gas, an MEC-MFC coupled system was designed [13]. The MFCs were linked both in series and through parallel connections employed as a power supply. Also, the power inputs were regulated by connecting the resistors across the circuit in series. In both MEC and MFC, the electrodes at the anodes and cathodes were manufactured of plain carbon paper (not wet proofed) and carbon paper with Pt on it, respectively. An anaerobic sludge had been applied as inoculums and acetate substrate in order to cultivate the anode biofilms of the coupled reactor over a ten-month period. By increasing loading resistances, the rate of H_2 generation and the density of circuit current was decreased from 2.9 ± 0.2 ml/ld to 0.2 ± 0.0 ml/ld and from 78 ± 12 mA/m^2 to 9 ± 0 mA/m^2, respectively. In a similar trend, the hydrogen yield reduced as a result of resistance increasing. Both the R_{H2} and CE were decreased by increasing the resistance. Comparing the MFCs in a series connection with the parallel configuration indicated that in the former one, the hydrogen generation is tremendously increased, while the latter decreased the production. Using MFCs as a direct power supply for MECs has limitations due to the low voltage generated by MFCs. However, a high voltage can be provided by capacitor circuits and effectively transfer energy and enhance the utilized voltages from MFCs to MECs. Connecting MFCs in a parallel way which empowers the MECs using a capacitor-based circuit in series can enhance the voltage applied from MFCs to MECs, as depicted in Fig. 2a [25]. In order to boost the applied voltage, 0.01 F capacitors and DPDT relays which were programmed by a microcontroller were used to construct circuit for the MEC-MFC system. Adding two 0.02 F capacitors to the MEC-MFC system including two MFCs linked in parallel directly to an MEC increases the H_2 generation rate by approximately 212%. By increasing the number of capacitors to three in the system, a decrease in the H_2 production rate was obtained (Fig. 2b). The total coulombic efficiencies (CE) decreased when the capacitor circuit was added to the system because of circuit losses. Increasing the number of MFCs improved the CE due to providing capacitors' maximum charge and reduced overall energy leakage. Cathode with in-situ deposited copper through Cu(II) removal in an MFC test can enhance the H_2 generation in an MEC mode [26]. Different cathode materials of carbon cloth (CC), nickel foam (NF), and titanium sheet (TS) were chosen to manufacture the cathode. The highest H_2 production rate is estimated for the NF material with the amount of 0.80 ± 0.07 m^3/m^3 d. Increasing the copper value in the solution in the MFC test cannot increase the H_2 generation. It was concluded that by adding 150 mg/L Cu(II) to each catholyte, the highest values of H_2 production and current density were obtained. Also, the lowest values belong to the samples previous to adding Cu(II). However, hydrogen production was reduced when the two highest concentrations of Cu(II) (300 and 500 mg/L) were used. Tungsten (W) and molybdenum (Mo) transition metals are introduced as great electrode materials for H_2 evolution since they can modify the electrode characteristics like surface roughness, electrical conductivity, and resistance. BESs made of three parallel MFC units aligned with an MEC reactor in series were designed to generate H_2 along with the deposition of W and Mo [27]. Dual-chamber BESs with porous graphite felt anodes and stainless steel sheet cathodes were constructed in this study. An inoculum of suspended bacteria from acetate-fed MFCs was added to the anolyte. These connected systems spontaneously worked at a higher voltage than interconnected individual units (1#) because the internal resistance was reduced by adding a number of parallel connected units. It can create more negative cathodic potential in the 1# unit which is enough for the reduction of H^+. The H_2 production within the 1# unit is found to be 0.34 ± 0.01 m^3/m^3 d. The highest H_2 production is calculated for BES-4-1 (four parallel units connected to a single unit in series) and BES-3-2-1 (three parallel MFCs connected to two units in parallel serially stacked with a single unit) with 0.83 ± 0.20 m^3/m^3 d, and 0.82 ± 0.10 m^3/m^3 d, respectively. Although the BES-4-1 indicated the maximal hydrogen generation compared with BES-3-1, the BES-3-1 showed more metal deposition and separation. Therefore, the BES-3-1 module can be chosen for appraisal of multiple cycle operation. Sulfate-reducing bacteria (SRB) is a prominent microbe for preparing biocathode in the MEC. It was proposed to develop an MEC biocathode in an MEC-MFC system for H_2 generation [28]. The ex-situ and in-situ enrichment methodologies were used to enrich SRB. Firstly, the SRB was richened through an ex-situ procedure, in which a mixture of palm oil mill effluent (POME) in serum vials sulfate, CO_2, and H_2 were used. Two types of in-situ processes were applied to fabricate the biocathodes of MEC origin (MEC-O) and MEC-MFC system, which were conducted by transferring the ex-situ enriched culture to the MECs (Fig. 2c). As the H_2 production rate is higher in the MEC-O compared with the MEC-MFC system. In addition, H_2 gas was the major product in the MEC-O with insignificant methane, while methane was identified as the main byproduct in the MEC-MFC biocathode. It can be elucidated by a potential prevalence of methanogens over SRB via anodic enrichment attributed to the presence of acetate during the MFC mode. Increasing sulfate value increased methane generation in the MEC-MFC system. For hydrogen generation by MEC, an MFC reactor can be employed to prepare biofilm on the bioanode in double-chamber MEC units with two various ion exchange membranes constructed from FUMASEP-FBM and Nafion 117 [24]. Wastewater enriched with sodium acetate was used as an anolyte and fed to MFC, in order to cultivate the biofilm formation. When the anodes were fully colonized, the MFC reactor would be used in the MEC configuration. The H_2 generation rate for MFC-Nafion after 24 h was obtained about 90.03 μmoles, while there were not any methane or carbon dioxide detections. However, for MFC-BPM, the hydrogen detection was not possible. The H_2 production yield for the MEC-Nafion unit was measured as 10.3 mg H_2/g chemical oxygen demand (COD), while for MEC-BPM no generation was obtained. However, the H_2 recovery for both systems was relatively low when COD levels were high. Therefore, the bioelectrochemical systems can be used to generate hydrogen at low COD levels.

Molybdenum carbide (Mo_2C) is an impressive electrocatalyst for generating hydrogen in alkaline media that can be produced using different carbon sources. Graphene aerogel is a great candidate for hydrogen production due to its porous structure, leading to improve HER activity. Besides, adding the heteroatoms can increase electrocatalytic activity. Therefore, nitrogen-doped graphene aerogel-supported

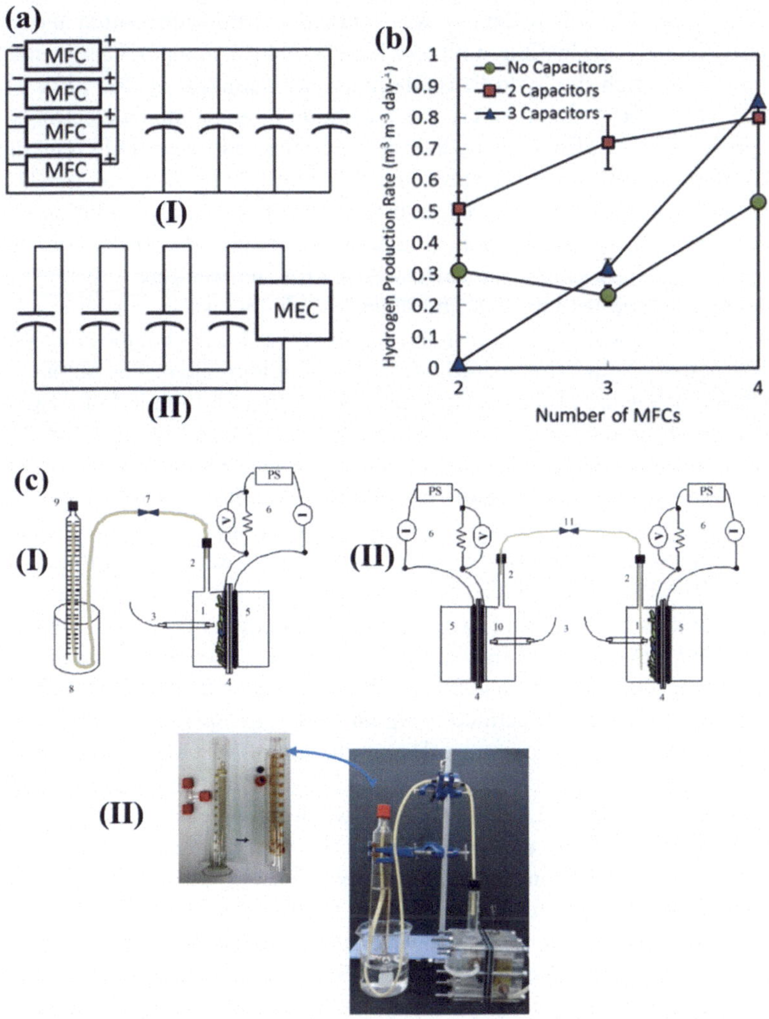

Fig. 2 **a** Circuit based on capacitors, where (I) involves charging capacitors in parallel and (II) discharging capacitors in series. **b** Displays varying H_2 production rates based on different numbers of MFCs and capacitors. Reprinted with permission from Ref. [25]. Copyright 2013, Elsevier. **c** Presents a schematic of (I) the investigated MEC; and (II) hydrogen generated by coupling the MEC Pt-cathode and MEC biocathode during the in-situ biocathode enrichment process. Components include: 1. Biocathode chamber; 2. Gas collection system; 3. Ag/AgCl reference electrode; 4. Gaskets and membrane; 5. Anode chamber; 6. MEC electrical circuit, PS, and multimeter connections; 7. Marprene tube connection and one-way valve; 8. Water displacement system; 9. Rubber stopper screw cap of Hungate tube for gas sampling; 10. Pt-cathode; and 11. Marprene tube connection with one-way valve. (III) Depicts the experimental setup of the MEC and a user-friendly, cost-effective, and simplified new design for the gas collection and sampling apparatus. Reprinted with permission from Ref. [28]. Copyright 2017, Elsevier

molybdenum carbide (Mo_2C/NGA) was fabricated from ammonium molybdate, poly (propylene glycol) bis(2-aminopropylether) (D400), and graphene oxide (GO) as precursors through the hydrothermal-annealing method (Fig. 3a) [29]. The as-fabricated Mo_2C/NGA was then used as a cathode in an ammonia electrolysis cell (AEC) powered with MFCs for hydrogen generation purposes. Different samples with different annealing temperatures and different GO/D400 ratios (no D400 = GA, 4:1, 1:1, and 1:4) were created, and their effect on the H_2 evolution reaction was investigated. Increasing the annealing temperature provided better conductivity. Thus, the Mo_2C/NGA sample annealed with 800 °C temperature illustrated higher HER electrocatalytic activity than the other samples with the annealing temperatures of 600, 400, and 0 °C, as shown in Fig. 3b. It was found that the NGA-800 has poor HER activity, indicating that Mo_2C is the main ingredient. The output voltage of MFCs was 1 V which was used to supply AEC. In a 24-h-period of AEC-MFCs operation, the H_2 production rate was measured about 198.0 mL/g Mo_2C/NGA which is almost three times higher than the obtained nitrogen production rate (66.0 mL/g Mo_2C/NGA). Compared with the Mo_2C/N-rGO/carbon felt-based AEC-MFCs system, the prepared system based on the Mo_2C/NGA indicated better performance which was due to its 3D porous structure and high conductivity. Sulfate-reducing bacterial (SRB) can be employed to bioreduction of heavy metals under anoxic conditions. It was reported that two types of SRB strains (BY7 and SR10) were utilized for the reduction of sulfate and Antimony (Sb) and the generation of H_2 in an MEC reactor [30]. The BY7 and SR10 were inoculated into the MEC anodes. SRB biocathode MEC was manufactured using an MFC that operated with a voltage of 0.8 V. Under this voltage, the average H_2 production of MEC-BY7 and MEC-SR10 were found to be about 67 ± 4.5 and 76 ± 2.3 mL of H_2/cycle, respectively. The H_2 production rate is enhanced after each cycle when the process of sulfate reduction accelerates with the current density rising. As shown in Fig. 3c, after adding heavy metal in SRB, the production rate reduced in both MEC-BY7 and MEC-SR10. This could happen due to the reduction of electron transfer rate as the Sb (V) to Sb (III) converts to other subsequent forms. However, the current density of MEC-SR10 and MEC-BY7 did not alter after the addition of Sb. In a novel design, zinc anode and inoculation of *Shewanella oneidensis* MR-1 were utilized for reducing number of power supply of MFCs for MEC unit [31]. This system was employed to remove oxalate and H_2 production in the MEC coupled with two MFCs which have been illustrated in Fig. 3d. A straight and a spiral MFC anodic compartments were proposed to enhance the highest H_2 production performance. It was found that after reaching a stable condition, the maximum values of OCPs for the spiral and straight microchannels were 1254 mV and 1300 mV, respectively. The OCPs were also stable during 17 h, suggesting that these MFCs are great candidates for operating microfluidic MEC. By comparing the two structures, it was obtained that the current density of the straight microchannel was higher than the spiral one. This could be related to the anodic surface area of the two designs, which holds significance in the biofilm growth. By increasing the flowrate of oxalate, the H_2 production rate increased in MEC-coupled with one MFC, contributing to the improved mass transfer of oxalate in the anolyte of cells and enhanced nutrition of the biofilm. In the zinc anode implemented hybrid system including microfluidic MEC unit coupled with two MFCs, the hydrogen production performance was increased. The maximum hydrogen production rate and power generation in this research were measured about 1.12 molH_2/mol substrate day and 694 mW/m^2.

The presence of Co sites with high adsorption of H^+ and H_2O can provide significant hydrogen evaluation activity for ZIF-67. In addition, a uniform charge carrier distribution can be achieved for ZIF-67 through ligand group or heteroatom of metal Co atom. Carbon cloth modified with ZIF-67 can provide a stronger attachment for sulfate-reducing bacteria (SRB) and increase the generation of hydrogen in a MEC-MFC coupled system [32]. ZIF-67 based biocathode was manufactured from a bioanode MFC which was operated with 0.8 V for inversing polarity. The prepared biocathode indicated lower onset potential compared with the control sample, suggesting fast electron transfer and high hydrogen generation. The charge transfer resistance of the ZIF-67 modified biocathode was lower than the control sample, confirming the rapid charge transfer efficiency of this biocathode. The H_2 production was also compared for both the experimental and control systems prior to and post Sb addition. Before adding Sb, the maximum H_2 production was 0.923 m^3H_2/m^3

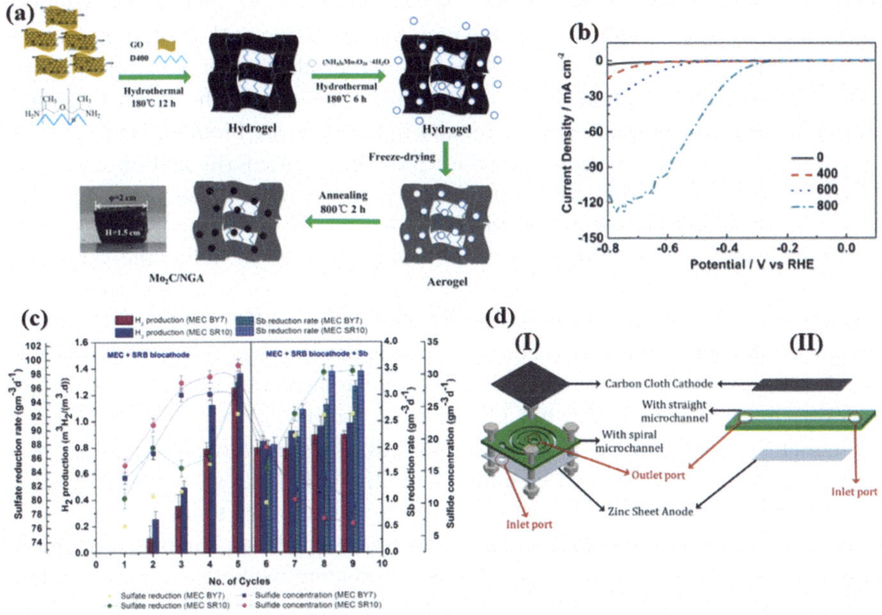

Fig. 3 **a** Depiction of the synthesis of the 3D Mo_2C/NGA material. **b** LSV plots illustrate the HER activity of various samples in KOH aqueous solution. Reprinted with permission from Ref. [29]. Copyright 2020, Elsevier. **c** Comparison of sulfate reduction with H_2 production in relation to Sb reduction and sulfide concentration. Reprinted with permission from Ref. [30]. Copyright 2021, Elsevier. **d** Schematic of a microfluidic MFC featuring (I) a spiral microchannel and (II) a straight microchannel. Reprinted with permission from Ref. [31]. Copyright 2023, Elsevier

and 0.755 m^3H_2/m^3 for the experimental and reference systems, respectively. Therefore, ZIF-67 increased the electron transfer between SR10 and electrode, moreover had a catalytic activity for HER. After adding Sb, the H_2 production was decreased in both systems.

Three different types of organic wastes, including mess food waste (FW), compost leachate (CL), and sewage sludge (SS) were used in dual-chamber H-type MFCs for evaluating bioelectricity generation potential in MFCs and hydrogen generation in MECs [33]. For the SS-, CL-, and FW-fed MFCs, open circuit voltage (OCV) peak value was found at 595 mV after 9 days, 505 mV after 12 days, and 690 mV after 14 days, in the same order. The peak values for power densities also appeared at 2.4 W/m^2, 1.3 W/m^2, and 1.8 W/m^2 for FW-, Cl-, and SS-fed MFCs, respectively. As the obtained results illustrated for the FW-fed MEC, the H_2 generation was higher than SS and CL integrated systems.

2.2 H_2 Production in a Combined Dark Fermentation-MEC-MFC System

Biological hydrogen generation can be achieved through three approaches including dark fermentation, direct photolysis, and photofermentation [34]. Dark fermentation is a facile approach for producing hydrogen from the valorization of various organic wastes. In dark fermentation, bacteria act on the substrate, which can be industrial wastewater, municipal solid waste, and lignocellulosic biomass, and generate hydrogen [35, 36]. A two-step procedure including dark fermentation and electrohydrogenesis was conducted to produce hydrogen from either corn stover or cellobiose. A stability in the produced H_2 gas was observed during multiple cycles when the combined inoculum from the MFCs acclimated to individual substrates was employed in the MEC (single substrate inocula and synthetic (SSI-Syn)) [37]. The applied anodes in MEC were firstly enriched through MFCs. A high H_2 generation rate (1.65 L H_2/L-day) was achieved when cellobiose was selected as substrate (Fig. 4a). Lignocellulose provided a lower H_2 generation rate of 0.25 L H_2/L-day. However, the ratio of hydrogen to CO_2 was 0.93 for the cellobiose substrate which was higher than the lignocellulose substrate (0.85). High hydrogen efficiency was also observed for the combined inoculum from the individual substrates compared with synthetic fermentation effluent (Fig. 4b). It was reported that the maximal H_2 production rate was 88% elevated when SSI was used. In a similar investigation, Wang et al., designed an integrated hydrogen producing method, combining a dark fermentation reactor with the MEC-MFC CS which has been shown in Fig. 4c [38]. Two MFCs which were applied as a power source for MECs, connected in series to an EMC. In addition, cellulose was preferred as a renewable starting substrate. The H_2 generation from cellulose fermentation was constant after 72 h flow operation. The MEC-MFC system with a large MEC and two MFCs was selected as the optimal system because one MFC connected to a large MEC created a high hydrogen production rate as well as three MFCs did not provide significant voltage compared with two MFCs. In a 28 h batch cycle, the mean rate of H_2 generation and the hydrogen yield in the two-reactor system is obtained about 0.28 m^3/m^3d and 9.8 mmol H_2/g COD, respectively. The hydrogen generation rate in the integrated fermentation with MEC-MFC systems was improved by about 41% compared to the dark fermentation.

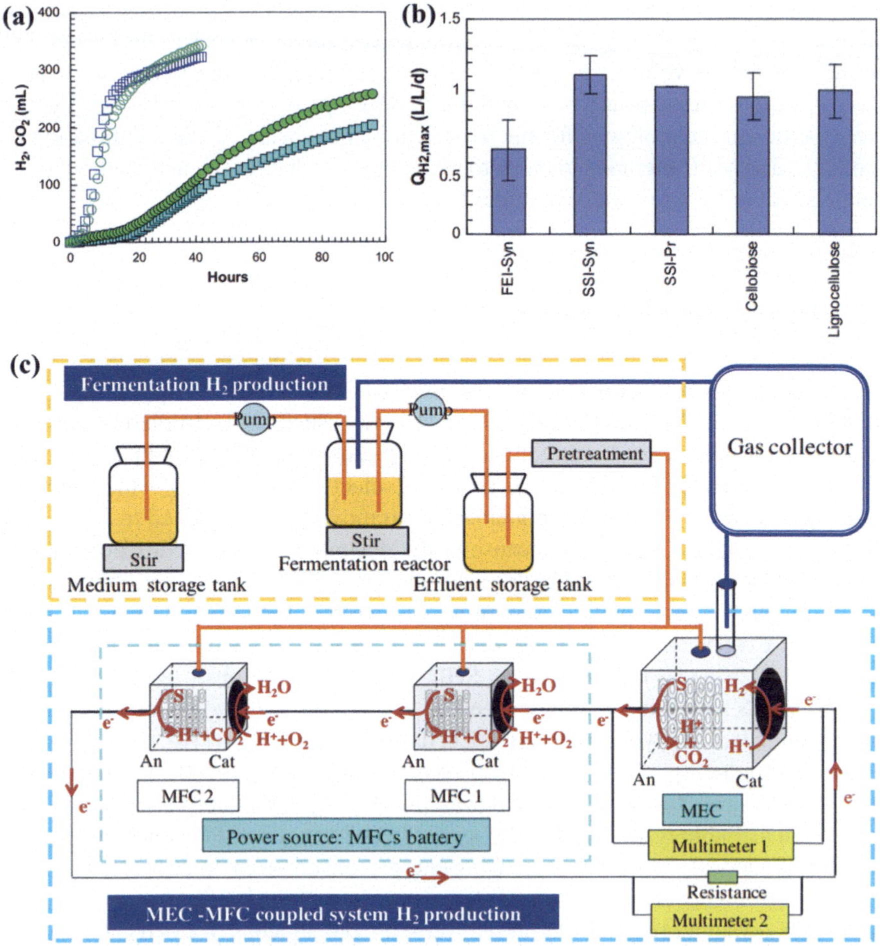

Fig. 4 **a** Hydrogen (squares) and CO_2 (circles) generation by Clostridium thermocellum using cellobiose (open symbols) and corn stover lignocellulose (filled symbols) as the substrate in batch fermentation. **b** Maximum H_2 production rate for fermentation effluents (Standard deviations are calculated for duplicate reactors over three cycles). Reprinted with permission from Ref. [37]. Copyright 2009, Elsevier. **c** Illustration of the integrated hydrogen production process comprising fermentation, an MEC, and MFCs (S: substrates; An: anode electrode; Cat: cathode electrode). Reprinted with permission from Ref. [38]. Copyright 2011, Elsevier

3 Wastewater Treatment by MEC-MFC Systems

The treatment of industrial wastewaters has received much attention in the world. The wastewater should be cleaned of toxic substances, heavy metals, dyes, and organic substances, and if they are not removed, they can strongly pollute soils, surface water, and groundwater [39]. The electrochemical procedure requires high energies and expensive catalysts for decreasing the overpotentials of the electrode. Conventional biological methods can be either replaced or entirely coupled with traditional chemical-physical processes, but they encounter certain issues such as the production of sludge with high volumes or organic carbon requirements as the continuous source [39–41]. Therefore, it would be beneficial to develop environmentally friendly and energy-efficient technologies for the treatment of wastewaters. The combination of two MFC and MEC methods in order to reduce the challenges of each method alone can be suitable for wastewater treatment.

3.1 Heavy Metals' Removal

Heavy metals such as cobalt (Co), chromium (Cr), lead (Pb), copper (Cu), and cadmium (Cd) can be found in industry wastewaters such as electroplating, mining, metal processing, leather tanning, battery, and pigment manufacturing [16, 39]. The new BESs such as MEC and MFC can be a good selection for recovery of these metals, e.g., Cu(II) and Cr(VI) can be spontaneously removed and converted to Cu(0) and Cr(III) on the MFCs' cathodes because of the desirable half-cell redox potential of 0.34 and 1.4 V (versus standard hydrogen electrode, SHE), respectively, compared to that of organic matter (standard conditions, ca. −0.30 V for acetate) [39, 42–44] and Cd(II) and Co(II) can be chemically diminished in MECs at a more negative potential of −0.52 V and −0.23 V (versus SHE, standard conditions), respectively [39, 45, 46]. MFCs have been widely utilized for the diverse reduction of spontaneously reacting metals, namely Au(III), Ag(I), Cu(II), V(V), and Cr(VI) [16], and MECs can be used for reducing of nonspontaneous-reacting metals like Pb(II), Cd(II), Ni(II), Sn(II), Ni(II), and Zn(II) utilizing external power [16]. An effective and promising self-driven technology, the MEC-MFC system, can be used for different metal reduction, simultaneously as well as decreasing energy consumption. In 2014, Huang et al. reported a full recovery of Co from lithium cobalt oxide ($LiCoO_2$) by a self-driven MEC-MFC system [14]. In order to recover Co from aqueous of Co(II), at first, Co should be leached from $LiCoO_2$ particles. Self-driven MEC-MFC system can entirely execute these two procedures. In the first step, Co(II) was obtained from particles of $LiCoO_2$ on the MFCs' cathodes and after that reduced on the MECs' cathodes powered by the Co leaching MFCs [14]. SEM images indicated the particle sizes on the surface of MECs' and MFCs' electrode before (Fig. 5a and b) and after (Fig. 5c and d) Co leaching such that a and c are in MFCs and b and d are in the during of Co(II) reduction in MECs. Smaller particles were observed with blurred borders (Fig. 5c) which was the reason for the apparent disappearance and dissolution of $LiCoO_2$ particles. The surface of MECs' electrode was evenly and densely dispensed with crystals after the reduction of Co(II) leading to a six-party crystal structure (Fig. 5d) in comparison with the bare electrode (Fig. 5b). By this system, energy was recovered from wastewaters using MFCs, and this energy was used to power MECs, and also, it evaded to the requirement of external power support for MECs which achieved complete and simultaneous Co recovery from $LiCoO_2$ particles.

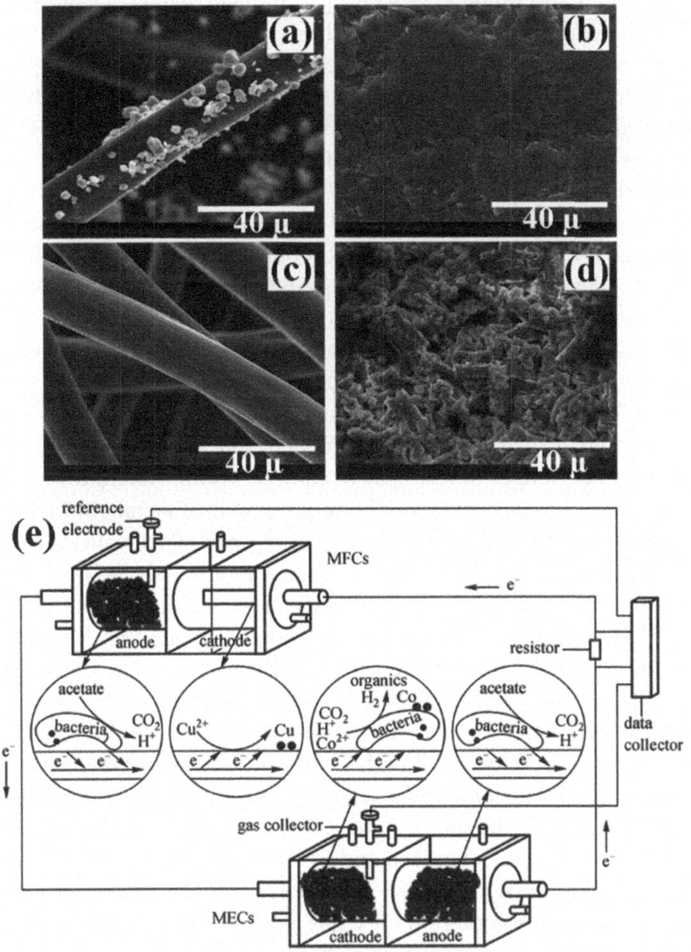

Fig. 5 Scanning electron microscope (SEM) images of the electrode surface prior to **a** and **b** and post **c** and **d** cobalt leaching, with **a** and **c** in MFCs and **b** and **d** during Co(II) reduction in MECs. Reproduced with permission from Ref. [14]. Copyright 2014, Elsevier. **e** Schematic depiction of MECs featuring biocathodes and powered by MFCs for concurrent removal of Co(II) and Cu(II). Reproduced with permission from Ref. [43]. Copyright 2015, Springer

There are effluents that contain different metal concentrations, so that they are very high to discharge into the environment [39, 47]. The recently integrated MEC-MFC systems exhibit promise for the removal of two or more metals under acidic conditions [39]. The employing of biocathodes in these coupled systems to decline electrode overpotentials prevents the requirement of costly inorganic catalysts in merely electrochemical methods; and the microorganisms can be regenerated themselves, and thus, their usage is sustainable [39]. In 2015, the self-driven MECs-MFCs using inexpensive stainless steel mesh (SSM) cathodes in MECs illustrated good wastewater treatment by complete separation of influent metals of Cu(II), Co(II), and Li(I) with low concentration [47]. It could directly be used for treating low-strength wastewater of spent Li-ion batteries treatment processes and chemical precipitation of effluents from spent Li-ion batteries wastewater treatment. It should be noted that in the Co(II)-reduced MEC (MEC_{Co}) driven with Cu(II)-reduced MFC (MFC_{Cu}), the cathode material of MEC_{Co} and cathode volumes in both MFC_{Cu} and MEC_{Co} played an important role in improving simultaneous recovery of Co and Cu with any external energy usage [48]. Either SSM or TS cathode could indicate the impressive reduction of Co(II), whereas carbon rod (CR) could not continue to do so. On the CR cathode with smaller cathode volumes, a significant Co(II) reduction was achieved in $MFC_{Cu,}$ but the highest reduction of Co(II) using SSM and TS was obtained in both MEC_{Co} and MFC_{Cu} with smaller cathode volumes. These sequential MFC_{Cu} and MEC_{Co} processes achieved a high reduction of Cu(II) and Co(II) with a value of 100% and 65.3–72.0% using either TS or SSM cathodes. This was primarily attributed to the higher resistance of TS compared to CR or even SSM, leading to a neglect of resistance changes in response to cathode volume alterations in either just MFC_{Cu} or both MEC_{Co} and MFC_{Cu}. The requirement of enhancing BESs' performance is a suitable power source to overcome overpotentials' electrode in MECs and harvest low power density of MFCs for the removal of metals [43]. In 2015, MECs driven by MFCs were introduced as an effective approach to prevent the individual challenges of MECs and MFCs for removal of metals such that MFCs worked as the external power of MECs and MECs worked as the external resistance for MFCs (seen Fig. 5e) [43]. Biocathodes of MECs which derived with Cu(II) reduction in MFCs could decline the Co(II) removal overpotentials on the MECs' cathodes and so achieved highly efficient removal of both Cu(II) and Co(II) in MFCs and connected MECs, respectively, without any external energy usage [43]. MECs with biocathodes driven by MFCs improved the removal rates of Cu(II) with a value of 6.0 ± 0.2 mg/L.h at 50 mg/L of initial concentration and 5.3 ± 0.4 mg/L.h for Co(II) at an initial concentration of 40 mg/L which were 3.3 and 1.7 folds more than those in MECs driven by MFCs and with abiotic cathodes. It is worth mentioning that a unique bacterial community plays an important role in the biocathodes' performance such that with a distinct constitution of the bacterial community, biocathodes of MECs using voltage from connected MFCs are developed for the removal of the metals [43]. The composition and diversity of the bacterial community on the MECs' cathodes were studied with high throughput sequencing of 16S rRNA gene and phylogenetic tests on biocathodes exhibited that *Proteobacteria* mainly constitute 67.9% of the entire reads and after it was *Firmicutes* (14.0%), *Bacteroidetes* (6.1%), *Tenericutes* (2.5%),

Lentisphaerae (1.4%), and *Synergistetes* (1.0%). Therefore, MEC-MFC CSs are an effective and cost-effective way for removing several heavy metals from wastewater at the same time.

In one more research [39], in order to improve multiple-reactor BESs' performance for reduction of mixed metals, at first the biocathodes were separately adapted to three different metals using MFCs with Cu(II) and Cr(VI) because these metals exhibit high redox potentials and MEC for reduction of Cd(II) has a more negative redox potential (Fig. 6a). The BESs were initially acclimated to a mixture of metals with low concentrations and subsequently transitioned to higher concentrations of those metals. This process resulted in a selective and complete reduction of metals at rates of 0.98 ± 0.01 mg/L h for Cd(II), 1.07 ± 0.01 mg/L h for Cu(II), and 1.24 ± 0.01 mg/L h for Cr(VI). Removal of metals in the abiotic systems was related to physical adsorption and chemical reduction, whereas in the biotic systems, the processes ascribed with the bacteria presence, bio-adsorption besides physical adsorption, and metals' biological reduction. Also, the biocathodes in the following reactors have compatibility with high mixed metal solutions. Community analysis indicated that the composition of bacteria was completely varied after increasing mixed metal influents, and overall they exhibited less diverse communities after exposure to solutions' mixed metal. The relative abundance comparison of bacterial reads recovered from the biocathodes which adaptively evolved with mixed and single metals and classified at the phylum and class levels' distribution of the most predominant phylum of Proteobacteria were indicated in Fig. 6b and c.

By applying various reaction potentials, a self-sustained MEC-MFC CS was used for reducing multiple metals simultaneously. Within three days, three heavy metals of Cr, Pb, and Ni were reduced in this kind of system with the efficiency of 95% [15]. The 189.4 mW/m^2 was the MFCs' power density, and it was enough to support MECs. The energy recovery, in relation to the energy storage circuit (ESC), amounted to approximately 450%. This system was anaerobically worked with metals as the acceptor of electrons, which significantly saved the operating cost and energy. Additionally, the self-sustained MEC-MFC CS could generate the flexible voltage required for heavy metals' removal with various redox potentials in MEC by controlling the wastewater pH, metal concentration as well as the number of electrodes and the capacitors in MEC and MFC. Through more investigations, it was achieved a high-effective Pb(II) reduction with a kind of MEC-MFC coupling setup in which the MEC was driven by the Cr(VI)-reduced MFC. Hence, the Pb(II) and Cr(VI) as heavy metal pollution were simultaneously reduced without external energy input as seen in Fig. 7a [16]. The performance of this coupling system depended on cathode materials, initial metal ions, and anions' concentration. In this order, three different electrodes incorporating NF, CC, and SSM were carried out as MEC cathode led to deposit Pb on cathode electrodes. Before and after Pb(II) reduction, the SEM images of CC, SSM, and NF are illustrated in Fig. 7b–g. The CC surface had dense and uniform fibers which after Pb(II) reduction, its surface was deposited by Pb particles. Many large cavities were structured on the NF and SSM electrodes in comparison with the CC electrode, and they resulted in more precipitation after Pb(II) reduction. NF with good properties namely high conductivity, low-cost, significant surface area, and excellent stability

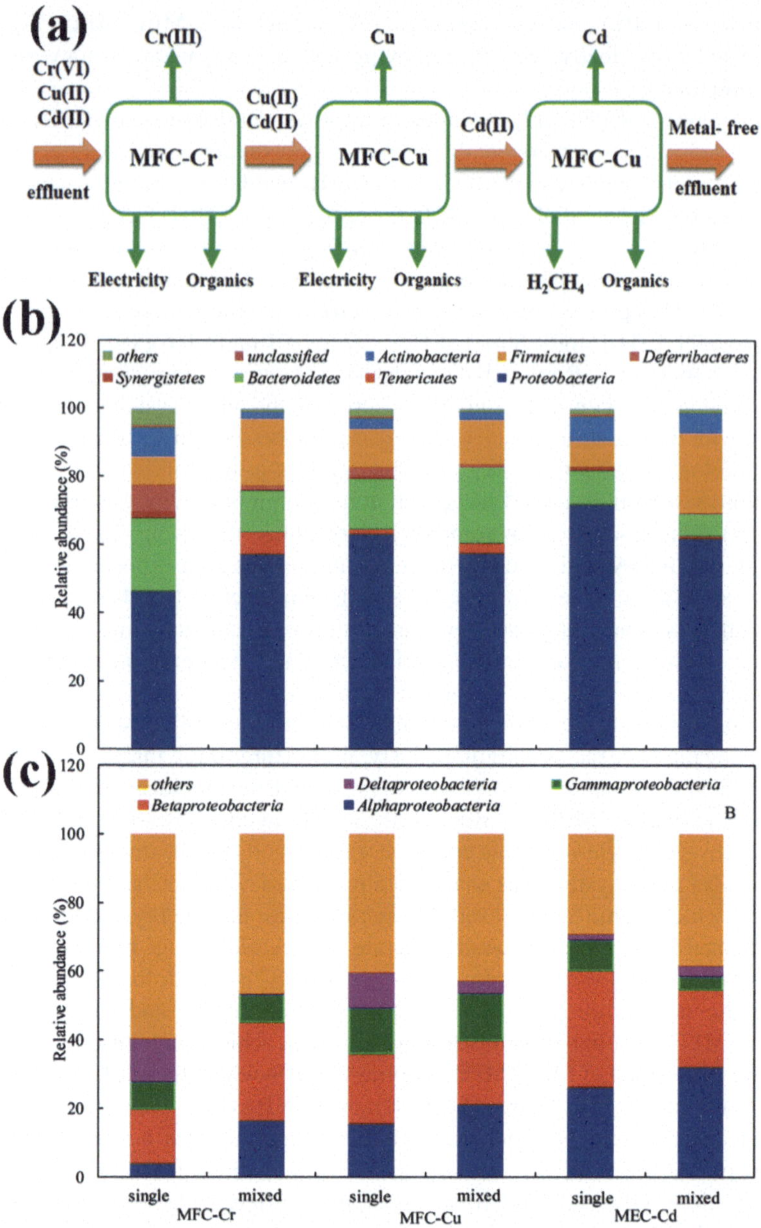

Fig. 6 **a** Operational mechanism of the MEC-MFC CS for the reduction of Cu(II), Cr(VI), and Cd(II). Comparative analysis of the proportional abundance of bacterial reads obtained from biocathodes adaptively evolving with single and mixed metals and categorized at **b** the phylum level and **c** the class level, highlighting the most prevalent phylum, Proteobacteria. Reprinted with permission from Ref. [39]. Copyright 2015, American Chemical Society

exhibited the best Pb(II) reduction when used as a cathode of MEC driven with the Cr(VI)-reduced MFC, and also it indicted the largest Cr(VI) reduction rate of about 1.72 g/m^3 h yielding maximum power output of 702.86 mW/m^2 compared with those of SSM (1.47 g/m^3 h) and CC (1.32 g/m^3 h) electrodes in MFCs. Continuous recovery and separation of multiple heavy metals can be performed utilizing a photo-assisted BESs consisting of one MFC_{Cu} (predominant for Cu), one MEC_{Ni} (primary for Ni), and one MEC_{Zn} (dominant for zinc (Zn)) feeding with etching terminal wastewater (ETW) [49]. Light radiation and circuital current had synergistic roles in heavy metal recovery. Every cathod of units was included in $WO_3/MoO_3/g\text{-}C_3N_4$ which had significant stability. This photo-assisted cathode provided photo-generated electrons and holes, leading to high circuital currents. Light radiation provided high open circuital voltage and power generation in MFC_{Cu} that led to high hydrogen production in MEC_{Ni} and MEC_{Zn}. A MFC could also generate the needed power for MEC in a photo-assisted BESs which feeds with etching terminal wastewater [50]. In MFCs, Ni(II), Zn(II), and Cu(II) can be simultaneously removed and hydrogen can also be generated in the coupled MEC. The heavy metal removal of Cu(II), Ni(II)), and Zn(II) was higher than that of uncoupled MEC. In the absence of light radiation, the heavy metal removal was better than control open circuital conditions, indicating a significant role of circuital current as opposed to light irradiation for removing heavy metals.

3.2 *Ammonium Ion (NH_4^+) Removal*

Although N as a basic life element has an important role in energy transferring and ecosystem substance circulation, thropogenic nitrogen such as NH_4^+ creates environmental pollutants. For example, aquatic NH_4^+ can provide widespread pollutants for rivers, underground water, and lakes. Accordingly, NH_4^+ removal is essential for retaining the building and function of aquatic ecosystems. It can be carried out through bioelectrochemical process which is a promising route for high efficiency removal of NH_4^+ utilizing stimulated microbial and approaches [51]. Anodic oxidation of NH_4^+ can also be performed in MFCs; however, a low external power needed to supply due to the low potential difference between the electrodes. The anodic oxidation of NH_4^+ to NO_2^- and NO_3^- produces electron which transfers to the cathode through the electric circuit and uses for reducing NO_2^- and NO_3^- to N_2. These processes can simultaneously be carried out in an MEC system with a small voltage supply which can be powered by a potentiostat and a double-chamber MFC [52]. It was found that the power supply by MFC can accelerate the oxidation of NH_4^+ in the MEC. The ammonium nitrogen (NH_4^+–N) removal rate was achieved at $E_{ap} > 0.8$ V conditions, as shown in Fig. 8a. Although no remarkable agglomeration was observed for NO_2^-, some agglomeration of NO_3^- appeared with a high NO_3^--N generation rate at E_{ap} of 0.8 V. It indicated that the reduction of NO_3^- is a limitation removing total nitrogen in this MEC-MFC. However, the agglomeration of NO_2^-

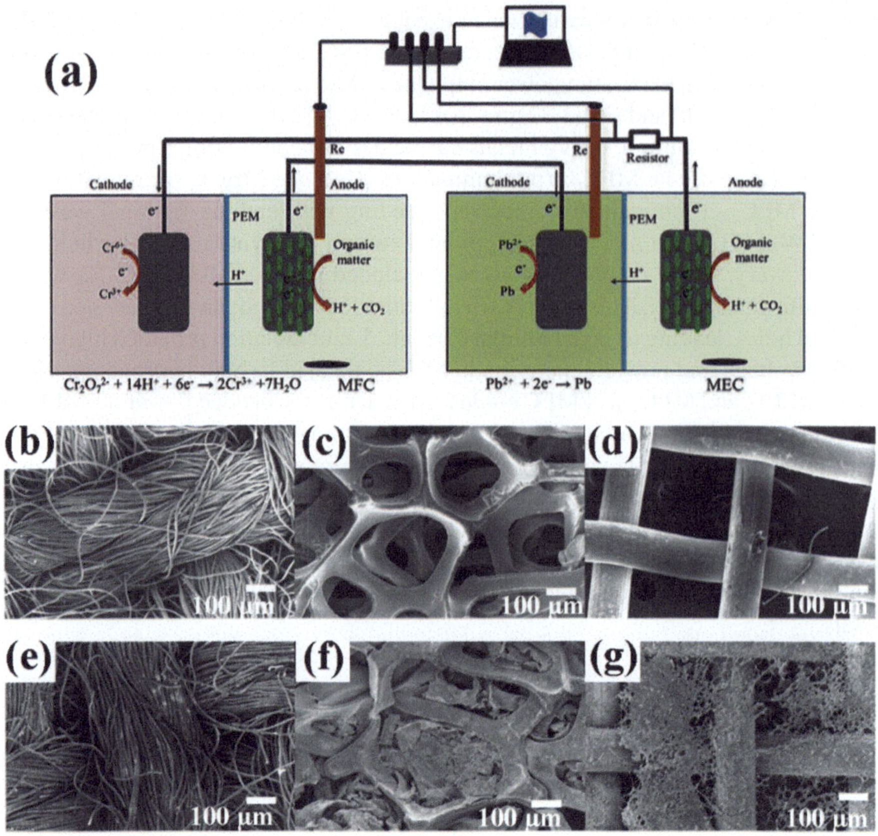

Fig. 7 **a** Depiction of the Pb(II) reduction process on diverse cathodes of MEC powered by chromium (Cr(VI))-reduced MFC. Scanning electron microscopy images on the carbon cloth cathode **b**, **e**, nickel foam cathode **c**, **f**, and stainless steel mesh cathode **d**, **g** of MEC before **b**, **c**, **d** and subsequent to **e**, **f**, **g** Pb(II) reduction. Reproduced with permission from Ref. [16]. Copyright 2019, Elsevier

and NO_3^- in MEC-potentiostat was more than MEC-MFC system. Figure 8b shows a summary of nitrogen elimination routes in a solitary chamber of MEC.

3.3 *Azo Dyes Removal*

A large number of azo dyes as one group of major pollutants that are toxic and very stable against biodegradation can be found in textile industries' wastewater. Various technologies developed for the treatment of azo dyes including absorption by organic and inorganic materials, mineralization and decolorization by photocatalysis and oxidation processes, membrane filtration, enzymatic or microbiological degradation,

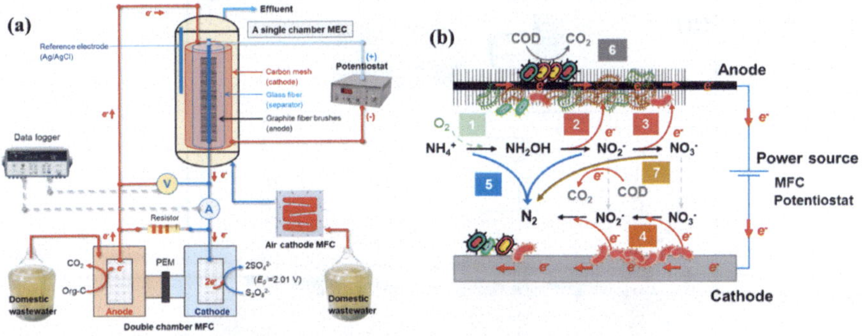

Fig. 8 **a** Diagram illustrating an MEC-MFC CS and an MEC-potentiostat system. The input voltage of the MEC was controlled by adjusting the external resistance. In the MEC-potentiostat system, a double-chamber MFC was substituted with a potentiostat. **b** Proposed pathways for removing nitrogen in a single-chamber MEC. (1) Oxidation of NH_4^+ to NH_2OH with ambient oxygen by ammonia-oxidizing bacteria (AOB), (2) anodic oxidation of NH_2OH to NO_2^- by AOB, (3) anodic oxidation of NO_2^- to NO_3^- by nitrite-oxidizing bacteria (NOB), (4) cathode-dependent denitrification, (5) anammox, (6) anodic oxidation of COD by anode-respiring bacteria (ARB), and (7) heterotrophic denitrification. Reproduced with permission from Ref. [52]. Copyright 2021, Elsevier

etc. [18, 53]. More recently, MFC as a promising and innovative method for azo dyes' decolorization has received considerable attention in wastewater treatment and energy recovery at the same time. For the first time, in 2015, to improve the removal of azo dye employing MFC, the MEC-MFC CS was introduced [18]. In this system, the MFC performed as a driving force, so it provided enough current flow in the MEC for reducing azo dyes. Furthermore, the electrons use in-situ with azo dye in the MEC cathode. This system can receive low energy generated with the MFC leading to prevent more energy losses due to the energy storage and transportation [18]. The MEC-MFC CS which its schematic diagram indicated in Fig. 9a was employed for methyl orange (MO) decolorization. It was reported that the MO decolorization rate in the MEC-MFC CS had a significant increase of 36.52% to 75.28% compared with the single MFC system. As seen in Fig. 9b and c, it was reported that the cathodic pH of the MEC and the MFC in a determined range were not so impressive on the removal of azo dyes in the MEC-MFC CS. It concluded that the dosage of buffer could be considerably saved in the MEC-MFC CS, which would significantly decrease the azo dye wastewater treatment cost. Zoua et al. reported a new approach using an electrolysis cell (EC) combined with MFC for methyl red (MR) removal and generating electricity at the same time [53]. The COD removal and MR decolorization efficiencies were 89.3% and 100%, respectively, with a cell voltage output generation of a 0.56 V.

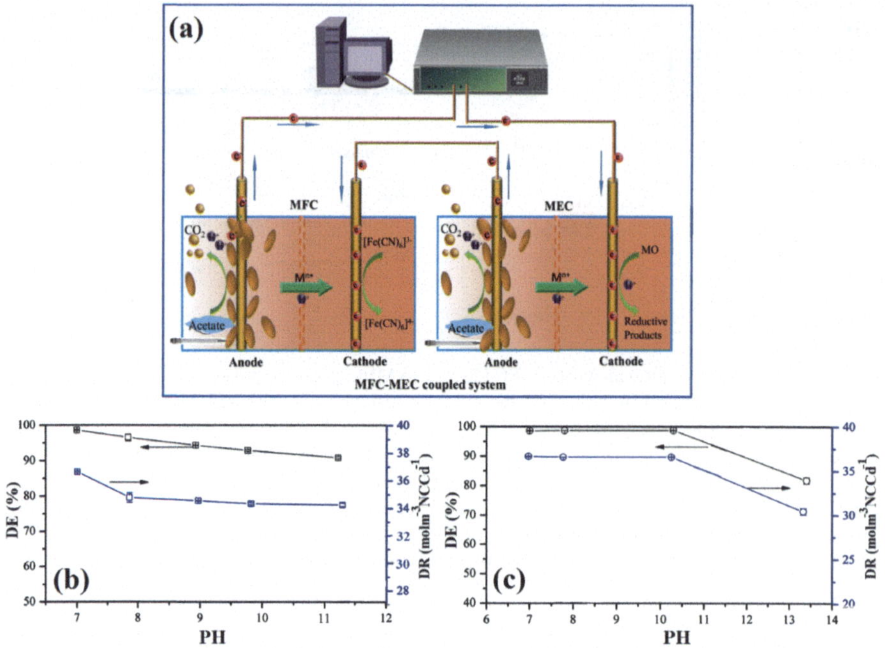

Fig. 9 Representation of the process in the MEC-MFC CS **a**, Impact of cathodic pH on **b** the MEC and **c** the MFC on the removal performance of methyl orange (MO) in the MEC-MFC CS. Reproduced with permission from Ref. [18]. Copyright 2016, Elsevier

3.4 Carbon Dioxide (CO_2) Reduction

Carbon dioxide (CO_2) as the main reason for global warming and therefore reducing its atmospheric concentration is an important issue. The CO_2 reduction to valuable chemical materials such as formic acid [54], methane [55], and methanol [56] using the electrochemical method is very momentous for CO_2 recycling [54]. In general, CO_2 reduction has a high overpotential in the electrochemical process. In this fact, when a common metal is used as the cathode, it requires a voltage greater than 4 V to reduce the potential of the cathode below −1.5 V (versus Ag/AgCl) toward highly efficient CO_2 reduction. MFC system can provide a low voltage of less than 0.8 V which cannot support the requirement potential for CO_2 reduction. Certainly, linking a series of MFCs can augment the overall voltage. However, persistent challenges like the elevated applied voltage, substantial overpotential, and heightened energy consumption continue to exist in the electrochemical CO_2 reduction process. MEC-MFC coupled system can be a good selection as high-effective technique for CO_2 reduction [54]. The multiwalled carbon nanotubes (MWCNT) and cobalt tetra-amino phthalocyanine (CoTAPc) composite electrode which was generated with the layer-by layer (LBL) self-assembly approach could enhance the CO_2 reduction in MEC driven by an MFC. It could decrease the overpotential of CO_2 reduction with the formic acid generation rate of up to 21.0 ± 0.2 mg/L h. The mechanism of MEC-MFC system is illustrated in Fig. 10a. This improvement in CO_2 reduction may be due to the two factors. First, the interaction between the CoTAPc and MWCNTs could decline the overpotential of CO_2 reduction to formic acid. Second, the MWCNTs have a good electrical conductivity which enhance the transferring of electrons between the active sites of CoTAPc and CO_2. As a result, the current was increased (by about 20%). As seen in the SEM image of MWCNT/CoTAPc (Fig. 10b), the MWCNTs were constantly distributed on the surface of the modified electrode. The MEC-MFC-coupled system was also developed to electrochemically CO_2 reduction to H_2 and CO. The mixture of H_2 + CO gas can be applied in some chemical catalyzed procedures to generate fuel or convert it into hydrocarbons [19]. Until now, different kinds of catalysts namely organometallics, metal catalysts, and ionic liquids have been developed for overcoming the high energy barrier of CO_2 electrochemical reduction. The 1-ethyl-3-methylimidazolium tetrafluoroborate (EMIM-BF_4) as a high practical ionic liquid could effectively diminish the CO_2 reduction' over-potential, but it still required more external voltage for coupled anodic H_2O oxidation. Hence, it was needed a more compatible process with various power sources to decrease its energy usage and the external voltage. The electrochemical CO_2 reduction integrated with MEC could extremely decrease the used external voltage. The optimum external voltage was −0.4 V for CO_2 reduction rate of 0.033 mmol/m^2 h in the MEC mode. In the MEC mode, the optimized external voltage was obtained −0.4 V for the CO_2 reduction rate of 0.033 mmol/m^2 h. If the voltage decreases more to about −0.5 V, the CO_2 reduction selectivity would considerably decrease. On the other hand, when the MFC powered the MEC by generating the maximal power density of 18.5 W/m^2, the CO_2 reduction to CO was achieved to be 0.06 mmol/m^2 h.

3.5 *H_2O_2 Production and Residual H_2O_2 Removal*

The Fenton process which consists of the in-situ production of hydroxyl radical (OH·) has been widely used as an efficient and promising process for contaminant removal. Sustainable supply of H_2O_2 and inexpensive removal of residual H_2O_2 are two main issues related to the prosperous field of the Fenton process for recalcitrant organic pollutants removal. The electro-Fenton process was used for in-situ generating H_2O_2, but its electric energy utilization was high. Therefore, MECs and MFCs was suggested as a promising technique for H_2O_2 generation [57]. The requirement of these methods is noble-metal-free cathode like graphite to form a two-electron reaction (Eq. (1)) and prevent more H_2O_2 oxidation. Despite the beneficial aspects of MFCs and MECs, their production rate of H_2O_2 falls significantly below the necessary threshold for achieving optimal treatment performance. MECs are better partners for Fenton reaction in an H_2O_2 generation capacity but require an electricity supply that could be powered with MFCs. Hence, MFCs could provide this requirement owing to their advantages in the cathode reduction and anode oxidation. It has been suggested that H_2O_2 is superseded as an electron acceptor to oxygen in the

MFCs' Pt-catalyzed cathode, where H_2O_2 is reduced to water (Eq. (2)).

$$O_2 + 2e^- + 2H^+ \rightarrow H_2O_2 (E = +0.450\ V_{SCE}) \tag{1}$$

$$H_2O_2 + 2e^- + 2H^+ \rightarrow H_2O\ (E = +1.534\ V_{SCE}) \tag{2}$$

As a result, a bioelectro-Fenton system, with the ability to switch alternately between MEC and MFC modes, was devised. In the MEC mode, the bioelectro-Fenton system generated hydrogen peroxide (H_2O_2), which subsequently reacted with Fenton's reagent (Fe II) to produce hydroxyl radicals (OH·). By transitioning the system to the MFC mode, any remaining unused H_2O_2 was eliminated as an electron acceptor (refer to Fig. 10c). To prove this claim, a synthetic wastewater including methylene blue (MB) was used in the MEC mode, and after 8 h, 97% of MB was decolorized. Subsequently, by 39 h switching on MFC mode, the residual H_2O_2 of 180 mg/L from the MEC was entirely removed with the removal rate of 4.61 mg/L h. The MB degradation and H_2O_2 production in the BES were developed by using several MFC reactors as power supply in the MEC mode. It should be noted that, in this kind of system, the residual H_2O_2 removal and MB degradation were influenced by cathode pH, initial MB concentration, and external resistance.

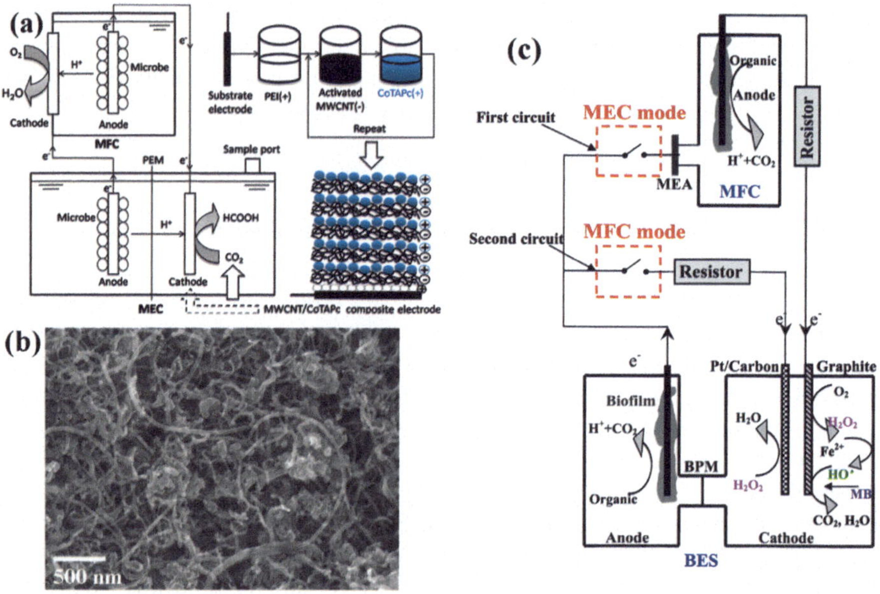

Fig. 10 **a** Mechanism of CO_2 reduction in an MEC-MFC system, **b** SEM image of MWCNT/CoTAPc electrode. Reproduced with permission from Ref. [54]. Copyright 2012, American Chemical Society. **c** Schematic representation of the bioelectro-Fenton setup. Reproduced with permission from Ref. [57]. Copyright 2015, Elsevier

4 Conclusion

MEC integration with other bioelectrochemical fuel cells is a new and successful technology for sustainable applications like CO_2 reduction, production of H_2 and H_2O_2 or other products as well as wastewater treatment by heavy metal, azo dyes, and NH_4^+ removal. Hybrid systems of this nature exhibit greater promise when compared to stand-alone systems. MFCs are able to recover the potential energy in wastewater and convert it directly into electricity and also can remove different kinds of pollutants. By using MFC as the power source of MEC, the necessary energy for the MEC can be provided. The remarkable advantages of the MEC-MFC system are better overall treatment efficacy, more stable generated voltage, and energy storage potential. Many factors can affect the performance of MEC-MFC systems namely reactor configurations, the materials of electrode and membrane, microorganisms, etc., which can improve the industrial application.

References

1. Arun J, SundarRajan P, Pavithra KG, Priyadharsini P, Shyam S, Goutham R, Le QH, Pugazhendhi A (2024) New insights into microbial electrolysis cells (MEC) and microbial fuel cells (MFC) for simultaneous wastewater treatment and green fuel (hydrogen) generation. Fuel 355:129530
2. Goren AY, Dincer I, Khalvati A (2023) A comprehensive review on environmental and economic impacts of hydrogen production from traditional and cleaner resources. J Environ Chem Eng 111187
3. Jaleh B, Nasrollahzadeh M, Eslamipanah M, Nasri A, Shabanlou E, Manwar NR, Zboril R, Fornasiero P, Gawande MB (2022) The role of carbon-based materials for fuel cells performance. Carbon 198:301–352
4. Tanzil AH, Brandt K, Wolcott M, Zhang X, Garcia-Perez M (2021) Strategic assessment of sustainable aviation fuel production technologies: yield improvement and cost reduction opportunities. Biomass Bioenerg 145:105942
5. Morya R, Andrianantenaina FH, Singh S, Pandey AK, Kim G-B, Verma JP, Kumar G, Raj T, Kim S-H (2023) Exploring rice straw as substrate for hydrogen production: critical challenges and opportunities. Environ Technol Innov 103153
6. Gude VG (2018) Integrating bioelectrochemical systems for sustainable wastewater treatment. Clean Technol Environ Policy 20:911–924
7. Ndayisenga F, Yu Z, Wang B, Wu G, Zhang H, Phulpoto IA, Zhao J, Yang J (2022) Thermophilic-operating environment promotes hydrogen-producing microbial growth in a lignocellulose-fed DF-MEC system for enhanced biohydrogen evolution. Process Saf Environ Prot 167:213–224
8. Heidrich ES, Edwards SR, Dolfing J, Cotterill SE, Curtis TP (2014) Performance of a pilot scale microbial electrolysis cell fed on domestic wastewater at ambient temperatures for a 12 month period. Biores Technol 173:87–95
9. Winfield J, Ieropoulos I, Greenman J (2012) Investigating a cascade of seven hydraulically connected microbial fuel cells. Biores Technol 110:245–250
10. Malik S, Kishore S, Dhasmana A, Kumari P, Mitra T, Chaudhary V, Kumari R, Bora J, Ranjan A, Minkina T (2023) A perspective review on microbial fuel cells in treatment and product recovery from wastewater. Water 15:316

11. Rabaey K, Rozendal RA (2010) Microbial electrosynthesis—revisiting the electrical route for microbial production. Nat Rev Microbiol 8:706–716
12. Chorbadzhiyska E, Apostolova D, Bardarov I, Mitov M, Hubenova Y (2020) Hybrid MFC-MEC systems: principle and applications. Bulg Chem Commun 52:30–34
13. Sun M, Sheng G-P, Mu Z-X, Liu X-W, Chen Y-Z, Wang H-L, Yu H-Q (2009) Manipulating the hydrogen production from acetate in a microbial electrolysis cell–microbial fuel cell-coupled system. J Power Sources 191:338–343
14. Huang L, Yao B, Wu D, Quan X (2014) Complete cobalt recovery from lithium cobalt oxide in self-driven microbial fuel cell–microbial electrolysis cell systems. J Power Sources 259:54–64
15. Li Y, Wu Y, Liu B, Luan H, Vadas T, Guo W, Ding J, Li B (2015) Self-sustained reduction of multiple metals in a microbial fuel cell–microbial electrolysis cell hybrid system. Biores Technol 192:238–246
16. Li M, Zhou S, Xu Y (2019) Performance of Pb (II) reduction on different cathodes of microbial electrolysis cell driven by Cr (VI)-reduced microbial fuel cell. J Power Sources 418:1–10
17. Zhang Q, Yang H, Zhou T, Chen X, Li W, Pang H (2022) Metal-organic frameworks and their composites for environmental applications. Adv Sci 9:2204141
18. Li Y, Yang H-Y, Shen J-Y, Mu Y, Yu H-Q (2016) Enhancement of azo dye decolourization in a MFC–MEC coupled system. Biores Technol 202:93–100
19. Li Z, Wang Q, Liu D, Yang B, Zhang X, Lei L (2013) Ionic liquid-mediated electrochemical CO_2 reduction in a microbial electrolysis cell. Electrochem Commun 35:91–93
20. Fudge T, Bulmer I, Bowman K, Pathmakanthan S, Gambier W, Dehouche Z, Al-Salem SM, Constantinou A (2021) Microbial electrolysis cells for decentralised wastewater treatment: the next steps. Water 13:445
21. Aboelela D, Soliman MA (2022) Hydrogen production from microbial electrolysis cells powered with microbial fuel cells. J King Saud Univ-Eng Sci
22. Logan BE (2004) Peer reviewed: extracting hydrogen and electricity from renewable resources. Environ Sci Technol 38:160A-167A
23. Sivaramakrishnan R, Shanmugam S, Sekar M, Mathimani T, Incharoensakdi A, Kim S-H, Parthiban A, Geo VE, Brindhadevi K, Pugazhendhi A (2021) Insights on biological hydrogen production routes and potential microorganisms for high hydrogen yield. Fuel 291:120136
24. Chacón-Carrera R, López-Ortiz A, Collins-Martínez V, Meléndez-Zaragoza M, Salinas-Gutiérrez J, Espinoza-Hicks J, Ramos-Sánchez V (2019) Assessment of two ionic exchange membranes in a bioelectrochemical system for wastewater treatment and hydrogen production. Int J Hydrogen Energy 44:12339–12345
25. Hatzell MC, Kim Y, Logan BE (2013) Powering microbial electrolysis cells by capacitor circuits charged using microbial fuel cell. J Power Sources 229:198–202
26. Wang Q, Huang L, Pan Y, Zhou P, Quan X, Logan BE, Chen H (2016) Cooperative cathode electrode and in situ deposited copper for subsequent enhanced Cd (II) removal and hydrogen evolution in bioelectrochemical systems. Biores Technol 200:565–571
27. Huang L, Li M, Pan Y, Shi Y, Quan X, Puma GL (2017) Efficient W and Mo deposition and separation with simultaneous hydrogen production in stacked bioelectrochemical systems. Chem Eng J 327:584–596
28. Jafary T, Daud WRW, Ghasemi M, Kim BH, Carmona-Martínez AA, Bakar MHA, Jahim JM, Ismail M (2017) A comprehensive study on development of a biocathode for cleaner production of hydrogen in a microbial electrolysis cell. J Clean Prod 164:1135–1144
29. Zhou Y, Niu J, Zhang G, Yu M, Yang F (2020) A three-dimensional self-standing Mo_2C/nitrogen-doped graphene aerogel: enhancement hydrogen production from landfill leachate wastewater in MFCs-AEC coupled system. Environ Res 184:109283
30. Arulmani SRB, Dai J, Li H, Chen Z, Zhang H, Yan J, Xiao T, Sun W (2021) Efficient reduction of antimony by sulfate-reducer enriched bio-cathode with hydrogen production in a microbial electrolysis cell. Sci Total Environ 774:145733
31. Gele MY, Kachooei ASS, Yaghmaei S, Mardanpour MM (2023) Developing a self-powered microfluidic microbial electrolysis cell (MEC) for converting oxalate into hydrogen. J Environ Chem Eng 11:109373

32. Dai J, Huang Z, Zhang H, Shi H, Arulmani SRB, Liu X, Huang L, Yan J, Xiao T (2023) Promoted Sb removal with hydrogen production in microbial electrolysis cell by ZIF-67-derived modified sulfate-reducing bacteria bio-cathode. Sci Total Environ 856:158839
33. Gautam R, Nayak JK, Talapatra KN, Ghosh UK (2023) Assessment of different organic substrates for bio-electricity and bio-hydrogen generation in an integrated bio-electrochemical system. Mater Today: Proc 80:2255–2259
34. Lee H-S, Salerno MB, Rittmann BE (2008) Thermodynamic evaluation on H_2 production in glucose fermentation. Environ Sci Technol 42:2401–2407
35. Kamran M (2021) Bioenergy, renewable energy conversion systems 243
36. Koutra E, Tsafrakidou P, Sakarika M, Kornaros M (2020) Microalgal biorefinery, Microalgae cultivation for biofuels production. Elsevier, pp 163–185
37. Lalaurette E, Thammannagowda S, Mohagheghi A, Maness P-C, Logan BE (2009) Hydrogen production from cellulose in a two-stage process combining fermentation and electrohydrogenesis. Int J Hydrogen Energy 34:6201–6210
38. Wang A, Sun D, Cao G, Wang H, Ren N, Wu W-M, Logan BE (2011) Integrated hydrogen production process from cellulose by combining dark fermentation, microbial fuel cells, and a microbial electrolysis cell. Biores Technol 102:4137–4143
39. Huang L, Wang Q, Jiang L, Zhou P, Quan X, Logan BE (2015) Adaptively evolving bacterial communities for complete and selective reduction of Cr(VI), Cu(II), and Cd(II) in biocathode bioelectrochemical systems. Environ Sci Technol 49:9914–9924
40. Heinzel E, Hedrich S, Janneck E, Glombitza F, Seifert J, Schlömann M (2009) Bacterial diversity in a mine water treatment plant. Appl Environ Microbiol 75:858–861
41. Ňancucheo I, Hedrich S, Johnson D (2012) New microbiological strategies that enable the selective recovery and recycling of metals from acid mine drainage and mine process waters. Mineral Mag 76:2683–2692
42. Jiang L, Huang L, Sun Y (2014) Recovery of flakey cobalt from aqueous Co(II) with simultaneous hydrogen production in microbial electrolysis cells. Int J Hydrogen Energy 39:654–663
43. Shen J, Sun Y, Huang L, Yang J (2015) Microbial electrolysis cells with biocathodes and driven by microbial fuel cells for simultaneous enhanced Co(II) and Cu(II) removal. Front Environ Sci Eng 9:1084–1095
44. Tao H-C, Liang M, Li W, Zhang L-J, Ni J-R, Wu W-M (2011) Removal of copper from aqueous solution by electrodeposition in cathode chamber of microbial fuel cell. J Hazard Mater 189:186–192
45. Choi C, Hu N, Lim B (2014) Cadmium recovery by coupling double microbial fuel cells. Biores Technol 170:361–369
46. Zhang Y, Yu L, Wu D, Huang L, Zhou P, Quan X, Chen G (2015) Dependency of simultaneous Cr(VI), Cu(II) and Cd(II) reduction on the cathodes of microbial electrolysis cells self-driven by microbial fuel cells. J Power Sources 273:1103–1113
47. Wu D, Pan Y, Huang L, Zhou P, Quan X, Chen H (2015) Complete separation of Cu(II), Co(II) and Li(I) using self-driven MFCs–MECs with stainless steel mesh cathodes under continuous flow conditions. Sep Purif Technol 147:114–124
48. Wu D, Pan Y, Huang L, Quan X, Yang J (2015) Comparison of Co (II) reduction on three different cathodes of microbial electrolysis cells driven by Cu (II)-reduced microbial fuel cells under various cathode volume conditions. Chem Eng J 266:121–132.
49. Sun S, Kong W, Huang L, Wang Q, Zhang G, Zhou P (2022) Synergistic light irradiation and circuital current for efficient mineralization of recalcitrant organics and sequential recovery of heavy metals from etching terminal wastewater using photo-assisted bioelectrochemical systems. J Power Sources 522:230991
50. Huang L, Kong W, Sun S, Wan H, Shi Y, Zhou P (2023) Photo-assisted self-driven bioelectrochemical systems for simultaneous enhanced treatment of etching terminal wastewater and selective recovery of heavy metals. J Power Sources 558:23262
51. Yu B, Liu C, Wang S, Wang W, Zhao S, Zhu G (2020) Applying constructed wetland-microbial electrochemical system to enhance NH^{4+} removal at low temperature. Sci Total Environ 724:138017

52. Okabe S (2021) Bioelectrochemical anoxic ammonium nitrogen removal by an MFC driven single chamber microbial electrolysis cell. Chemosphere 274:129715
53. Zou H, Wang Y (2017) Azo dyes wastewater treatment and simultaneous electricity generation in a novel process of electrolysis cell combined with microbial fuel cell. Biores Technol 235:167–175
54. Zhao H, Zhang Y, Zhao B, Chang Y, Li Z (2012) Electrochemical reduction of carbon dioxide in an MFC–MEC system with a layer-by-layer self-assembly carbon nanotube/cobalt phthalocyanine modified electrode. Environ Sci Technol 46:5198–5204
55. Villano M, Aulenta F, Ciucci C, Ferri T, Giuliano A, Majone M (2010) Bioelectrochemical reduction of CO_2 to CH_4 via direct and indirect extracellular electron transfer by a hydrogenophilic methanogenic culture. Biores Technol 101:3085–3090
56. Shironita S, Karasuda K, Sato M, Umeda M (2013) Feasibility investigation of methanol generation by CO_2 reduction using Pt/C-based membrane electrode assembly for a reversible fuel cell. J Power Sources 228:68–74
57. Zhang Y, Wang Y, Angelidaki I (2015) Alternate switching between microbial fuel cell and microbial electrolysis cell operation as a new method to control H_2O_2 level in Bioelectro-Fenton system. J Power Sources 291:108-116

Effect of Design and Other Parameters of Microbial Electrolysis Cell for Industrial Wastewater Treatment

Mahalakshmi Velrajan, L. Growther, and N. Susan Silvia

Abstract Waste is considered as a useful resource. Microbial electrolysis cells (MECs) as a novel carbon–neutral technology have replaced the conventional methods in treating industrial wastewater loaded with organic matters that could be converted to value-added products. MEC has proved to have diverse applications in industrial effluent treatment, biosynthesis of compounds such as hydrogen peroxide, methane, hydrogen, nutrient recovery, pollutant removal, and biosensing. However, this technology to be used in the circular industrial wastewater treatment for meeting the huge global energy demand is still in its infancy. The various parameters such as architectural design of the reactor and optimizing its working parameters such as feedstock, inoculum, electrode material and design, pH, temperature, and applied potential need to be standardized for the best output. Designing reactors operated using solar power or less electricity would be a real breakthrough. Similarly, choosing the apt electrode materials and its size to volume ratio of the reactor, reducing internal resistance, and experimenting with the inoculum type and time may reduce the current hindrances in the progression of this technology for industrial applications. The huge capital cost is an impediment, which if addressed with less capital expenditure and improved treatment efficiency and thereby an increased energy yield, then this fascinating technology may serve as a futuristic technology to solve the universal pollution problem and energy requirements and thereby ensuring sustainability.

Keywords Microbial electrolysis cell · Design of reactor · Operating conditions · Industrial wastewater treatment · Parameters

---

M. Velrajan (✉) · N. S. Silvia
Department of Microbiology, Madras Christian College, Chennai, India
e-mail: mahalakshmi@mcc.edu.in

L. Growther
Department of Microbiology, Hindusthan College of Arts and Science, Coimbatore, India
e-mail: laligrowther@hicas.ac.in

A. A. Yaqoob and A. Ahmad (eds.), *Microbial Electrolysis Cell Technology*, Sustainable Materials and Technology, https://doi.org/10.1007/978-981-97-3356-9_9

1 Introduction

Population explosion has depleted global resources, and industrialization has drastically changed the climatic conditions. The two vital items that determine the edification of the human race are water and energy. By 2050, the estimated world population is to touch 9.7 million, and the energy utility may surpass 736 quadrillion of British thermal units (BTUs) by 2040 [1]. Water is an essential commodity for all lifeforms, since it ascertains the standard of life [2]. Normally, 80% of untreated global wastewater from industrial, agricultural, and municipal sectors is simply dumped into the environment [3], and hence, there is a huge scarcity for clean and fresh water. Water contamination occurs due to varying determinants such as nutrients, surface run off, depletion of oxygen, microorganisms, suspended organic matter, chemicals, and oil spillage.

Therefore, nutrient-rich wastewater needs to be attended before being expelled to prevent eutrophication. Also, harmful chemicals and pathogens, if not eliminated, would cause gastroenteritis, skin infections, and leptospirosis-like diseases, and this certainly will pose a health risk for the scavengers [4]. Up to the minute access are needed to treat industrial wastewater than to depend on the traditional huge sewerage network to shift the effluent to a central processing unit, which is an inconvenient process too. Thus, microbial electrolysis cell serves as a decentralized wastewater treatment technology, where the sewage nutrients are trapped and its biomass is converted to hydrogen, methane (bioenergy), and other value-added chemicals [5]. The treated water in addition would solve the problem of water shortage of the future generations.

2 Microbial Electrolysis Cell Technology—A Novel Carbon–Neutral Technology

2.1 Constraints of Conventional Wastewater Treatment Facility

In traditional treatment machinery, the industrial effluent is taken by vast sewerage network to a centralized processing unit and the sewage is treated by a linear end of pipe technology. Due to urbanization and population explosion, this system is getting outdated and is also considered difficult to install and is expensive too. A switch over to an efficient, decentralized treatment facility is found to be more flexible and economical [6].

2.2 *MEC as Carbon–Neutral and Diverse Technology*

Microbial electrolysis cell is carbon–neutral and diverse technology that could get rid of organic matters present in the industrial effluent with concurrent biohydrogen and biomethane production. Thus, countless industrial plan of actions is imperative to mass-produce biohydrogen in a clarified and balanced state. This is because hydrogen has exceptional thermal efficiency (141.9 $MJ{\cdot}kg^{-1}$ higher heating value, HHV; 119.9 $MJ{\cdot}kg^{-1}$ lower heating value, LHV) in comparison with the list of other fuels available (e.g., gasoline: 47.5 $MJ{\cdot}kg^{-1}$ (HHV); 44.5 $MJ{\cdot}kg^{-1}$LHV) and also hydrogen can be conserved for a stretched time span. Biohydrogen when produced from the industrial wastewater is a green initiative making the activity more feasible [7]. Hydrogen an exceptional fuel for the time to come to replace global ultimatum for energy requirement, because of great energy denseness and environmentally safe flammable nature it can be combusted at favorable environmental conditions [8]. Currently, hydrogen is produced through conventional electrolytic, thermochemical methods, which emits greenhouse gases and hence detrimental to the atmosphere. These traditional practices require plenty of energy and so are costly. Therefore, novel biological and biotechnological practices using microbial electrolysis cell to produce green biofuels such as biohydrogen are the need of the hour. Biomethane is being generated deploying a process termed anaerobic digestion (AD), where organic matters in wastewater sludge is broken down to CO_2 and CH_4. Due to considerable production of CO_2, the energy efficiency of biomethane is decreased and chemical method such as cryogenic segregation is required to get rid of CO_2 [9]. Hence, MECs are an excellent carbon–neutral technology to replace biological anaerobic digestion method with simultaneous production of biohydrogen and biomethane. It is a diverse technology since microbial fuel cell (MFC) that employed microorganisms to generate current and used for wastewater treatment, evolved into MEC when it used electricity.

3 Process of MEC

A MEC device constitutes two compartments, one for the positive electrode and another for the negative electrode partitioned by an ion exchange membrane connected to electric power. During the operation of MEC, the microorganisms in the wastewater degrade the organic matters in the sludge and populate the positive electrode as a bioaggregate supplying electron. These electrons reach the negative electrode through an extraneous orbit and coalesce with protons, thus generating biohydrogen. Here, microorganisms serve as a biocatalyst in transforming chemical energy through degradation of organic matters in the sewage to electrical energy to generate valuable products. In addition, these biocatalysts diminish the energizing overcapacity of a redox event, thereby escalating the voltage effectiveness besides furnishing more byproducts [10, 11].

MECs that are constructed for biohydrogen generation need a partition in the middle of negative and positive electrode in order to prevent the transformation of hydrogen to methane, and this is achieved when microorganisms cohere only to the positive electrode as represented schematically in Fig. 1. In contrast, MECs constructed for methane generation is with a negative bioelectrode for harboring microbes to convert hydrogen and carbon dioxide to biomethane, and these systems are termed as methanogenic microbial electrolysis cell (MMEC) or electro methanogenic reactor [12] (Fig. 2).

In an oxygen-independent setup, when the positive and negative electrodes are coupled to energy source, the sewage go into the system, and by the influence of

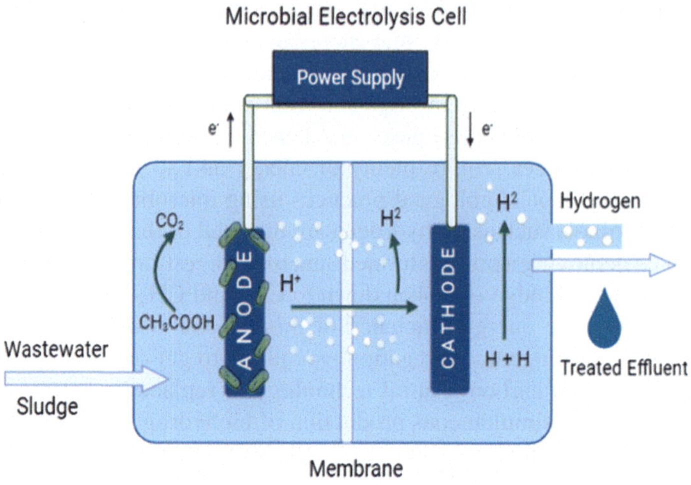

Fig. 1 MEC for generating hydrogen

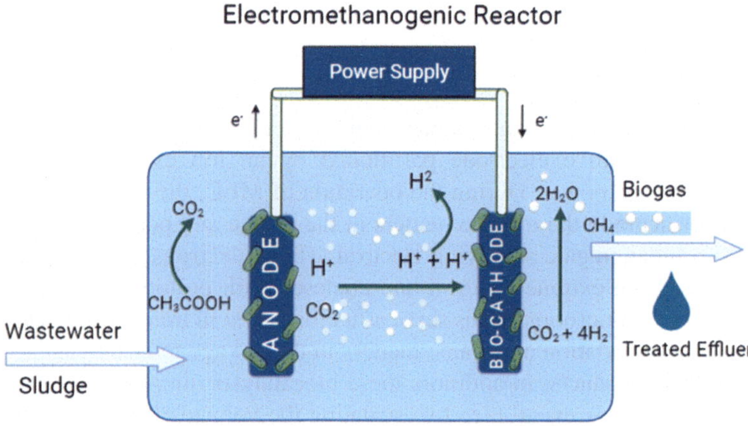

Fig. 2 MMEC for generating methane

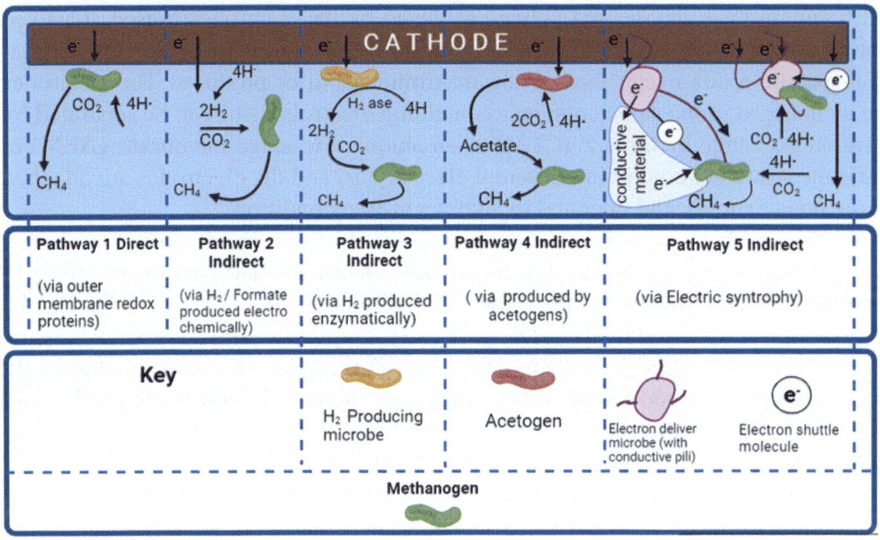

Fig. 3 Different pathways (1, 2, 3, 4, and 5) for methane generation

energy at controlled conditions, the organic pollutants in the wastewater are degraded by microorganisms to generate hydrogen and methane. The reactions happening at the negative and positive electrodes are positive electrode: $C_2H_4O_2 + 2H_2O \rightarrow 2CO_2 + 8H^+ + 8e^-$ [13].

For methane generation at negative electrode, various routes are based on the direct or indirect passing over of electrons. In direct passing over of electrons through the outer membrane, CO_2 is converted to methane, whereas in indirect passing over of electrons methane is generated via intermediates such as formate, acetate, and hydrogen (Fig. 3). The production of methane from hydrogen is termed hydrogenotrophic methanogenesis and from formate or acetate is termed acetogenic methanogenesis. When microorganisms serve as a channel for transmitting electrons with their pili as electron carriers, then it is called electric syntrophy. It was observed that considerable amount of biomethane was generated by hydrogenotrophic methanogenesis than by direct passing over of electrons [14, 15] and [16].

4 Architectural Design of Microbial Electrolysis Cell

The architectural design of the microbial electrolysis cell (MEC) system mainly consists of four components: anode, cathode, membrane for separation, and power supply. Each component takes part in a vital function in the bioelectrochemical system, which activates the organic matter in the industrial effluent into byproducts.

MEC functions at the desired voltage achieved by the electrodes poised at a fixed potential using microbial cultures as biocatalysts to lessen the activation overpotential of a specific redox event, thus giving maximum yield of products. The electrodes are submerged in the same container containing electrolytes or can be separated by a cation exchange membrane (CEM) or an anionic exchange membrane (AEM) in the same container containing different electrolytes, and the electrodes are attached to the power supply and operated in an oxygen-free condition.

At the anodic compartment, the electroactive bacteria (EAB) grow, produce a bio-mat that serves as a biocatalyst to generate electricity, and pass the electrons to the positive electrode where the protons are liberated. At the negative electrode, the electrons get transferred by the external circuit by applying a small current where the protons get reduced and produce biohydrogen. This process does not spontaneously occur but needs an additional voltage supply to increase biohydrogen production. The prime applied voltage ranges from 0.5 -1.0 V which is a small voltage used, and this serves as an advantage over the voltage used in water electrolysis.

$$\text{At positive electrode}: CH_3COO^- + 4H_2O \rightarrow 2HCO_3^- + 9H^+ + 8e^-$$

$$\text{At negative electrode}: 8H^+ + 8e^- \rightarrow 4H_2$$

There are several configuration possibilities and materials available for the MEC's composition. Various MEC setups can produce varying current densities and biohydrogen production when they use the same materials and operate in the same circumstances. This is brought on by different combinations of setup, electrode spacing, membranes, and inoculum forms.

4.1 Different Types of MECs Reactor Configurations and Their Effects in Industrial Effluent Treatment:

4.1.1 Two-Chamber MECs

A two-chamber reactor configuration was first developed and is the typical reactor design that constitutes the anode and cathode chamber which is partitioned by the membranous film. All types of two-chamber MECs have the membrane inserted since it plays two important roles.

1. It prevents the crossover of the bacteria and fuels from the anode to the cathode, thus producing a pure biohydrogen without any impurities like CO_2, CH_4, and H_2S.
2. It helps to prevent internal resistance.

The advantage of two-chamber MECs is that two different electrolytes can be used in the same container where a membrane separator is used, which increases the efficiency and production of biohydrogen. There are varied types of partitions present in two-chamber MECs, and the most common type is the proton exchange membrane (PEM) that utilizes the SO_3-functional group which permits free passage of protons [17]. The other membranes as substitute include anion exchange membranes (AEM), such as AMI7001, bipolar membranes, and charge mosaic membranes (CMM) [18]. Different configurations of two chambers MEC were designed, such as H-type, plate type, tubular type, cruse type, barrel-shaped, circular like a disc, concentric tube-shaped, and flat type to decrease the input resistance.

4.1.2 *Bioelectrochemically Assisted Microbial Reactor* (BEAMR)

In a bioelectrochemically assisted microbial reactor (BEAMR), cathode evolves the hydrogen by removing oxygen at the negative electrode by applying a minimum voltage to the circuit. The small voltage that is applied to generate hydrogen through acetate bioelectrolysis is ~ 0.25 V which is still less than 1.8–0.2 V needed for water electrolysis to produce hydrogen. Hydrogen may be lost during the BEAMR process as a result of it diffusing into the anode chamber from the cathode chamber via the cation exchange membrane (CEM). There is no chance of substrate loss in the BEAMR process since oxygen diffuses from the cathode chamber into the anode chamber, preventing aerobic bacterial growth. The lowest applied voltage for hydrogen production in a BEAMR process for various feedstocks neither can be predicted solely on MFC testing, nor can it be determined how the applied voltage may impact current density in these reactors. As a novel process, in BEAMR acetate alone has been investigated as a potential feedstock for the synthesis of biohydrogen. Since both acetate and wastewater may be used to create electricity in an MFC, it is plausible to assume that household wastewater might be utilized to make hydrogen in the BEAMR process. (Fig. 4).

4.1.3 Concentric Tubular MEC

Kyazze et al., in [19] built a dual-chamber MEC reactor, where two concentric tubular chambers made up the structure. Here, the inner tube was radially punctured at one end, and it was thrust into the larger outer tube. A positive electrode assembly wrapped around a plastic inner rod several times was housed inside the inner tube. To create a barrier in the middle of the interior contents of the two tubes, a cation exchange membrane (CEM) was enfolded on all outer area of the inner tube and kept over the perforations. The CEM was encircled by the cathode assembly. The highest hydrogen production rate (HPR) was achieved at 0.85 V, with 60% CE and 45% cathodic hydrogen recovery, and a hydrogen yield of up to 1.1 mol per mole of acetate converted, indicating a 30.5% reduction in chemical oxygen demand.

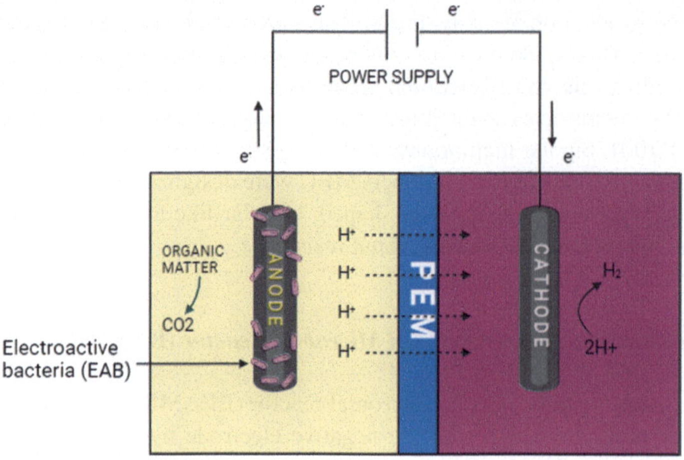

Fig. 4 Two-chamber MEC system

4.1.4 Enrichment of MEC Biocathodes from Sediment MFC Bioanodes

Electrotrophic biocathodes (that accept electrons) were synthesized using enriched graphite fiber brush electrodes from marine sediment, which are present in sediment-type microbial fuel cells (sMFCs) (Fig. 5). These electrodes are allowed to operate as negative electrodes, when inverted thus creating a two-chamber bioelectrochemical systems (BESs). Electron consumption occurred at set potentials of −439 mV and −539 mV, but not at −339 mV in minimal media lacking organic energy sources. MFC positive electrodes initially not adapted at a set potential produced electrotrophic biocathodes, whereas positive electrodes operating at a set potential did not. The electrotrophic biocathodes were autotrophic, as evidenced by the removal of CO_2 from the cathode headspace. One sample's synthesis of methane and subsequent loss of hydrogen gas showed hydrogenotrophic methanogenesis. A rise in turbidness due to microorganisms on the surface of the cathode indicated a luxuriant microbial growth in the cathode chamber [20].

4.1.5 Quantification of Soluble Microbial Products (SMP) in MEC

An and Lee created a two-chamber MEC that kept the distance between the anode and the stainless-steel mesh cathode low while offering a sizeable anode surface area versus the membrane surface area. Under situations of substrate deficiency, the MEC optimized to maintain a high concentration of anode-respiring bacteria (ARB) per membrane surface area yielded 8.3 A m^{-2} of membrane area, of which 57% was produced by endogenous degradation of ARB. For the acetate-utilizing ARB in the MEC, soluble microbial products (SMP) were measured; the SMP proportion of

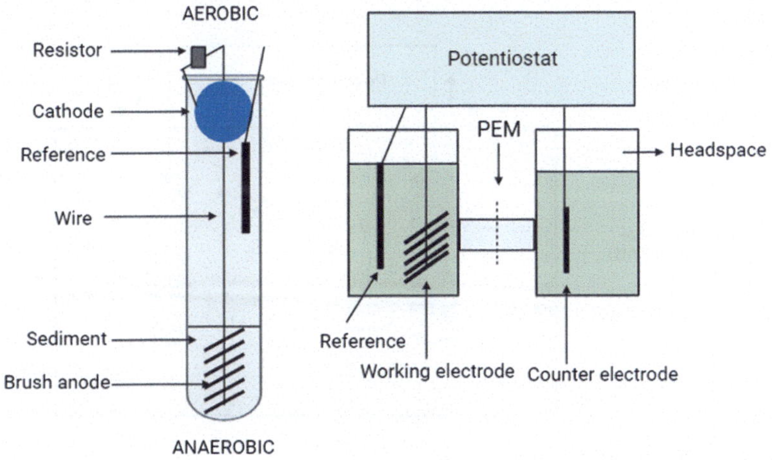

Fig. 5 Enrichment of MEC biocathodes from sediment MFC bioanodes

influent COD was 23%. At an applied voltage of 0–1.2 V, HPR varied from 66.4 ± 18.0 to 137.2 ± 14.4 L H_2/m^3 d, with a maximum hydrogen recovery of 93 ± 25%.

4.2 *Single-Chamber MECs*

Singe chamber MECs work without the membrane component, making it less expensive and providing a streamlined architectural design. This structure was created to optimize volumetric hydrogen production with the standard quality of gas production. Advantages of single-chamber MECs include the membrane-less concept of reduced internal resistance, making it more suitable for field applications, energy loss, pH gradient, and system cost.

Due to the lack of membrane between the anode and cathode, many transformations occur internally, which results in major drawbacks in this configuration. The most important disadvantage is that membrane-less MECs might encourage the growth of hydrogenotrophic methanogens that use hydrogen to create methane, lowering the off-gas purity. Another significant performance is the hydrogen recycling phenomenon, where the cathodic hydrogen is re-oxidized on the anode, which increases electrical current and decreases the energy efficiency of the system, and the hydrogen can be converted into acetate when there is no restriction of homoacetogenic microorganisms, which is then consumed by the anode similar to hydrogen recycling. Thus, there are important weaknesses in single-chamber MECs. (Fig. 6).

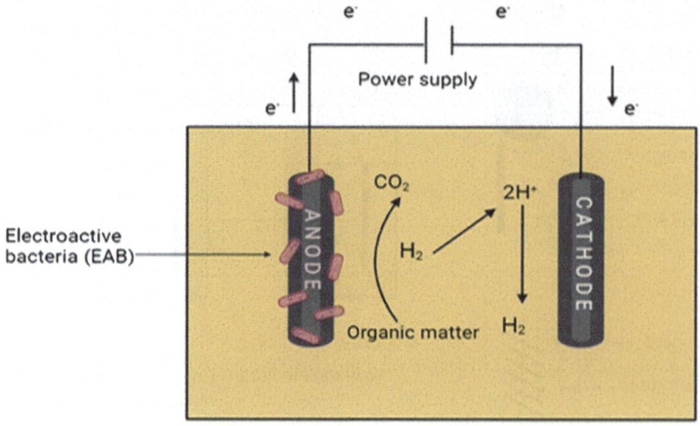

Fig. 6 Single-chamber MEC system

4.2.1 A Single-Chamber MEC with a Brush Anode and a Flat Carbon Cathode

A single-chamber MEC was created by in 2008 [21], a polycarbonate cylindrical chamber where the anode was an ammonia-treated graphite brush placed at the center of the chamber and the cathode was wet-proofed carbon cloth kept at the side of the compartment. Cathodic hydrogen recoveries of 78 ± 1% to 96 ± 1% were attained using an MEC at applied voltages of 0.3–0.8 V and 7.5 mS/cm of solution conductivity. An applied voltage of 0.8 V resulted in a maximum hydrogen generation rate of 3.12 ± 0.02 m^3 H_2/m^3 per day due to the membrane-less system, tight electrode spacing, and graphite fiber brush anode.

4.2.2 An Upflow Single-Chamber MEC

In the upflow single-chamber MEC, set up by Lee et al., the cathode was placed on top of the glass cylinder that served as the reactor. Long graphite rod fragments were crammed into the anode bed, and the cathode was carbon felt without the need for a metal catalyst. Graphite rods were placed into the upper portions of the granule anode and used to puncture the carbon felt to link the two electrodes. After 32 h of batch assessment studies, it was discovered that at an initial acetate concentration was 10 mM, the Coulombic efficiency (CE) was 60 ± 1%, the H_2 output was 59 ± 2%, and the formation of methane was negligible. However, methane buildup caused a considerable loss of H_2 despite a longer batch reaction period (about seven days) and greater accumulating acetate concentrations (30 or 80 mM).

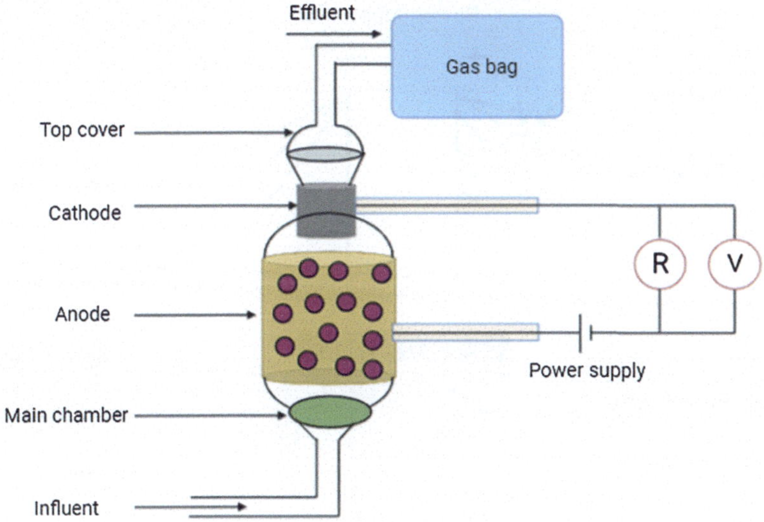

Fig. 7 Cathode on top single-chamber MEC system

4.2.3 A Cathode on Top Single-Chamber MEC

In this design, the cathode was placed above the anode, and the reactor was separated into two parts, the main chamber and the top part, both made of glass. The positive electrode synthesized from graphite granules was kept at the base of the main compartment, and the negative electrode made of minor titanium tube coated with platinum was kept at the top of the system. A power supply given to provide voltage, and the voltage across the resistor was monitored to calculate the current. The substrate and electrolyte were liberated from the vent into the chamber, and the applied voltage was from 0.2 V to 1.0 V with a 0.1 V interval. A comprehensive energy efficiency of 86.78% with an applied voltage of 0.6 V was achieved. Thus, this constructional design of the cathode on top of the anode provided increased hydrogen recovery and a high hydrogen production rate during industrial wastewater treatment. (Fig. 7).

4.2.4 The Smallest-Scale MEC

The smallest-scale MECs were developed by Call and Logan in 2011 with small 5 ml crimp-top serum bottles with graphite plate anode and stainless-steel cathode. All the anodes were polished, soaked, and cleaned, and the cathode mesh was placed and operated using a single power source. The maximum volumetric current density of 240 A/m^3 was got using stainless-steel mesh cathode and a feedstock inoculum at an applied voltage of 0.7 V. (Fig. 8).

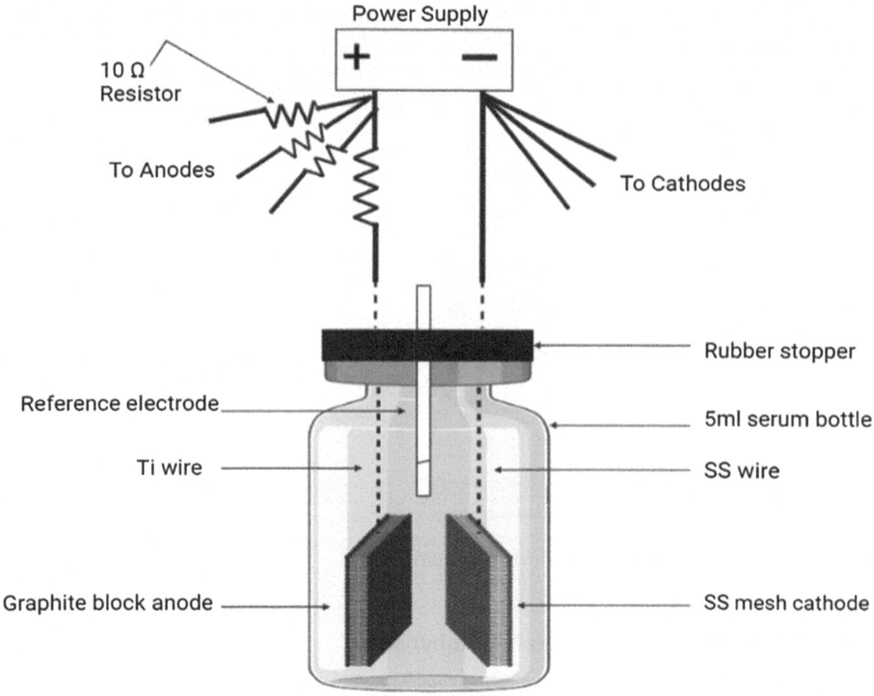

Fig. 8 Smallest-scale MEC system

4.2.5 Single-Chamber Glass Tubular MEC Using Non-Precious Metal Cathode

Tubular membrane-free MECs consisting of newly developed cathodes for hydrogen production are designed using serum tubes made of glass. Electrodepositing of NiMo and NiW on a carbon fiber-weaved cloth material, used a non-precious metal as negative electrode. In MECs, NiMo electrode performance was greater than NiW electrode as negative electrode. The MECs with NiMo cathode produced 2.0 m^3/day/m^3 of hydrogen at a current density of 270 A/m3 (12 A/m^2) at an applied voltage of 0.6 V. This was 33% more than the NiW MECs and somewhat less than the Pt catalyst-equipped MECs (2.3 m^3/day/m^3). The NiMo MECs' energy efficiency based on electrical energy input achieved 240% at an applied voltage of 0.4 V. These findings showed how carbon cloth combined with Ni-alloy catalysts has a lot of potential as a cathode material for MECs [22].

4.3 *Continuous Flow MECs*

4.3.1 Continuous Flow MEC Fed with Wastewater

The setup by Escapa et al. of continuous membrane-less single-chambered reactor was with a Ni-based cathode operating at hydraulic retention times (HRTs) ranging from 3 to 48 h, with an emphasis on the impact of applied voltage (Vapp) and organic loading rate (OLR) on energy consumption, hydrogen production rates, and the efficiency of the treatment (removal of COD) when treating full-strength, organics loaded wastewater. This resulted in a COD reduction of 76%, energy consumption was less or lower than the aerobic treatments. The hydrogen production rate peaked at 0.3 L L_r $^{-1}$ d^{-1} with a hydraulic retention time of 3–6 h.

4.3.2 A Semi-Pilot Tubular MEC in Wastewater Treatment

The impact of a semi-pilot modular microbial electrolysis cell (MEC) structure designed by Gil-Carrera et al. and the organic loading rate on energy consumption during the treatment of wastewater was investigated. The MEC reactor, that was designed from two hydraulically linked tubular units in succession (M_A and M_B), could lower the influent's chemical oxygen demand (COD) concentration by up to 85% at a comparatively low energy consumption of 1.6 kW h kg-COD1. The lower quantities of organic matter fed into the reactor and the cathode's subpar efficiency limited the quantity of hydrogen that could be produced (Fig. 9).

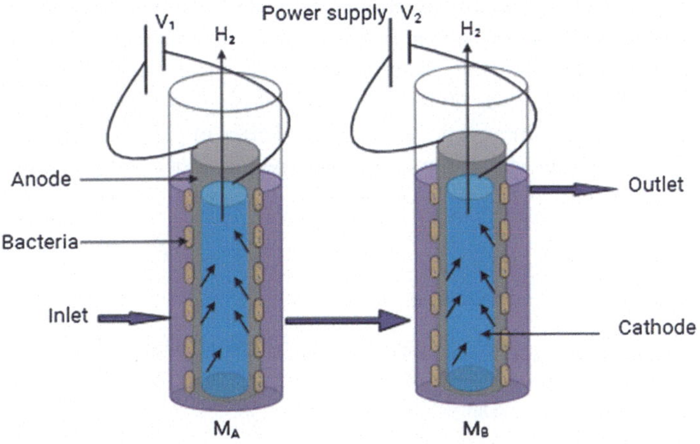

Fig. 9 Semi-pilot tubular MEC system

5 Parameters that Influence Design of Reactors for Industrial Wastewater Treatment

The following parameters are important since they influence not only the execution of the reactor's production but also improvise the commercial feasibility of MEC technology, thus popularizing it in industrial wastewater treatment. Therefore, MEC technology of future should have large wastewater with organic matters loading capacity, low electrode cost with long shelf-life and less requirements for maintenance [23]. Research in these lines will speed-up the technology to be transformed by the industry for the best performance.

5.1 Substrates

The type of substrates has a great influence on the working efficiency of MEC, since it denotes the strength of organic matters, energy needed to treat the substrate, time required for treatment, reactor configuration, and byproducts generated. The more the organic matter of the effluent (organic loading rate, OLR), more will be the hydraulic retention times (HRT). Studies reveled that adjusting OLR, HRT help in decreasing the size of the reactor and energy generation (Table 1).

By improvising the design of MECs, an organic load of 800 to 1400 mg/L of industrial effluent could be treated [23]. Low organic matter loaded domestic wastewater having a COD of 300–500 mg/L needed an HRT of 5–9 h for treatment, while heavily organic matter loaded piggery effluent with more COD of 18,300 mg/L needed a lengthy HRT of 314 h for treatment. Simultaneously heavily organic matter loaded effluents could generate increased energy, since they have increased energy potential and hence economically feasible in terms of energy production. Thus, generally a yield of 76.2 kWh/m^3/day was attained for 16 L of piggery wastes and an electrode surface area of 11.25 m^2/m as reported by San-Martín in [25]. But HRTs for industrial effluents with higher strength may be an impediment for industrial use, thereby increasing the size and cost of the reactor.

Table 1 List of industrial wastewater with respective chemical oxygen demand

Industrial effluent	COD g/L
Alcohol refining beer & malt	5–22
Coffee	3–15
Dairy	1.5–5.2
Meat & Poultry	2–7
Fish	2.5
Swine or piggery	18.3

HRT to treat various industrial effluents calculated using the formula, OLR = COD (mg/L) × Flow rate (m^3/d) [24]

But the research done by Gil-Carrera et al. revealed that a MEC, workable for hydrogen production by treating industrial effluent having an OLR of 1000–2000 mg COD/L/day, may replace the activated sludge treatment even with high HRT [23]. In conventional methods of treating industrial wastewater, especially activated sludge treatment, required huge energy for degradation of sludge, whereas MECs integrated with anaerobic digesters speeded the process of organic matter degradation with more energy yield (nearly 1.7 times more than traditional methods) [26]. Many studies are focusing on the efficiency of MECs in treating various industrial effluents and the means of substrates removal by microbial consortium and their marketing potentiality, thereby creating the futuristic MEC technology.

5.2 *Inoculum*

Inoculum represents the type of microorganisms that survive, multiply, and form a mat on the surface of the electrodes in MECs, and based on the type of inoculum, the method of degradation of organic matter in the effluent will occur [27]. Hence, it is crucial to choose the correct inoculum to kick start the process in generating energy and thereby increasing the productivity of the reactor. Studies were done to improvise the ways to energize the reactor for action. It was found that the positive electrode at beginning with −0.25 V versus Ag/AgC and when raised to −0.5 V versus Ag/AgCl ignited the system immediately and was found to harbor *Geobacter* than the anodes at −0.36 and −0.42 V versus Ag/AgCl as experimented by Lim et al. in [28]. They also observed that mat formation and activation of microorganisms happened faster at the positive electrode that the negative electrode and gradual raise in mat growth occurred between 0.3 and 1.0 V. Methanogenic bacteria from the effluent grew on the negative electrode as a thin film. Compared to single cultures, consortium of microorganisms obtained by mixing 1:4 activated sludge and municipal.

Sewage elevated methane generation by 18.5%, thus proving the significance of different microbial cultures involved in the degradation process. Consortium also accelerates survival of hydrogenotrophic methanogenic bacteria for biomethane generation [29]. Thus, in bulkier operation, to speed-up the process, habituating the electrodes with the recommendable inoculum showed desirable results in kick starting the reactor, thereby increasing the commercial prospects of the MEC technology for industrial wastewater treatment [30].

5.3 Electrode Material

5.3.1 Positive Electrode Material

Negative electrode material plays a very important role as electrode material in bioelectrochemical systems as it is the final acceptor for the electroactive microorganisms (EAB) to form a biofilm which directly influences the hydrogen recovery in MEC systems. Since the EAB adheres to the anode and CO_2 gets oxidized by the organic species, EAB is an important factor in CO_2 production. Essential characteristics of a model anode material are biocompatibility, chemical stability, non-corrosiveness, high electrical conductivity, low resistance, affordability, and porosity. The measurements of current density, COD reduction, and Coulombic efficiency (CE) can be used to assess the anode and biofilm's efficiency. Potentially used anode materials are carbon-based materials, graphite, carbon fiber, carbon fiber brushes, carbon nanotubes (CNTs), molybdenum anodes, and stainless steel.

Currently, carbon materials are found to have wider applications as the electrode material due to their high surface area, strong EAB anchoring, and availability. It has been demonstrated that carbon compounds encourage interfacial microbial colonization, which speeds up the production of biofilms, and they are extremely homogenous and efficient. Cross-linked structures, which help to generate a strong connection between the particles and anodes, are what distinguish them. However, in comparison with metals, carbon compounds have a comparatively poor conductivity [31]. Graphite due to its affordability, abundance, and conductivity, is the most sorted out electrode materials for utility. Numerous types of graphite electrodes such as brushes, granules, rods, felts, and foams are being used. However, in contrast to other carbon materials, graphite has a planar molecular structure and shape because of its low surface porosity, which is necessary for bacterial adherence.

Carbon nanotubes are hollow nanostructures with remarkable mechanical and electrical capabilities that are surrounded by carbon sheets that are one atom thick. Their huge specific surface areas, good stability, and conductivity make them attractive candidates for MEC anodes; nevertheless, they also offer several major problems, chief among which is their bacterial toxicity. Since mesh-like anodes have better mass transfer, surface area, and biofilm development than flat or plate-shaped anodes, they often provide higher current densities (more current flows) [32]. Although carbon fiber brushes produce good test results, large-scale BES rarely utilize them due to their high cost. According to separate research that employed recycled carbon fiber anodes, discovered that using recycled carbon fiber electrodes had advantages over graphite felt anodes in generating greater outcomes at a lower cost.

Out of all the materials, molybdenum anodes—both treated and untreated—showed the highest maximum power densities. Additionally, molybdenum anodes showed good endurance and were neither decreasing nor corroding. Stainless steels are typical industrial materials with excellent mechanical qualities, a long lifespan against corrosion, having a wide range of commercially accessible compositions and morphologies. Stainless steel is worthy of study for its merits, much more research needed to construct microbial anodes, given that the relatively few attempts conducted with stainless-steel anodes yielded current densities of around 25% of the greatest values reported so far with carbon and graphite [33, 34]. As an outcome, stainless steel presents an up-and-coming option for creating an effective microbial anode.

5.3.2 Negative Electrode Material

In the cathode of MECs, hydrogen gas and other value-added compounds are produced. Since the cathode is where hydrogen evolves, choosing the cathode material is the most important component of MECs. The cathode increases electrode–microbe electron transfer to improve hydrogen evolution reactions (HER) by acting as an electrocatalyst and biocatalyst [35]. High specific surface area, great biocompatibility, good corrosion resistance, good conductivity, and good mechanical qualities are essential for cathode materials to be successful. Cathode materials also need to lower significant hydrogen evolution overpotentials. Cathode construction for large -scale applications should be inexpensive, utilizing easily obtained materials and established manufacturing processes for industrial level implementation. Platinum due to its extremely low overpotential, corrosion resistance, and lack of biological activity, is frequently promoted as the ideal electrode material. Additionally, platinum has the strongest HER activity, which improves H_2 evolution. But there are a lot of drawbacks to mining platinum, including its high cost and significant environmental effect [12, 36].

Stainless steel is a regularly used metal for electrodes and is also reasonably priced. For the generation of H_2, stainless steel with a high specific surface area can function similarly to a platinum catalytic electrode incorporating carbon. Iron, carbon, chromium, and nickel combined with stainless steel have also been widely used in possible applications. Nickel is one of the non-platinum metals with the best HER activity and corrosion resistance. Additionally, nickel is more resilient than stainless steel. An electrode's capacity to withstand economic hardship is crucial. The enhanced performance is attributed to hydrogenotrophic methanogens, suggesting that nickel's high HER activity in comparison with the other materials enhanced its performance [37]. As an alternative to Pt in MECs for the creation of hydrogen, biocathodes have been employed. It has been stated that because of their low cost, sustainable operation, and HER overpotential limitation, biocathodes are a preferable substitute for metal-based catalysts.

5.4 pH

The pH of the electrolyte determines the HER at the negative electrode cathode and hence plays a vital role in the general execution of MECs. A variance in oxidation–reduction potential connecting the positive and negative compartments may result in large overpotentials causing a change in protons and electrons traversing the electrodes, thereby making the cathode alkaline and the anode acidic. Practically, a voltage deficiency of 59 mV is resulted because of pH variance in the middle of positive and negative electrode. Since microorganisms' survival and their processes are directly related to pH, the pH variance may lead to changes in microbial communities and their activities leading to decreased byproducts production. A favorable situation occurs at alkaline pH when protons accumulate and simultaneously improve electrons formation, and studies have revealed that at regular intervals polarity change will equilibrate the pH in the anodic and cathodic compartments of MECs [38]. Merrill et al. has suggested the use of an electrolyte, where the PO_4 groups of a weak acid will serve as a transporter for charge, thus increasing electrical conductivity, which in turn will have a positive influence on the cathode material by lowering its over potential. This also will increase the efficiency of the reactor. Different types of electrolytes in cathodic compartment are suggested like NaCl solution, deionized/ acidified/ tap water, and PO_4 buffer to yield the highest H_2 production.

5.5 *Temperature*

The other important parameter in the effective functioning of MECs is temperature since it favors the selectivity and generation of exoelectrogen. In general, mesophilic microorganism are the most active at a temperature range of 35–40 °C and hence multiply and form a mat that degrades the organic matters and produce the energy. The best operating temperature as studied by Omidi et al. was 31 °C in COD removal. However, Lu et al., in 2011 [39] had revealed that operating a single-compartmentalized MEC at low temperature of 4 °C resulted in hydrogen yield but lowering biomethane production.

5.6 *Applied Potential*

High overpotentials will hamper the working efficiency of MECs, and so, a favorable potential range of 0.2–0.8 V is mandatory for scaling up the process for industrial applications. Studies showed that though high applied potential benefitted H_2 generation at the negative electrode, the lowest applied potential of 0.2 V is needed to interrupt the thermodynamic obstacle to generate H_2. In addition, altering the applied potential will reduce metabolic activities of microorganisms and promote the autolysis of cells [40, 41].

6 Diverse Applications of MEC

6.1 *Electrosynthesis of Compounds*

6.1.1 Electrosynthesis of Hydrogen Peroxide

Though both MFCs and MECs are deployed in the synthesis of hydrogen peroxide during electron reduction of oxygen in the negative chamber of the reactor, MFCs usage in treating large industrial effluent is restricted due to reduced hydrogen peroxide production. Therefore, investigators are making use of MECs to speed-up the H_2O_2 synthesis simultaneously while these systems are used for treating industrial effluents [42, 43]. The other lead of this technology over MFCs is that MECs do no not utilize any deleterious constituent and could be operated with negligible power supply, thereby enabling the process congruous for sustainability.

Similarly, MFCs successfully can separate organic matters from the industrial effluent, still to take out the residual pollutants, harmful microorganisms, and other recalcitrant compounds, a third-order therapy is mandatory. Nevertheless, when second-order and third-order therapies are done at negative and positive compartments, either of the demands are met with by MECs—the effluent after organic matter removal in bulk at positive electrode reaches the negative electrode for hydrogen peroxide generation [44, 45]. In addition, the hydrogen peroxide when generated at the negative compartment can take out colorant and recalcitrant chemicals, thus contributing a wholesome wastewater treatment by MECs. Since both the processes can be done in a sole equipment, with compatible order, running cost and overhead charges can be drastically reduced and thus making MEC a structured system than MFC in the speed of hydrogen peroxide production [39]. Though MEC popped up as an operational equipment for sustainable hydrogen peroxide generation, some impediment exist in the large-scale operation of the reactor, viz the oxygen usage speed is less in the positive electrode and hence to increase aeration pumps are being used, which in turn lead to more input of energy. But if the design of the reactor is upgraded and new negative electrode such as graphite material, carbon cloth and carbon felt, that consumes less energy with enhanced oxygen transfer potential, can be used, the deficiency can be addressed in time ahead [46].

6.1.2 Electrosynthesis of Methane

Since methane is a fossil gas, it has wider use as a fuel and in creating a number of products. Globally, there is a huge demand for methane gas, and in the years to come this may go up to 55% as per the prediction by the IEA (International Energy Agency), amounting to an overall consumption of 53,330 TWh of methane [47]. The novel and carbon–neutral MEC technology is rapidly replacing the varied conventional processes of methane generation by photocatalysis, thermal, electrical, and biological means. This system utilizes the combined bioelectrochemical method to quicken the degradation of highly soaking organic matters, dangerous compounds, and undecaying matters, thereby amplifying methane production. The degradation commences when a low voltage of 0.2 V to 0.8 V is supplied to the system by which the excited bacteria break down organic matters by producing electrons at the anode, and these electrons then move toward the cathode forming methane. Equations (1) and (2) summarize the events happening in the closed electrical circuit [48, 21].

$$C_2H_4O_2 + H_2O \rightarrow 2CO_2 + 8H^+ + 8e^- \tag{1}$$

$$CO_2 + 8H^+ + 8e^- \rightarrow CH_4 + 2H_2O \tag{2}$$

The efficiency of methane generation by anaerobic digester and anaerobic digester integrated with MEC was compared by Park et al in [49]. Nickel-coated graphite carbon mesh electrode was utilized in the integrated model to strengthen the passage of electricity. The productivity rate was found to be the same in two of them, however, MEC integrated design reactor quickened the process, thus having an edge over anaerobic digester. With a novel chitosan carbon felt electrode and a steady voltage of 0.4 V, MEC had an ability to remove 67% of COD with a simultaneous production of 87 μmol/day of methane as reported by Seelajaroen et al. in [50]. He was also instrumental in designing a reactor that could metabolize organic matters at the positive electrode, while at the negative electrode with an anaerobic environment created with a mixed inoculum, conversion of carbon-di-oxide to methane took place.

6.1.3 Electrosynthesis of Hydrogen

Biohydrogen is considered as an environment-friendly fuel of the future transportation vehicles. Hydrogen production through fermentation and biocatalytic means is not economical due to poor outcomes in terms of supply and energy content. Efficient synthesis of hydrogen is achieved using MEC deploying a range of renewable organic substances. The MEC reactor degrades the organic substances at the anode, and the protons and electrons released during the degradation process result in the formation of hydrogen at the cathode. Still for large-scale operation of MEC for hydrogen generation, the system needs to be improvised. An attempt was made in building an integrated MEC with aerobic digester, where positive and negative electrodes were coupled making the reactor concise and portable, and in addition, enhancing the production efficiency of methane (30.12 ± 0.5 mL of hydrogen in conventional system to 40.05 ± 0.5 mL of hydrogen in the upgraded reactor) and better treatment of the industrial effluent was achieved (COD removal potential of wastewater found to be 77.5% and 75.6% in 70 h, by conventional and upgraded reactors, respectively) [51]. When a binary compartmentalized MEC was used to produce hydrogen from three different lignocellulosic substrates, an extreme yield of 28.67 L of hydrogen/kg of substrates was observed [52]. Thus, experimental results favor the use of MECs in remediating untreated industrial wastewater with loads of organic matters as well sewage with low content of organic matters.

6.2 *Remediation of Industrial Wastewater*

Industrial effluent is considered a "neglected wealth" since it is used to produce standard products, when treated in addition to generation of energy. In order to tap the energy from industrial effluents, mostly anaerobic treatment processes using microorganisms is the favored option due to their solidity and heterogeneity. Previously microbial fuel cell (MFC) is used to retrieve the energy from wastewater; however, currently MEC has replaced MFC in execution not only in tapping the energy content but also in producing biohydrogen, bioethanol, biomethane, and H_2O_2 from the effluent, thus proving to be the energy fount for the future generations. Kai Wang et al., in their research have proved that MEC in union with sulfate-reducing bacteria (SRB) is used to lessen sulfate content of effluent, which is otherwise difficult to treat since it is devoid of electron donors. Thus, MEC utilizing SRB achieved the highest removal of 14.9% that any other system in treating sulfate-rich effluent [53]. An instant removal of nitrogen from nitrates, salt contents, and metals from wastewater was reported by Yan Li et al., in 2017 [31] when microbial desalination cell (MDC) was integrated with MEC. To increase the denitrification of the nitrates and nitrites in the effluent at the anode, oxygen with carbon is supplied. MDC displayed an elimination efficacy of 62.9% nitrogen at a holding period of 48 h, which is higher than indigenous reverse osmosis and can also take out salt over an extra holding time. Thus, by combining denitrification of MEC and water recycling in MDC resulted in 95.1% nitrogen removal, 63.7% desalination, and 99.5% lead (II) in 48 h [31]. Thus, this research has future perspective in designing novel MEC integrated with anaerobic digester for treating beer industry effluent for the simultaneous methane generation.

6.3 *Recovery of Nutrients*

There may be a huge crisis for food due to uncontrolled proliferation of human race, and hence, to meet this global demand for food, the need for manure to grow crops is also increasing. It is predicted that need for N_2 and P based fertilizers will increase substantially in the coming years. These nutrients are abundant in the industrial wastewater and hence can be recovered from the industrial effluent and used in the manufacture of fertilizers. Hui Guo et al. in 2019 [54] created a MEC-1, MEC-5, and MEC-10 reactors which contained one, five, and ten electrode pairs, respectively, each containing 28, 35 and 40 mL relative volumes of effluent. The anodic material was a surfactant treated activated carbon, and the cathodic electrode was a net of stainless steel wedged by two rubber gaskets and one plastic net, each electrode with its separate controls. The effluent was allowed to pass through the wedged electrodes, P removal was found to be from 20 to 40% than reactors with flat electrodes. Using a plastic tube attached to the electrodes, the removal of methane and nitrogen from wastewater was achieved; however, the nitrogen removal in the system was found to be higher than met hane from anaerobic digester due to microaeration [54]. A microbial electrolysis desalination cell (MEDC) was designed by in 2020 [55] to eliminate salts and organic contents from industrial wastewater with simultaneous nutrient removal, with a removal efficiency of 66.7 ± 4.7% and 66.3%, of N_2 and P, respectively [55].

7 Conclusion and Future Perspectives

Thus, it is evident that the architectural design of the reactor with other parameters such as substrates that make up the industrial effluents and the electrode material determine the efficacy and economic feasibility of MECs for large-scale application by the industry. The reactor architecture being critical in scaling up the process needs optimization to maximize energy efficiency and output by reducing internal resistance and increase substrate mixing. The existence of a membrane separating the anode and the cathode compartments largely dictates the architecture of the system. Studies conducted so far are encouraging with different reactor's configuration in utilizing less electricity in remediating the industrial wastewater with simultaneous production of energy. Still, designing reactors operating in solar power can transform it to a total carbon–neutral technology and will help in its speedy commercialization globally. Currently carbon anodes and stainless-steel cathodes have proved to be the desired choice for electrodes, and further their resistance to corrosion, good biocompatibility, electrical conductivity, and large surface area to reactor volume may add value for the future state-of-the art MECs technology. A knowledge gap occurs in the electrode material and design since bench top reactors performing well with the electrode material, fail to do so while scaling up to pilot scale, thus posing an obstacle in taking the technology forward. A cumulative scientific data on other operational parameters such as such as HRT, COD levels, inoculum, pH, temperature, and applied potential may determine exactly which parameter can improvise the performance of the system. Though there are a plenty means by which the efficiency of the technology could be increased, yet it should not result in complexity of the system and consecutively increase its cost. There needs to be a striking balance between the capital expenditure and operational expenditure to carry the technology forward for industrial applications. MECs is advantageous than water electrolysis in not producing oxygen during the process and the anode producing chloride at low potential and operating at neutral pH, all these factors adding to the safety aspects of the technology. Similarly, MEC/ MEC integrated with AD has yielded promising results in industrial effluent treatment, and scaling up is feasible with increased volume of effluent with high organic matter/COD level or by gradually increasing the flow-rate of the wastewater into the reactor.

Since MEC technology has an edge over the other conventional methods for industrial wastewater treatment and in the generation of value-added products, thereby decarbonizing the current energy network, industries may adapt this shortly. Promising new researches may find a lasting solution to the obstacles in commercializing it for large-scale operations and ensure sustainability.

References

1. Baruah J, Nath BK, Sharma R, Kumar S, Deka RC, Baruah DC, Kalita E (2018) Recent trends in the pretreatment of lignocellulosic biomass for value-added products. Front Energy Res 6:141 CrossRef
2. Savla N, Pandit S, Khanna N, Mathuriya AS, Jung SP (2020) Microbially powered electrochemical systems coupled with membrane-based technology for sustainable desalination and efficient wastewater treatment. J Korean Soc Environ Eng 42:360–380 CrossRef
3. WWAP (United Nations World Water Assessment Programme) (2017) The United Nations world water development report 2017. Wastewater: The Untapped Resource. UNESCO, Paris, France
4. Tiwari RR (2008) Occupational health hazards in sewage and sanitary workers. Indian J Occup Environ Med 12:112–115 CrossRef
5. Zhang Y, Merrill MD, Logan BE (2010) The use and optimization of stainless steel mesh cathodes in microbial electrolysis cells. Int J Hydrogen Energy 35:2020–12028 CrossRef
6. Starkl M, Brunner N, Feil M, Hauser A (2015) Ensuring sustainability of non-networked sanitation technologies: an approach to standardization. Environ Sci Technol 49:6411–6418 CrossRef
7. Nikolaidis P, Poullikkas A (2017) A comparative overview of hydrogen production processes. Renew Sustain Energy Rev 67:597–611 CrossRef
8. Parkhey P, Gupta P (2017) Improvisations in structural features of microbial electrolytic cell and process parameters of electrohydrogenesis for efficient biohydrogen production: a review. Renew Sustain Energy Rev 69:1085–1099 CrossRef
9. Awe OW, Zhao Y, Nzihou A, Minh DP, Lyczko N (2017) A review of biogas utilisation, purification and upgrading technologies. Waste Biomass Valor 8:267–283 CrossRef
10. Kitching M, Butler R, Marsili E (2017) Microbial bioelectrosynthesis of hydrogen: current challenges and scale-up. Enzym Microb Technol 96:1–13 CrossRef
11. Pant D, Singh A, Van Bogaert G, Olsen SI, Nigam PS, Diels L, Vanbroekhoven K (2011) Bioelectrochemical systems (BES) forsustainable energy production and product recovery from organic wastes and industrial wastewaters. R Soc Chem Adv 2:1248–1263 CrossRef
12. Siegert M, Yates MD, Spormann AM, Logan BE (2015) Methanobacterium dominates biocathodic archaeal communities in methanogenic microbial electrolysis cells. ACS Sustain Chem Eng 3:1668–1676 CrossRef
13. Kadier A, Simayi Y, Abdeshahian P, Azman NF, Chandrasekhar K, Kalil MS (2016) A comprehensive review of microbial electrolysis cells (MEC) reactor designs and configurations for sustainable hydrogen gas production. Alex Eng J 55:427–443 CrossRef
14. Blasco-Gómez R, Batlle-Vilanova P, Villano M, Balaguer MD, Colprim J, Puig S (2017) On the edge of research and technological application: a critical review of electromethanogenesis. Int J Mol Sci 18:874. [CrossRef] [PubMed]
15. Villano M, Aulenta F, Ciucci C, Ferri T, Giuliano A, Majone M (2010) Bioelectrochemical reduction of CO_2 to CH_4 via directand indirect extracellular electron transfer by a hydrogenophilic methanogenic culture. Bioresour Technol 101:3085–3090 CrossRef
16. Nevin KP, Hensley SA, Franks AE, Summers ZM, Ou J, Woodard TL, Snoeyenbos-West OL, Lovley DR (2011) Electrosynthesis of organic compounds from carbon dioxide is catalyzed by a diversity of acetogenic microorganisms. Appl Environ Microbiol 77:2882–2886 CrossRef

17. Selembo PA, Merrill MD, Logan BE (2010) Hydrogen production with nickel powder cathode catalysts in microbial electrolysis cells. Int J Hydrog Energy 35:428–437 CrossRef
18. Rozendal RA, Sleutels THJA, Hamelers HVM, Buisman CJN (2008) Effect of the type of ion exchange membrane on performance, ion transport, and PH in biocatalyzed electrolysis of wastewater. Water Sci Technol 57:1757–1762. [CrossRef] [PubMed]
19. Kyazze G, Popov AL, Dinsdale RM, Esteves S, Hawkes FR, Premier GC, Guwy AJ (2010) Influence of catholyte pH and temperature on hydrogen production from acetate using a two-chamber concentric tubular microbial electrolysis cell. Int J Hydrogen Energy 35(15):7716–7722. https://doi.org/10.1016/j.ijhydene.2010.05.036
20. Pisciotta JM, Zaybak Z, Call DF, Nam JY, Logan BE (2012) Enrichment of microbial electrolysis cell biocathodes from sediment microbial fuel cell bioanodes. Appl Environ Microbiol 78(15):5212–5219. https://doi.org/10.1128/aem.00480-12
21. Logan BE, Call D, Cheng S, Hamelers H, Sleutels T, Jeremiasse AW, Rozendal RA (2008) Microbial electrolysis cells for highyield hydrogen gas production from organic matter. Environ Sci Technol 42:8630–8640. [CrossRef] [PubMed]
22. Hu H, Fan Y, Liu H (2009) Hydrogen production in single-chamber tubular microbial electrolysis cells using non-precious-metal catalysts. Int J Hydrogen Energy 34(20):8535–8542. https://doi.org/10.1016/j.ijhydene.2009.08.011
23. Aiken DC, Curtis TP, Heidrich ES (2019) Avenues to the financial viability of microbial electrolysis cells [MEC] for domestic wastewater treatment and hydrogen production. Int J Hydrogen Energy 44:2426–2434 CrossRef
24. IPCC (2006) Wastewater treatment and discharge. In: 2006 IPCC guidelines for national greenhouse gas inventories, vol 5. IPCC: Geneva, Switzerland, pp 6.1–6.28
25. San-Martín MI, Sotres A, Alonso RM, Díaz-Marcos J, Morán A, Escapa-González A (2019) Assessing anodic microbial populations and membrane ageing in a pilot microbial electrolysis cell. Int J Hydrogen Energy 44:17304–17315 CrossRef
26. Bo T, Zhu X, Zhang L, Tao Y, He X, Li D, Yan Z (2014) A new upgraded biogas production process: coupling microbial electrolysis cell and anaerobic digestion in single-chamber, barrel-shape stainless steel reactor. Electrochem Commun 45:67–70 CrossRef
27. Achinas S, Euverink GJW (2019) Effect of combined inoculation on biogas production from hardly degradable material. Energies 12:217 CrossRef
28. Lim SS, Fontmorin J-M, Izadi P, Daud WRW, Scott K, Yu EH (2020) Impact of applied cell voltage on the performance of a microbial electrolysis cell fully catalysed by microorganisms. Int J Hydrogen Energy 45:2557–2568 CrossRef
29. Escapa-González A, San-Martín M, Mateos R, Morán A (2015) Scaling-up of membraneless microbial electrolysis cells (MECs) for domestic wastewater treatment: bottlenecks and limitations. Bioresour Technol 180:72–78 CrossRef
30. Fudge T, Bulmer I, Bowman K, Pathmakanthan S, Gambier W, Dehouche Z, Al-Salem SM, Constantinou A (2021) Microbial electrolysis cells for decentralised wastewater treatment: the next steps. Water 13(4):445. https://doi.org/10.3390/w13040445
31. Li Y, Styczynski J, Huang Y, Xu Z, McCutcheon J, Li B (2017) Energy-positive wastewater treatment and desalination in an integrated microbial desalination cell (MDC)-microbial electrolysis cell (MEC). J Power Sources 356:529–538 CrossRef
32. Rasheed T, Hassan AA, Kausar F, Sher F, Bilal M, Iqbal HMN (2020) Carbon nanotubes assisted analytical detection—sensing/delivery cues for environmental and biomedical monitoring. Trends Anal Chem 132:116066 CrossRef
33. Pocaznoi D, Calmet A, Etcheverry L, Erable B, Bergel A (2012) Stainless steel is a promising electrode material for anodes of microbial fuel cells. Energy Environ Sci 5:9645–9652 CrossRef
34. Guo K, Donose BC, Soeriyadi AH, Prévoteau A, Patil SA, Freguia S, Gooding JJ, Rabaey K (2014) Flame oxidation of stainless steel felt enhances anodic biofilm formation and current output in bioelectrochemical systems. Environ Sci Technol 48:7151–7156. [CrossRef] [PubMed]
35. Zhang T, Nie H, Bain TS, Lu H, Cui M, Snoeyenbos-West OL, Franks AE, Nevin KP, Russell TP, Lovley DR (2012) Improved cathode materials for microbial electrosynthesis. Energy Environ Sci 6:217–224 CrossRef

36. Sangeetha T, Guo Z, Liu W, Cui M, Yang C, Wang L, Wang A (2016) Cathode material as an influencing factor on beer wastewater treatment and methane production in a novel integrated upflow microbial electrolysis cell (Upflow-MEC). Int J Hydrog Energy 41:2189–2196 CrossRef
37. Siegert M, Yates MD, Call DF, Zhu X, Spormann A, Logan BE (2014) Comparison of nonprecious metal cathode materials for methane production by electromethanogenesis. ACS Sustain Chem Eng 2:910–917. [CrossRef] [PubMed]
38. Borole AP, Mielenz JR (2011) Estimating hydrogen production potential in biorefineries using microbial electrolysis cell technology. Int J Hydrog Energy 36:14787–14795 CrossRef
39. Hua T, Li S, Li F, Zhou Q, Ondon BS (2019) Microbial electrolysis cell as an emerging versatile technology: a review on its potential application, advance and challenge. J Chem Technol Biotechnol 94:1697–1711. [Google Scholar] [CrossRef]
40. Nam J-Y, Tokash JC, Logan BE (2011) Comparison of microbial electrolysis cells operated with added voltage or by setting the anode potential. Int J Hydrog Energy 36:10550–10556 CrossRef
41. Chae K-J, Choi M-J, Kim K-Y, Ajayi FF, Chang I-S, Kim IS (2009) A solar-powered microbial electrolysis cell with a platinum catalyst-free cathode to produce hydrogen. Environ Sci Technol 43:9525–9530. [CrossRef] [PubMed]
42. Munoz-Cupa C, Hu Y, Xu C, Bassi A (2021) An overview of microbial fuel cell usage in wastewater treatment, resource recovery and energy production. Sci Total Environ 754:142429 [Google Scholar] [CrossRef]
43. Chung TH, Meshref MN, Hai FI, Al-Mamun A, Dhar BR (2020) Microbial electrochemical systems for hydrogen peroxide synthesis: critical review of process optimization, prospective environmental applications, and challenges. Bioresour Technol 313:123727 [Google Scholar] [CrossRef]
44. Ghangrekar MM, Chakraborty I (2019) Exploiting bioelectrochemical systems for wastewater treatment and value-added product recovery. In: Post treatments of anaerobically treated effluents. IWA Publishing, London, UK, pp 389–408. [Google Scholar] [CrossRef]
45. Das S, Mishra A, Ghangrekar MM (2020) Production of hydrogen peroxide using various metal-based catalysts in electrochemical and bioelectrochemical systems: mini review. J. Hazardous Toxic Radioact. Waste 24:06020001. [Google Scholar] [CrossRef]
46. Radhika D, Shivakumar A, Kasai DR, Koutavarapu R, Peera SG (2022) Microbial electrolysis cell as a diverse technology: overview of prospective applications, advancements, and challenges. Energies 15(7):2611. https://doi.org/10.3390/en15072611
47. Palhares DDDF, Vieira L, Damasceno JJR (2018) Hydrogen production by a low-cost electrolyzer developed through the combination of alkaline water electrolysis and solar energy use. Int J Hydrogen Energy 43:4265–4275. [Google Scholar] [CrossRef]
48. Capuano L (2018) International energy outlook 2018 (IEO2018). US energy information administration
49. Park J, Lee B, Tian D, Jun H (2018) Bioelectrochemical enhancement of methane production from highly concentrated food waste in a combined anaerobic digester and microbial electrolysis cell. Bioresour Technol 247:226–233 CrossRef
50. Seelajaroen H, Spiess S, Haberbauer M, Hassel MM, Aljabour A, Thallner S, Guebitz GM, Sariciftci NS (2020) Enhancedmethane producing microbial electrolysis cells for wastewater treatment using poly(neutral red) and chitosan modified electrodes. Sustain Energy Fuels 4:4238–4248 CrossRef
51. Nainamohamed S, Spurgeon J, Matheswaran M, Satyavolu J (2020) Simultaneous biohydrogen production with distillery wastewater treatment using modified microbial electrolysis cell. Int J Hydrogen Energy 45:18266–18274 CrossRef
52. Zhang L, Wang Y-Z, Zhao T, Xu T (2019) Hydrogen production from simultaneous saccharification and fermentation of lignocellulosic materials in a dual-chamber microbial electrolysis cell. Int J Hydrogen Energy 44:30024–30030 CrossRef
53. Wang K, Sheng Y, Cao H, Yan K, Zhang Y (2017) Impact of applied current on sulfate-rich wastewater treatment and microbial biodiversity in the cathode chamber of microbial electrolysis cell (MEC) reactor. Chem Eng J 307:150–158 CrossRef

54. Guo H, Kim Y (2019) Stacked multi-electrode design of microbial electrolysis cells for rapid and low-sludge treatment of municipal wastewater. Biotechnol Biofuels 12:23 CrossRef
55. Li J, Liu R, Zhao S, Wang S, Wang Y (2020) Simultaneous desalination and nutrient recovery during municipal wastewater treatment using microbial electrolysis desalination cell. J Clean Prod 261:121248 CrossRef
56. An J, Lee H (2013) Implication of endogenous decay current and quantification of soluble microbial products (SMP) in microbial electrolysis cells. RSC Adv 3(33):14021. https://doi.org/10.1039/c3ra41116h
57. Call D, Logan BE (2008) Hydrogen production in a single chamber microbial electrolysis cell lacking a membrane. Environ Sci Technol 42(9):3401–3406. https://doi.org/10.1021/es8001822
58. Lee HS, Torres CI, Parameswaran P, Rittmann BE (2009) Fate of H2 in an up-flow single-chamber microbial electrolysis cell using a metal-catalyst-free cathode. Environ Sci Technol 43(20):7971–7976. https://doi.org/10.1021/es900204j
59. Guo K, Tang X, Du Z, Li H (2010) Hydrogen production from acetate in a cathode-on-top single-chamber microbial electrolysis cell with a mipor cathode. Biochem Eng J 51(1–2):48–52. https://doi.org/10.1016/j.bej.2010.05.001
60. Escapa A, Gil-Carrera L, García V, Morán A (2012) Performance of a continuous flow microbial electrolysis cell (MEC) fed with domestic wastewater. Biores Technol 117:55–62. https://doi.org/10.1016/j.biortech.2012.04.060
61. Gil-Carrera L, Escapa A, Moreno RM, Morán A (2013) Reduced energy consumption during low-strength domestic wastewater treatment in a semi-pilot tubular microbial electrolysis cell. J Environ Manage 122:1–7. https://doi.org/10.1016/j.jenvman.2013.03.001
62. Li S, Cheng C, Thomas A (2017) Carbon-based microbial-fuel-cell electrodes: from conductive supports to active catalysts. Adv Mater 29:1602547. [CrossRef] [PubMed]
63. Merrill MD, Logan BE (2009) Electrolyte effects on hydrogen evolution and solution resistance in microbial electrolysis cells. J Power Sources 191:203–208 CrossRef
64. Omidi H, Sathasivan A (2013) Optimal temperature for microbes in an acetate fed microbial electrolysis cell (MEC). Int Biodeterior Biodegrad 85:688–692 CrossRef
65. Lu L, Ren N, Zhao X, Wang H, Wu D, Xing D (2011) Hydrogen production, methanogen inhibition and microbial community structures in psychrophilic single-chamber microbial electrolysis cells. Energy Environ Sci 4:1329–1336 CrossRef

Recent Advances in Microbial Electrolysis Cells for High Production of Hydrogen

Irwan Ibrahim, Krishan Balachandran, Ryan Yeo Yow Zhong, Kee Shyuan Loh, Mimi Hani Binti Abu Bakar, Bor Chyan Jong, Jamaliah Binti Md. Jahim, and Swee Su Lim

Abstract The generation of hydrogen gas through microbial electrolysis (MEC), which is one of the bioelectrochemical system (BES) approaches, has been extensively explored in recent years. The biohydrogen produced in this process is considered green hydrogen and thus is an alternative to other conventional methods that mostly produce blue or gray hydrogen which is unsustainable. MEC harnesses electrogenic microbes, which can metabolize organic matter and generate electricity to drive the recombination of protons to hydrogen gas, which can metabolize organic matter and generate electricity to drive the recombination of protons to hydrogen gas. Albeit its' potential as a green technology for green fuel synthesis, the process is still far from realization, mainly due to the low productivity and the low rate of reaction which is still low. Plus, the feasibility of MEC on a large scale is challenging because expensive electrode catalysts usually consist of noble metals. However, recent revelations and breakthroughs have provided insight into the accelerating potential of MEC in terms of electrode materials, system configurations, microbiology aspects of the microbes used, modeling, and recent scale-up attempts at MEC that have shown promising results. This review will discuss recent achievements relating to the MEC system and highlight future recommendations.

Keywords Nanomaterials · 3D printing · Photoelectrode · Thermophiles · Macroalgae · Thermoelectric · Modeling · Scale-up

---

I. Ibrahim · K. Balachandran · R. Y. Y. Zhong · K. S. Loh · M. H. B. A. Bakar · S. S. Lim (✉)
Fuel Cell Institute, Universiti Kebangsaan Malaysia, UKM, 43600 Bangi, Malaysia
e-mail: limss@ukm.edu.my

J. B. Md. Jahim
Department of Chemical and Process Engineering, Faculty of Engineering and Built Environment, Universiti Kebangsaan Malaysia, UKM, 43600 Bangi, Malaysia

B. C. Jong
Agrotechnology and Bioscience Division, Malaysian Nuclear Agency, 43000 Kajang, Malaysia

A. A. Yaqoob and A. Ahmad (eds.), *Microbial Electrolysis Cell Technology*, Sustainable Materials and Technology, https://doi.org/10.1007/978-981-97-3356-9_10

1 Recent Advances in Electrode Materials in Increasing the HER Efficiency

1.1 Nanomaterials as the Research Paradigm in MEC

Both anode and cathode play a crucial role as an intermediate for redox reaction in synthesizing hydrogen gas in MEC, from oxidation of substrate to produce electron and proton in anode, to combining the same precursor at the cathode to produce hydrogen. Thus, the material selection of both electrodes is the key aspect as the properties of the materials allow the reaction rate to occur, in terms of morphology and electrochemical properties. In recent years, several promising types of materials and fabrication methods have been studied and concluded that can significantly impact the hydrogen generation process and be applied in real-world applications. Examples such as nanomaterials and fabrication methods using 3D printing will be further discussed in this topic to highlight recent breakthroughs of MEC's electrode materials in producing high yields of hydrogen.

One such promising material is nano-electrocatalysts; unlike conventional electrocatalysts, nano-electrocatalysts offer enormous surface area which acts as an active site for electrochemical activity in a nanometer scale. Compared to normal 2D or 3D structures based on normal electrodes that have a smooth surface, nanomaterials have microscopic structures that increase the surface area exponentially. Conventionally, surface modification is used to achieve this morphology, either through chemical methods such as chemical pretreatment and electrodeposition; or physical methods such as pyrolysis, to name a few.

One such example is by using electrodeposition method; this method has been used for surface modification for decades and has been proven to increase the morphology of electrodes. It is also demonstrated that the electrodeposition process can be done in-situ bioelectrochemically, as opposed to normal electrochemical deposition. One such example is using electrogenic microbes such as *Shewanella oneidensis* to deposit palladium (Pd) catalyst on the surface of carbon cloth substrate. This approach works by reducing Pd (II) to Pd metal, which is carried out by the microbes. *S. Oneidensis* supplies the electron needed to reduce Pd (II) to Pd nanoparticle on the surface of its cell membrane or inside periplasmic space. Then, a Nafion binder is added to the surface to increase the stability of Pd nanoparticle attachment on the cathode surface. By implementing this method in double-chamber MEC, the hydrogen yield obtained was 37.70% better as compared with the system with the absence of electrogenic microbes [1]. This shows that the nanoparticle cathode coupled with electrogens has better effectiveness in generating HER and offers good biocompatibility for microbial attachment. Aside from Pd metal, several other nanomaterials that are used in MEC are summarized in Table 1.

Another one of the emerging materials that has been extensively studied is graphene, a 3D carbon element structure that is made entirely of carbon atom in a single layer in a hexagonal arrangement but is one atom thick and exhibits stronger hardness than steel but also flexible [2]. Owing to this elemental arrangement, the

Table 1 Nanomaterials used in previous studies

Anode	Cathode	Microbes	Applied voltage (V)	Current density (A/m^2)	Hydrogen yield	Reference
Graphite brush	Ni_2P/C	Enriched MFC	0.9	5.7	0.29 L H_2/L d	[8]
Graphite	Ni–Co–P/SS	Enriched MFC	0.6	4.28	0.16 m^3 H_2/m^3 d	[9]
Graphite plate	NiO · rGO	Sugar industrial wastewater	1.0	2.01×10^{-6}	4.38 mmol/L d	[10]
Graphite	NiMoO4–NF	Sugar wastewater	1.0	8.1	0.12 L/L d	[11]
Graphite plate	$NiCo_2O_4$–graphene	Sugar wastewater	1.0	1.25	0.14 L/L dl	[5]

properties of graphene exceed any of its carbon element counterparts such as carbon, graphite, and activated carbon. Synthesizing graphene requires several reaction steps in series from the precursor material, which is graphite. This process starts with transforming graphite into graphite oxide through the oxidation process, then graphite oxide is converted into graphene oxide via ultrasonication in solution before it is finally transformed into graphene through reduction to remove functional groups [3]. This process is called Hummer's method, and it is a widely used procedure in synthesizing graphene as illustrated in Fig. 1.

Through the application of graphene as cathode, using stainless-steel mesh (SSM) as substrate, and polyaniline (PANI) as conductive polymer binder, the hydrogen production was at 0.805 m^3H_2/m^3 anolyte daily under 1.0 V applied potential. This difference is 20% lower compared to the control SSM/Pt electrode [4]. Another study was done using a nanocomposite of $NiCo_2O_4$/Graphene on a nickel foam-supported cathode, which is performed through a modified Hummer's method. This process involved the addition of $CoCl_2 \cdot 6H_2O$ and $NiCl_2 \cdot 6H_2O$ into GO dispersion, after which the ammonia and sodium borohydride are added to reduce the GO into rGO. The MEC indicated that the hydrogen production rate increased at a rate more than three times higher compared to the control cell at 0.14 L/L d [5]. This demonstrates the potential improvement that graphene can bring as a new substitute for carbon-based electrodes.

Plus, another aspect of interest in graphene is that it has the potential to become an alternative material in replacing noble metals such as platinum. This is one of the technical challenges that researchers have been trying to overcome in finding a catalyst that has the same catalytic properties as platinum, but at a fraction of the cost and can be sourced sustainably. This can be demonstrated by using graphene-coated nickel foam-supported graphene cathode (NF–G) in the study and comparing it with Pt/C based cathode at Pt loading of 0.5 mg Pt/cm^2 as control [6]. The modified NF–G is made by coating the NF with graphene ink, which is simple and easily reproducible

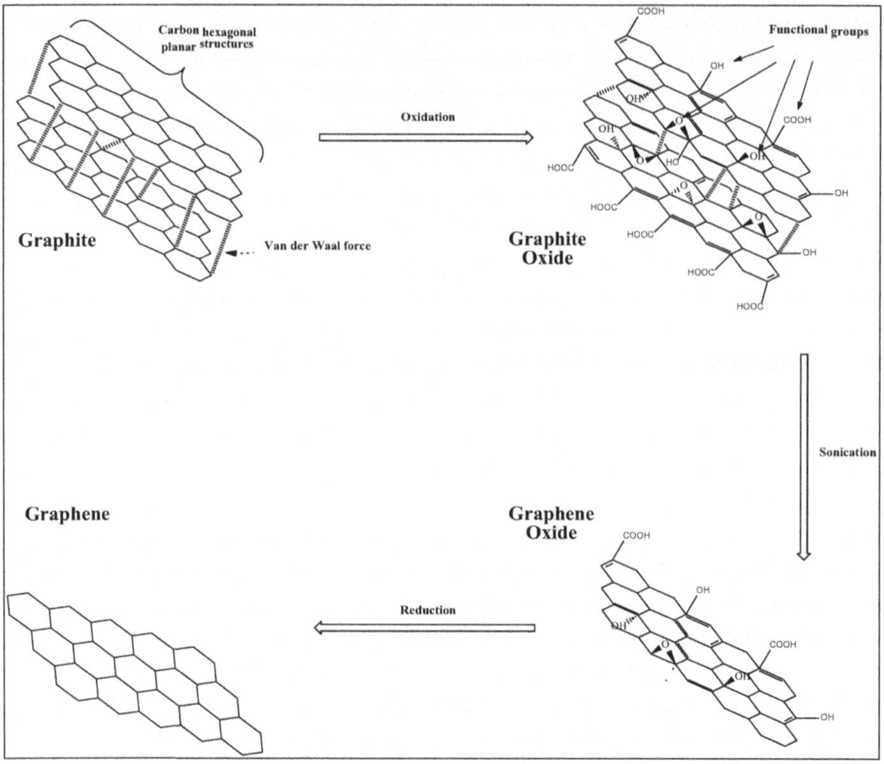

Fig. 1 Reaction steps in Hummer's method

at low cost and mass-produced in a short amount of time. The MEC operation analysis showed that the hydrogen production of NF–G and Pt/C is almost similar at 1.31 mL H_2/mL d and 1.32 mL H_2/mL d, respectively. This highlights the graphene potential as a candidate material to substitute noble metal-free catalysts; thus, more studies could be done to improve the graphene to make performance better than noble metals. One such example, like Pt/C based catalyst, is to manipulate the loading rate of graphene in graphene ink to determine the maximum amount of loading rate it could achieve to maximize the HER performance in MEC while still maintaining the cost optimization or exploring other means of the graphene ink coating, such as spray printing which can minimize the cost of coating while achieving the same result.

Graphene structure also could be further exploited in a variety of other modifications. For example, two materials that exhibit almost similar resemblance to graphene structure, called graphene oxide (GO) and reduced graphene oxide (rGO), are also used as electrode material in MEC; GO has a similar structure as graphene, but with the additional functional groups such as carboxyl, epoxy, and carbonyl group. Additionally, rGO also shares this similar configuration, but in a significantly smaller quantity compared to GO, and it also has the closest structural arrangement to graphene. To demonstrate the idea, a study was done on incorporating Pt nanoparticles into rGO (Pt/rGO). This is done by sonicating the rGO solution with K_2PtCl_4 and then $NaBH_4$ before it is washed and dried. This material is then turned into ink, brushed onto carbon cloth, and made as a cathode catalyst for MEC. The result shows that the current density of Pt/rGO was 56% higher compared to commercial Pt/C ink with Coulombic efficiency (CE) of 116% with a hydrogen production rate of 28% higher than that of Pt/C cathode at 1.1 m^3/m^3 d [7]. This is achieved as the Pt/rGO exhibits the lowest nanoparticle size at 8–32 nm with the largest surface area of 24.40 m^2/g and low ionic resistance. Additionally, several other graphene-based cathodes are listed in Table 1 to summarize the electrode's proficiency.

1.2 Application of Photoelectrode-Initiated MEC

The use of photoanode has been popular in fuel cell technology as both anode and cathode, photoelectrode harnesses semiconductor on its electrode surface to generate holes on the photoanode side which is where the oxidation of water to oxygen takes place, and electron generation on the photoanode would result in the reduction of CO_2 into hydrogen, formate, carbon monoxide, alcohol, and other hydrocarbons [12]. This process is theoretically practical in converting CO_2 such as methanol or hydrogen production, especially as it virtually uses free energy derived from sunlight for photolysis and electrical power from renewable sources. The application of this system is still limited because of the complex system of the photoreactor, lack of charge separation, and limited light absorption of the photoelectrode [14].

The mechanisms of photolysis can be divided into five step processes [13] that is shown in Fig. 2, Solar energy is absorbed by the semiconductor known as light-harvesting step where the energy of solar radiation should be equal or higher than the semiconductor's band gap energy. (1) The electrons in the valence band are excited to the conductance band, thus resulting in proton generation in the conductance band and holes in the valence band. (2) The electrons move toward the outer region of the semiconductor, thus confined by the active catalyst sites present at the surface. (3) Volume recombination occurs which is the recombination of electrons and holes. (4) Surface recombination occurs due to the recombination in the previous step. (5) Oxidation and reduction reactions occur by utilizing the holes and electrons generated.

Currently, the application of photocathode is proven to eliminate heavy metals in water [15] although the hydrogen production is not high, at only 0.1138 m^3H_2/m^3 d using semiconductor carbon dots/graphitic carbon nitride (CDs/g–C_3N_4). However, in other materials such as $ZnFe_2O_4$/g-C_3N_4, a high yield of hydrogen is achieved up to 0.55 m^3 H_2/m^3 d [16] over a long cycle. Plus, the use of photoanode also reported high production of hydrogen through the use of both cerium oxide (CeO_2) and rGO photoanode, showing the H_2 recovery at 93%, equivalent to 5 m^3 H_2/m^3 d and cathodic recovery at 95% [17] while simultaneously reduce COD in wastewater up to 95%.

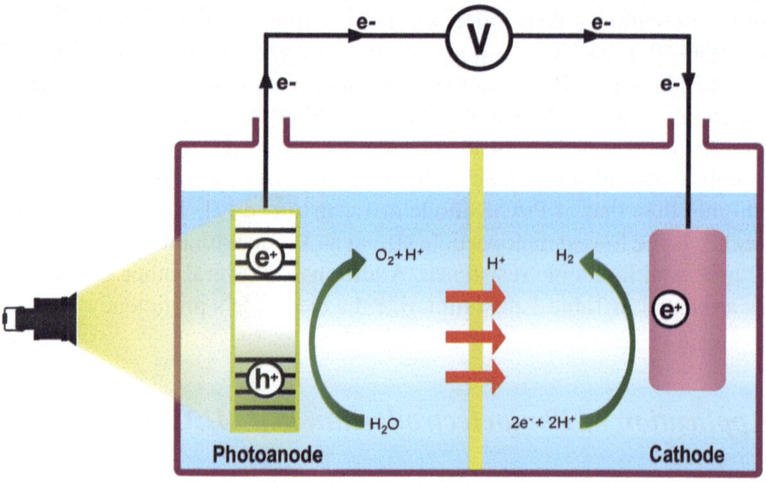

Fig. 2 Photoelectrode application in MEC

1.3 3D Printing as the New Frontier in Electrode Fabrication

Aside from the material selection, the mean of production has also seen some advancements in the fabrication of electrodes at high speed, lower cost, and greater efficiency. One such method that has gained attention is the 3D printing method, also known as additive manufacturing, where an object is created by adding material layer by layer. This manufacturing approach allows the fabrication of complex parts, which are already being used in the medical, fashion, transportation, construction, and even food industries applications and have paved the way for a myriad of new innovative research. 3D printing also allows for more cost and time-effective fabrication of materials as the process is fully automated and can be adjusted by using modeling software, thus eliminating human errors. The substrate of 3D printers usually ranges from polymers such as plastics and resins, to metals and ceramics. This process starts by pulling the materials through a tube in malleable form and depositing it to the plate, where it cools. The 3D object is created through layering, where the printer will add one layer of the object at a time until a fully formed structure is achieved.

Although the prospect of a 3D printer is promising, the application of it in MEC studies for hydrogen production is still very limited. There is only one notable example of 3D, which is applied in the fabrication of anode material [18]. This is done by using Electrifi conductive filament, a highly conductive 3D printer material that has high electrical conductivity with a low resistance of 0.006 Ω cm and fashioned into a spiral-shaped anode with a base diameter of 5 mm and 953 mm^3 active surface area, but with different height and pitch numbers as the manipulated variable. The results suggested that the best anode design exhibits high current density at 1.5 mA/ cm^2 at 2.0 V applied potential and produced 5 ml of hydrogen only in 70 min at best. This demonstrates the viability of 3D printing as a new approach in electrode fabrication, not to mention the possible modifications and potential opportunities that are yet to be explored. For example, modifying the 3D printed electrodes with other materials such as metals and carbon materials to enhance the performance is one of the possibilities that could be exploited in future studies.

2 New Variation of Microorganisms Used in MEC and Application of Genetic Engineering

The microbial aspect is the core element that separates MEC from conventional electrolysis systems. It is one of the prominent aspects that researchers have been extensively involved in to make the MEC viable. These include trying different variations of electrogenic microbes' strain, culture condition of microbes used, enrichment of electrodes with different electrolytes, and other operating parameters used to study the response behavior of electroautotrophs. In recent years, there have been several interesting discoveries being made in MEC in terms of microbiology elements that could be harnessed to further ameliorate hydrogen production.

2.1 Thermophiles

Historically, the majority of MECs used mesophilic inoculum as a biocatalyst to facilitate the HER; mesophilic microbes thrive at mild or room temperature and require minimal heat regulation requirement, which equates to low energy consumption. However, recent studies focused on different kinds of microbes, specifically those that thrive in higher temperatures, called thermophiles. Thermophiles are a type of extremophiles, in which organisms that live naturally in harsh environmental conditions where normal organisms would perish. Naturally, thermophiles are commonly found in hot springs, geysers, and underwater hydrothermal vents where temperatures exceed 50 °C consistently.

Moreover, aside from the unique living conditions, the metabolisms in most thermophiles also differ from their mesophilic counterpart. This is due to the lack or absence of oxygen as their electron donor and acceptor. Instead, they use hydrogen and sulfur-derived compounds as they live in limited oxygen environments where there is a lack of organic compounds. Another interest the thermophilic microbes possess that intrigues the researcher recently is the ability to produce hydrogen via the Embden–Meyerhoff pathway for sugar catabolism where it generates hydrogen using NADH-dependent hydrogenase, such as genera *Thermoanaerobacter* and *Caldicellulosiruptor* [19].

A few notable examples of the use of thermophiles within the MEC system include using hyperthermophilic *Geoglobus acetivorans*, which thrive in temperatures above 80 °C and can live up to 100 °C. By implementing *G. acetivorans* in single-chamber MEC with an applied potential of 0.7 V, the hydrogen production rate managed to attain 0.57 m^3H_2/m^3 d with peak current density at 1.53 A/m^2 which operates at 80 °C, one of the highest reported current produced in pure culture hyperthermophilic microbes [20]. Additionally, another study was done to compare the performance of thermophiles in converting lignocellulosic waste into biohydrogen in mesophilic and thermophilic conditions using dark fermentation-MEC-based systems [21]. It was found that the system ran in thermophilic conditions outperformed its mesophilic counterpart by more than 6 folds, which equates to 5.416 mmol H_2/g-lignocellulosic waste. Plus, microbial communal analysis verified that the thermophilic system at 55 °C was dominated by hydrogen-producing thermophiles such as phyla of *Proteobacteria*, *Thermotogota*, and *Coprothermobacteria* while the mesophilic system consisted mainly of methanogens as the high-temperature inhibits the growth of them.

2.2 *Macroalgae*

Aside from archaea and bacteria, another type of microorganism that has gained attention recently to facilitate hydrogen production in MEC comes from photosynthetic organisms named macroalgae. Macroalgae have a unique classification where it can be categorized into two main kingdoms of taxology, namely Plantae with 4400 different species and Chromista with 700 different species [22]. Plus, it has been documented that macroalgae are partly responsible for recycling CO_2 from the atmosphere globally and converting it to energy in the form of carbohydrates and oil at a higher rate compared to other alga types. Thus, previous research harnesses its ability to convert algae into biofuel, pharmaceuticals, and feedstock-based products for animals. Algae also are more productive than terrestrial plants as they can grow up to 100 times more per acre on land and require significantly less time to harvest as they multiply exponentially, which is up to 30 times faster than conventional crops [23], which proves their sustainability.

The photosynthetic quality of macroalgae makes the organism differ from electrogenic organisms because it can generate energy directly via sunlight. Owing to these properties, macroalgae use two different autotrophic processes, called light-dependent photolysis and light-independent dark fermentation. Biohydrogen production of macroalgae happens specifically during dark fermentation, when protein and carbohydrates are consumed during metabolism until oxygen depletion occurs. This in turn, creates anaerobic conditions, thereby activating oxygen-sensitive hydrogenase that creates hydrogen from the electrons generated by the fermentation pathway [24].

Macroalgae also can show promising results when compared with platinum-based cathode; it is demonstrated by using *Saccharina japonica* type macroalgae as biocatalyst coupled with anodized stainless-steel mesh cathode. The study indicates that the system operates at up to 75% efficiency as the Pt/C-based control, considering the cost of material that is significantly less than the respective noble metal [25]. Another research example uses different species of macroalgae to compare the biohydrogen production performance, which includes *Gracilaria tenuistipitata* (Rhodophyta), *Ulva lactuca* (Chlorophyta), and *Sargassum polycyctum* (Phaephyta) [26]. The outcome indicates that *S. polycyctum* showed the best performance producing biohydrogen at 472.4 mL/g VS while reducing the methane byproduct using inhibiter in the form of chloroform, which corresponds to higher H_2 yield and overall energy recovery by around 150 and 170%.

3 Operating Condition of MEC

The operating condition of MEC plays one of the vital aspects aside from the electrode and the biocatalyst as it provides the environment for the hydrogen evolution process to occur. Generally, conventional MEC bioreactors consist mainly of two different sets, which are double-chamber and single-chamber MEC. However, recent progress has extrapolated this mainstream bioreactor by manipulating certain fundamental parameters or hybridization with an assortment of systems to explore new findings of MEC in real-world applications, instead of confining it to limited applications. As previously mentioned in another chapter, MEC can be coupled with other BES, which are proven to increase the functionality and versatility of MEC, or used in various aqueous waste treatments. This chapter will discuss the new recent applications of MEC and how these pioneered breakthroughs could be harnessed and further explored in the future.

3.1 Integration of Thermoelectric Generator with MEC

The majority of industries use heat exchangers as a heating system to supply heat energy for operations or to release excess heat from the system. However, most of the heat energy is wasted as it dissipates to the surrounding environment without being utilized strategically. While some engineering approaches practice pinch analysis to reduce the loss of heat energy by funneling the energy to other parts of operations that require the heat, thus optimize, and minimize the waste energy; another approach can also be used to harness this energy to power the MEC system. This prospect is promising as the energy is virtually free and could be recycled to power electrical equipment, which contributes to net zero emission as the majority of energy comes from fossil fuel.

A thermoelectric generator (TEG) is one of the studied models that has been used to illustrate this approach. TEG converts heat directly into electricity with no moving parts, as heat moves through the thermoelectric module, it causes an electric current to flow creating DC electricity. This is because there are hot and cold regions of TEG, which create temperature differential and result in current flow that is directed to the load. The system is cost-effective and reliable as there are no moving parts and has already been deployed in space missions, transportation to increase system efficiency by recycling excess heat into energy, and isolated places with no power grid. Aside from normal convection and conduction heat, solar energy can also be used as a heat source to power TEG [27], making this system more viable by integrating TEG with renewable resources.

TEG-MEC integrated system example is done through hydraulic and electrical means. For electrical connection, the positive side of the TEG was connected to the anode of the MEC. The negative side was connected to the cathode of the MEC with a resistor in between. The heat transfer is initiated through wastewater pumped from the cold region generator at a variety of flows, where it is being heated and circulated to the anode side, and the cycle is completed as the anode effluent is redirected back to the cold region of TEG as illustrated in Fig. 3, with manipulated hydraulic retention time (HRT). The TEG-MEC operation comes together after the hydrogen generation becomes stable after being supplied with an applied voltage of 1.0 V. The research concludes that high hydrogen generation is improved as the flow rate and HRT increase at a certain point up to 0.36 m^3 H_2/m^3 d [28]. This creates good potential in using TEG to power MEC as TEG generates a viable amount of energy to power MEC independently, and more research could tap into using thermophilic microbes which could proliferate hydrogen yield as the current research use mesophilic microbes. Another similar study focused on the temperature difference of TEG and compared it with a constant supply load together with a medium of heat transfer. The experiment indicates that large temperature differences affect the TEG-MEC positively while water is a preferable medium compared to air because of less susceptibility to fluctuation [29] and the fact that fundamentally water has more heat capacity than air.

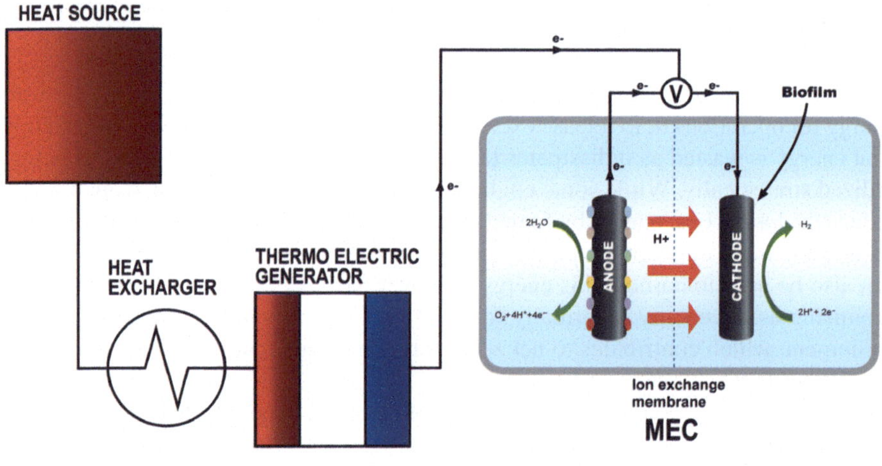

Fig. 3 Configuration of TEG-MEC

3.2 Modeling

The complex behavior of the MEC system happens as a result of biological components in the system, either in the form of biofilm or suspension. Microbes, being living things, make the system fluctuate constantly as compared to an abiotic system. Thus, the computational approach also called computer modeling is one of the methods to predict the nonlinear and dynamic behavior of MEC and simplify it for easier maneuverability and piloting [30]. Alternatively, computer modeling and simulation are more cost-effective and faster approach in determining the parameters needed to effectively proliferate the MEC performance with rigorous experimenting that takes many iterations, time, and cost.

Response surface methodology (RSM) is one of the orthodox and established ways to predict the performance of a system. This method relies on the data taken from a preset study and the use of the second-order polynomial model to generate possible outcomes that may happen as the variable changes. One such use of this approach is the treatment of palm oil mill effluent (POME) to biohydrogen conversion via MEC with three variables manipulation, namely temperature, pH, and dilution rate, utilizing central composite design and several experiments. Using RSM modeling suggests the highest production rate of 1.16 m^3 H_2/m^3 d H_2 with determined manipulated variables given and the actual experiment result showed the approximation to the model [31]. Aside from the electrolyte variable, RSM is also used to optimize the HRT and applied potential of MEC in treating dairy effluent [32]. Using the parameters optimized via RSM, the H_2 rate increased by more than 15%.

Another insight that will be conventional in the future is the use of artificial intelligence (AI) in predicting the MEC model. This approach is new and has not been fully implemented in MEC for hydrogen production. However, the use of AI in methane production has started recently. It is done to predict the AD-MEC model fed with alkaline-pretreated waste-activated sludge [33]. A two-hidden-layer artificial neural network (ANN) with topology 2–25–34–6, was obtained from the response surface methodology. The result suggested that net energy output and net monetary value were significantly enhanced by up to 160% and 300%, respectively, compared to the AD alone.

3.3 Scale-Up

The prospect of realizing the MEC on a larger scale and producing high hydrogen yield was initiated years ago in the search for the most realistic and best possible outcome MEC can be deployed on a commercial scale. The first stage of MEC scale-up is mostly done in a pilot scale plan as opposed to bench scale design to study the parameters that will deviate from the small-scale experiment setup, as the size becomes larger. The fundamental principle that must be considered is the engineering challenges that must be overcome in replicating similar results while changing the design aspects is paramount in achieving the endeavor. Furthermore, the consideration of both capital expenditure (CAPEX) and operating expenditure (OPEX) are other important components to consider, as the cost usually rises exponentially with the increase in scale. Thus, several creative solutions should be used in making the scale-up MEC cost-effective and competitive with other systems.

The carbon-based electrode is one of the most popular materials in scaling the MEC as the material is highly effective and relatively cheap to acquire compared to metal-based electrodes. However, the cost can further be reduced by reusing and recycling the carbon-based material. Recycled carbon fiber fabric (rCF) is one of the creative proposals that has been studied and implemented in larger-scale MEC [34]. rCF with a dimension of 1 m^2 is modified by coating the surface with manganese oxide (MnO_x) through electrodeposition before being installed into a large MEC reactor with an anodic volume of 175 L with surface area-to-volume ratio of 34 m^2/m^3 and 0.8 m^2 cathode with cathodic volume of 7.2 L and specific area of 13 m^2/m^3. By using the rCF, the hydrogen yield outperforms graphite felt control by more than 18-fold, which is equivalent to 66.77 L H_2/d while graphite felt produced only 3.65 L H_2/d per 1 m^3 reactor. The cost estimation is also significantly low, with cost saving up to 93% by using the rCF which would lead to lower CAPEX and OPEX consideration. Another remarkable progress was also made using carbon cloth modified with molybdenum phosphide and stacked electrode assembly in 10-L flow hybrid fermentation, and the MEC reactor was able to achieve an H_2 generation rate of 5.9 L/L/D with a current density of 970 A/m^2 due to vertically oriented electrodes with solution recirculation [35].

Substrate and raw materials needed for conversion of hydrogen is another aspect that should be considered. While most of the previous substrates used in MEC studies heavily relied on wastewater for hydrogen recovery, the search for alternative materials led to the use of other waste materials valorization. Agriculture waste, especially lignocellulose waste, consists mostly of the elements carbon, oxygen, and most importantly hydrogen. Oftentimes, these wastes were burned openly across the world, contributing to air pollution and degradation of air quality. The untapped potential of these lignocellulosic materials has been used as substrates in generating considerable amounts of hydrogen while reducing polluting materials. The illustration of this approach is verified by using lignocellulosic hydrolysate-fed substrate in 10-L single-chamber MEC with stacked bioanodes and biocathodes. The model produced 0.71 L/L/D of hydrogen with 91% efficiency while achieving anode and cathode current densities of 13.7 and 16.5 A/m^2, respectively [36]. Additionally, another model used leftover corn stover fermentation waste, a complex waste composition substrate rich with sugars and organic acids which produced an impressive hydrogen generation rate of up to 20 L/L/d and a maximum current density of up to 27.2 A/m^2. This indicates that substrate composition plays an important role in producing a high yield of hydrogen while being sustainable at the same time [37].

The high productivity of large-scale MECs is also largely responsible for the operation mode. Ideally, the continuous mode is the main choice in industries as compared to batch and fed-batch because MEC must able to accommodate large quantities of substrate in a short amount of time. A comparison between the continuous and fed-batch indicated that the fed-batch feeding regime is 46% inferior with only 4.27 L/L/D of hydrogen and 5.1 A/m^2 performance while the continuous regime outperformed with 7.9 L/L/D of hydrogen and 9.9 A/m^2 in a three week time duration of operation [38]. However, more studies should be done in a bigger setup of experiments to correlate the relation as the study focused in bench scale setup.

References

1. Wang W, Zhang B, He Z (2019) Bioelectrochemical deposition of palladium nanoparticles as catalysts by Shewanella oneidensis MR-1 towards enhanced hydrogen production in microbial electrolysis cells. Electrochim Acta 318:794–800. https://doi.org/10.1016/j.electacta.2019.06.038
2. Ibrahim I, Salehmin MNI, Balachandran K, Me MFH, Loh KS, Loh KS, Jong BC, Lim SS (2023) Role of microbial electrosynthesis system in CO_2 capture and conversion: a recent advancement toward cathode development. Front Microbiol 14. https://doi.org/10.3389/fmicb.2023.1192187
3. Zhou A, Yu T, Liang X, Yin S (2023) H_2O_2-free strategy derived from Hummers method for preparing graphene oxide with high oxidation degree. FlatChem 38:100487. https://doi.org/10.1016/j.flatc.2023.100487
4. Ghasemi B, Yaghmaei S, Abdi K, Mardanpour MM, Haddadi SA (2020) Introducing an affordable catalyst for biohydrogen production in microbial electrolysis cells. J Biosci Bioeng 129(1):67–76. https://doi.org/10.1016/j.jbiosc.2019.07.001
5. Jayabalan T, Matheswaran M, Mohamed SN (2020) $NiCo_2O_4$-graphene nanocomposites in sugar industry wastewater fed microbial electrolysis cell for enhanced biohydrogen production. Renewable Energy 154:1144–1152. https://doi.org/10.1016/j.renene.2020.03.071
6. Cai W, Liu W, Han J, Wang A (2016) Enhanced hydrogen production in microbial electrolysis cell with 3D self-assembly nickel foam-graphene cathode. Biosens Bioelectron 80:118–122. https://doi.org/10.1016/j.bios.2016.01.008
7. Sánchez-Peña P, Rodriguez JO, Gabriel D, Baeza JA, Guisasola A, Baeza M (2022) Graphene functionalization with metallic Pt nanoparticles: a path to cost-efficient H_2 production in microbial electrolysis cells. Int J Hydrogen Energy 47(34):15397–15409. https://doi.org/10.1016/j.ijhydene.2022.03.078
8. Kim K, Habas SE, Schaidle JA, Logan BE (2019) Application of phase-pure nickel phosphide nanoparticles as cathode catalysts for hydrogen production in microbial electrolysis cells. Biores Technol 293:122067. https://doi.org/10.1016/j.biortech.2019.122067
9. Chaurasia AK, Shankar R, Mondal P (2021) Effects of nickle, nickle-cobalt and nickle-cobalt-phosphorus nanocatalysts for enhancing biohydrogen production in microbial electrolysis cells using paper industry wastewater. J Environ Manage 298:113542. https://doi.org/10.1016/j.jenvman.2021.113542
10. Jayabalan T, Matheswaran M, Preethi V, Mohamed SN (2020) Enhancing biohydrogen production from sugar industry wastewater using metal oxide/graphene nanocomposite catalysts in microbial electrolysis cell. Int J Hydrogen Energy 45(13):7647–7655. https://doi.org/10.1016/j.ijhydene.2019.09.068
11. Jayabalan T, Matheswaran M, Radhakrishnan TK, Mohamed SN (2021) Influence of Nickel molybdate nanocatalyst for enhancing biohydrogen production in microbial electrolysis cell utilizing sugar industrial effluent. Biores Technol 320:124284. https://doi.org/10.1016/j.biortech.2020.124284

12. Xie S, Zhang Q, Liu G, Wang Y (2016) Photocatalytic and photoelectrocatalytic reduction of CO_2 using heterogeneous catalysts with controlled nanostructures. Chem Commun 52(1):35–59. https://doi.org/10.1039/c5cc07613g
13. Wu J, Huang Y, Ye W, Li Y (2017) CO_2 reduction: From the electrochemical to photochemical approach. Adv Sci 4(11):1700194. https://doi.org/10.1002/advs.201700194
14. Minggu LJ, Salehmin MNI, Mohamed M, Arifin K, Yunus RM, Kassim MB (2020) A low overpotential photoelectrochemical reduction of carbon dioxide to methanol with highly photoactive hierarchical structured cuprous oxide. Ceram Int 46(16):26004–26016. https://doi.org/10.1016/j.ceramint.2020.07.091
15. Zhang J, Wang Q, Wan H, Shi Y, Huang L (2023) Enhanced etching terminal wastewater treatment and H_2 production by in-situ deposited heavy metals on carbon dots/g-C_3N_4 photocathode microbial electrolysis cells. J Hazard Mater 459:132178. https://doi.org/10.1016/j.jhazmat.2023.132178
16. Huang L, Kong W, Song S, Quan X, Puma GL (2023) Treatment of industrial etching terminal wastewater using $ZnFe_2O_4$/g-C_3N_4 heterojunctions photo-assisted cathodes in single-chamber microbial electrolysis cells. Appl Catal B-Environ 335:122849. https://doi.org/10.1016/j.apcatb.2023.122849
17. Pophali A, Singh S, Verma N (2020) Simultaneous hydrogen generation and COD reduction in a photoanode-based microbial electrolysis cell. Int J Hydrogen Energy 45(48):25985–25995. https://doi.org/10.1016/j.ijhydene.2020.01.053
18. Baş F, Kaya MF (2022) 3D printed anode electrodes for microbial electrolysis cells. Fuel 317:123560. https://doi.org/10.1016/j.fuel.2022.123560
19. Topçuoğlu BD, Holden JF (2019) Extremophiles: hot environments ☆. Elsevier eBooks. https://doi.org/10.1016/b978-0-12-809633-8.90683-6
20. Kas A, Yılmazel YD (2022) High current density via direct electron transfer by hyperthermophilic archaeon, *Geoglobus acetivorans*, in microbial electrolysis cells operated at 80 °C. Bioelectrochemistry 145:108072. https://doi.org/10.1016/j.bioelechem.2022.108072
21. Ndayisenga F, Yu Z, Wang B, Wu G, Zhang H, Phulpoto IA, Zhao J, Yang J (2022) Thermophilic-operating environment promotes hydrogen-producing microbial growth in a lignocellulose-fed DF-MEC system for enhanced biohydrogen evolution. Process Saf Environ Prot 167:213–224. https://doi.org/10.1016/j.psep.2022.09.026
22. Dawes C (2016) Macroalgae systematics. *Seaweed in Health and Disease Prevention*, 107–148. https://doi.org/10.1016/B978-0-12-802772-1.00004-X
23. Ziolkowska JR, Simon LK (2014) Recent developments and prospects for algae-based fuels in the US. Renew Sustain Energy Rev 29:847–853. https://doi.org/10.1016/j.rser.2013.09.021
24. Nagarajan D, Lee DH, Kondo A, Chang JS (2017) Recent insights into biohydrogen production by microalgae—from biophotolysis to dark fermentation. Biores Technol 227:373–387. https://doi.org/10.1016/j.biortech.2016.12.104
25. Yun WH, Yoon YS, Yoon HH, Nguyen PKT, Hur J (2021) Hydrogen production from macroalgae by simultaneous dark fermentation and microbial electrolysis cell with surface-modified stainless steel mesh cathode. Int J Hydrogen Energy 46(79):39136–39145. https://doi.org/10.1016/j.ijhydene.2021.09.168
26. Phan PT, Nguyen TL, Nguyen PKT (2023) Evaluation and enhancement of hydrogen production from Vietnamese macroalgae in the single-stage of dark fermentation and microbial electrolysis cell. Biomass Bioenerg 175:106885. https://doi.org/10.1016/j.biombioe.2023.106885
27. Riyadi TWB, Effendy M, Utomo BR, Wijayanta AT (2023) Performance of a photovoltaic-thermoelectric generator panel in combination with various solar tracking systems. Appl Therm Eng 235:121336. https://doi.org/10.1016/j.applthermaleng.2023.121336
28. Jain A, He Z (2020) Improving hydrogen production in microbial electrolysis cells through hydraulic connection with thermoelectric generators. Process Biochem 94:51–57. https://doi.org/10.1016/j.procbio.2020.04.008
29. Jain A, He Z (2020) Powering microbial electrolysis cells by electricity generation from simulated waste heat of anaerobic digesters using thermoelectric generators. Int J Hydrogen Energy 45(7):4065–4072. https://doi.org/10.1016/j.ijhydene.2019.12.073

30. Asrul MAM, Atan MF, Yun HAH, Lai JCH (2021) Mathematical model of biohydrogen production in microbial electrolysis cell: a review. Int J Hydrogen Energy 46(75):37174–37191. https://doi.org/10.1016/j.ijhydene.2021.09.021
31. Kadier A, Wang J, Chandrasekhar K, Abdeshahian P, Islam MA, Ghanbari F, Bajpai M, Katoch SS, Bhagawati PB, Li H, Kalil MS, Hamid AA, Hasan HA, Ma P (2022) Performance optimization of microbial electrolysis cell (MEC) for palm oil mill effluent (POME) wastewater treatment and sustainable Bio-H_2 production using response surface methodology (RSM). Int J Hydrogen Energy 47(34):15464–15479. https://doi.org/10.1016/j.ijhydene.2021.09.259
32. Rani G, Banu JR, Kumar G, Yogalakshmi K (2022) Statistical optimization of operating parameters of microbial electrolysis cell treating dairy industry wastewater using quadratic model to enhance energy generation. Int J Hydrogen Energy 47(88):37401–37414. https://doi.org/10.1016/j.ijhydene.2022.03.120
33. Nguyen VT, Ta THQ, Nguyen PKT (2022) Artificial intelligence-based modeling and optimization of microbial electrolysis cell-assisted anaerobic digestion fed with alkaline-pretreated waste-activated sludge. Biochem Eng J 187:108670. https://doi.org/10.1016/j.bej.2022.108670
34. Carlotta-Jones DI, Purdy KJ, Kirwan K, Stratford JP, Coles SR (2020) Improved hydrogen gas production in microbial electrolysis cells using inexpensive recycled carbon fibre fabrics. Biores Technol 304:122983. https://doi.org/10.1016/j.biortech.2020.122983
35. Singh L, Miller A, Wang L, Liu H (2021) Scaling-up up-flow microbial electrolysis cells with a compact electrode configuration for continuous hydrogen production. Biores Technol 331:125030. https://doi.org/10.1016/j.biortech.2021.125030
36. Wang L, Long F, Fan W, Xiao X, Hong L (2021) Hydrogen production from lignocellulosic hydrolysate in an up-scaled microbial electrolysis cell with stacked bio-electrodes. Biores Technol 320:124314. https://doi.org/10.1016/j.biortech.2020.124314
37. Satinover SJ, Schell DJ, Borole AP (2020) Achieving high hydrogen productivities of 20 L/L-day via microbial electrolysis of corn stover fermentation products. Appl Energy 259:114126. https://doi.org/10.1016/j.apenergy.2019.114126
38. Lewis AJ, Borole AP (2019) Microbial electrolysis cells using complex substrates achieve high performance via continuous feeding-based control of reactor concentrations and community structure. Appl Energy 240:608–616. https://doi.org/10.1016/j.apenergy.2019.02.048

Role of Microbial Electrolysis Cells in Nutrient Recovery Process

Thiruppathi Krithika and Jagannathan Anitha

Abstract Microbial electrolysis cells (MECs) are an exciting and cutting-edge technology that holds tremendous potential for energy and resource recovery during waste treatment. By harnessing the power of microorganisms, MECs have the ability to convert biodegradable waste into valuable products such as hydrogen (H_2), biofuels, and other high-value compounds. Through extensive research, MECs have demonstrated their versatility and wide range of applications, including the removal of stubborn pollutants, synthesis of chemicals, resource recovery, and even biosensors. One of the remarkable advantages of MECs is their ability to not only treat wastewater but also recover valuable resources simultaneously. Unlike conventional wastewater treatment methods, MECs offer a dual benefit of both remediation and recovery. This is a significant development in waste management, as it not only addresses the environmental concerns associated with wastewater but also offers a solution for resource scarcity and cost-effectiveness. While the potential of MEC technology is undeniably exciting, it is important to acknowledge that it is still in its early stages of development. As such, there are several challenges that need to be overcome before large-scale practical applications can be realized. These challenges include optimizing the recovery of nutrients, water, and energy from high-strength sidestream centrate. However, researchers in the field are optimistic about the future of MECs and believe that with further advancements, these challenges can be successfully addressed. One particular area of interest is the recovery of various valuable resources from industrial wastewaters. With the application of microbial electrolysis cells, it becomes possible to recover hydrogen (H_2), methane (CH_4), hydrogen peroxide (H_2O_2), phosphate ions (PO_4^{3+}), nitrogen (N), formaldehyde (CH), formic acid (HCOOH), and even a variety of metals. While these applications are still primarily confined to laboratory settings, there is a growing belief that MEC technology holds great promise for practical applications in the future. Microbial electrolysis cells represent a groundbreaking approach to waste treatment that not only addresses the environmental impact of wastewater but also offers the possibility of resource recovery. Through continuous research and development, the challenges

T. Krithika (✉) · J. Anitha
Department of Biochemistry, Kongunadu Arts & Science College, Coimbatore, Tamil Nadu 641029, India
e-mail: krithikat2294@gmail.com

A. A. Yaqoob and A. Ahmad (eds.), *Microbial Electrolysis Cell Technology*, Sustainable Materials and Technology, https://doi.org/10.1007/978-981-97-3356-9_11

associated with MEC technology can be overcome, paving the way for large-scale applications that align with sustainability goals. As we strive for a more sustainable future, MECs undoubtedly offer a path to transform waste into valuable resources, ensuring a cleaner and more efficient world for generations to come.

Keywords Microbes · Bio fuels · Organic materials · Electrolysis

1 Introduction

There's a big problem that the planet is facing right now. Fossil fuels are non-renewable resources that have the potential to run out entirely, and they currently provide an astounding 80% of our energy needs. Furthermore, worries regarding these resources' detrimental effects on the world's ecosystems, human health, and climate are growing. To satisfy our ever-increasing energy needs, scientists are continuously looking for greener options.

Hydrogen is one such highly promising substitute. Since hydrogen's oxidation result is just water, it does not contribute to greenhouse gas emissions, acid rain, or ozone depletion like burning fossil fuels does. Moreover, hydrogen has the largest energy content per unit weight among gaseous fuels, demonstrating its remarkable efficiency: H_2 at 120 MJ/kg versus 44 MJ/kg for fuel. In addition, hydrogen is inexpensive, clean, renewable, and sustainable because it can be generated from a range of sources, including organic wastes. It is important to remember, nevertheless, that non-renewable fossil fuels like natural gas continue to provide the majority of the hydrogen produced today [13].

Recent years have seen tremendous advancements in the creation of microbial electrolysis cells (MECs), which produce hydrogen from organic materials. This strategy outperforms conventional water-based hydrogen production techniques in terms of efficiency. A microbial fuel cell (MFC) and a MEC system are similar in that they both have an external voltage and a sealed cathode. In MECs, protons, CO_2, and electrons are produced from organic molecules by electrochemically active microorganisms that flourish near the anode. Through an electrical circuit, these microbes then move the produced electrons to the anode and ultimately to the cathode, where they produce hydrogen. According to Thiruppathi et al. [35] biodegradation is a promising method for removing organic contaminants from a wide range of industrial wastewaters. Since it has been demonstrated that microorganisms are capable of decolorizing and metabolizing complex compounds, bioremediation-based technologies are currently the most popular methods for treating textile wastewater [23]. An essential component of living systems for the breakdown of certain pollutants is the microbial community. Anion exchange membrane and cation exchange membrane, two ion-selective membranes similar to those used in water electrolyzers, are used to stop the unwanted interaction between oxygen and the produced hydrogen. Because of its reduced internal resistance, which makes it easier for ions to pass through the membrane, the anion exchange membrane is usually regarded as being more efficient of the two [40].

Because of its potential to be used in the treatment of various wastewater types from a variety of industries, MECs have drawn the interest of researchers all over the world. Among them are the following: domestic or municipal wastewater; food processing; industrial protein food practices; acidogenic food waste; beverages; wineries; breweries; confectioneries; dairy; yogurt production; agro-processing; cassava mills; palm oil mills; mustard tuber industries; livestock; disposal of animal carcasses; swine industries; mining and related industries; recalcitrant pharmaceutical industries; production of steroidal drugs; paper recycling industries; and petrochemical industries. These materials are being developed at the vanguard of a new discipline that combines material science, biology, and nanotechnology. According to Kritika et al. [16], this multidisciplinary discipline is at the forefront of technological innovation.

This chapter explores the pivotal role of MECs in the recovery of various nutrients and the bioremediation of commonly used organic pollutants. Specifically, it provides a comprehensive review of MEC technologies employed in the removal and/or recovery of phosphorus (P), nitrogen (N), organics (in terms of chemical oxygen demand), and bioenergy (in the form of electricity) from diverse wastewater sources. The aim of this review is to offer valuable insights into the current state of MECs, with a focus on recent advancements in nutrient removal/recovery and power generation.

2 Microbial Electrolysis Cell (MEC)

Microbial electrolysis cells (MECs) have emerged as a groundbreaking technology since their discovery in 2005. By introducing a small amount of electricity at the cathode, a microbial fuel cell can be transformed into an MEC, enabling the production of hydrogen gas and other valuable products [12]. Unlike traditional fuel cells, which rely on chemical reactions, MECs utilize the biochemical reactions of microorganisms, specifically bacteria, to generate electrical power.

Synchronous bioremediation and power generation are the proper uses of MFC. The process of managing polluted media, such as mud, water, and subterranean debris, is known as bioremediation. It involves changing the surrounding environment to encourage the growth of microorganisms and reduce the pollutants [15]. The anode chamber and the cathode chamber, which are divided by proton exchange membranes (PEM), make up the microbial fuel cell. Microbes in the anode chamber of the MFC use the nutrients in their medium to grow, and some of that energy is released as electricity. In this process, organic material is oxidized by bacteria at the anode and oxygen is reduced at the cathode [17].

Microbial biofilms in the anodic chamber of MECs electrochemically oxidize organic materials. Protons and electrons are produced by this process and are subsequently used in a reduction reaction to produce useful chemicals like methane and hydrogen. The anodic reaction is accelerated by the electrogenic biofilm on the anode, which functions as a biocatalyst. Through an external electrical connection, electrons provided to the anode by the anodic biofilm proceed to the cathode, where they convert protons and H_2O to produce OH^- and H_2, which are then discharged from the cathode compartment. Since this linked redox process is thermodynamically unfavorable, an external voltage supply is required to enable it. However, because of the energy provided by the organic carbon decomposition in a MEC, this process requires less power than typical water electrolysis. Moreover, Zhou et al. [46] report that additional products, such as methane, may be produced on the cathode if a substance other than water or protons is reduced.

3 MEC for Nutrient Recovery

Aquatic habitats are seriously threatened by the release of wastewater with high nutrient and organic content, which can result in problems like eutrophication and hypoxia. In order to solve this issue, phosphate (PO_4^{3+}) and ammonium (NH_4^+) nutrients are removed or recovered by wastewater treatment facilities (WWTPs) using energy-intensive processes in order to meet discharge standards [6].

All forms of life require phosphorus (P), which is also essential to crop yields. Phosphorus cannot currently be substituted, and its availability is critical to the security of food worldwide. The deposits of commercial phosphate rock, however, are expected to run out in a century. Phosphorus recovery from wastewater has therefore attracted a lot of attention. With an estimated yearly potential of 3 million tons of phosphate worldwide, WWTPs provide an unexplored source of phosphorus. Should it be completely restored, the total phosphorus present in sewage may meet between 15 and 20% of the world's phosphorus needs, providing a feasible alternative to phosphate rock mines [28].

In addition to phosphorus, nitrogen (N) is a crucial nutrient for agricultural production. With the growing global population, relying solely on natural nitrogen-fixation processes is insufficient. Currently, approximately half of the world's food production depends on synthetic fertilizer, which is predominantly manufactured through energy-intensive processes like the Haber–Bosch process. This process requires high temperature, pressure, and non-renewable fossil sources for hydrogen gas, leading to the release of greenhouse gases. Furthermore, industrial and municipal wastewaters contain high concentrations of ammonium, which further contributes to the demand for ammonia production. The ammonium in domestic wastewater alone is projected to increase to 35 million tons annually by 2050. Hence, finding sustainable solutions for nitrogen removal and recovery is of utmost importance [2].

4 PO_4^{3-}–P and NH_4^+–N Removal

An MEC and an algal bioreactor were used in a study by Jiang et al. to treat wastewater from homes. Significant gains were shown by this hybrid system in the elimination of nitrogen, organics, and total phosphorus. Wang et al. [37] investigated the removal of nutrients (nitrogen and phosphorus) and the recovery of bioenergy through the use of an immobilized microalgal-based photoautotrophic microbial fuel cell (PMFC) system. High removal efficiencies were attained by the PMFC system for phosphate (PO_4^{3+}–P), ammonium (NH_4^+–N), and soluble chemical oxygen demand (COD). The incorporation of phosphate and nitrogen into the microalgal biomass was the main cause of the decrease in nutrient levels in the cathode compartment. Even though the manufacture of biodiesel, lighting, and the pumping of anolyte and catholyte all required energy consumption, the PMFC system demonstrated its capacity to recover energy. The power produced accounted for a significant portion of the energy consumed [37]. These studies demonstrate how MECs combined with other technologies can enhance nutrient removal through microbial assimilation.

5 NH_4^+–N Removal

Ammonification, nitrification, and denitrification are three techniques that can be used to remove organic nitrogen from synthetic wastewater. Ammonia volatilization is the main method for NH_4^+–N elimination in a single-chamber MEC. In contrast, the majority of the organic nitrogen in a dual-chamber MEC is converted into $NH4^+$–N in the anode chamber. The main mechanism for removing NH_4^+–N is the migration of NH_4^+–N via a proton exchange membrane (PEM) from the anode chamber to the cathode chamber and volatilization in the cathode chamber. NH_4^+–N can build up to high concentrations in the cathode chamber if it is not volatilized. This NH_4^+–N migration can be understood as a proton shuttle. According to earlier research, volatilization is important for NH_4^+–N removal in single and dual-chamber MECs [28].

6 Production of H_2

H_2 gas holds enormous potential in terms of eco-friendly transporter of energy. It is seen as a crucial alternative to lessen the dependence of the globe on fossil fuels. Here are a few unbeatable features of H_2:

- It is a renewable energy source that is non-toxic.
- The generation of H_2 does not release any additional toxic gases.
- With a high calorific value of 142.9 kJ g^{-1}, it is an extremely flammable gas.

In recent laboratory-scale experiments, microbial electrolysis cells (MECs) have successfully produced H_2 from various substrates, proving to be more efficient than traditional water electrolysis. However, raw organic sewage has been the primary substrate employed in the majority of MEC experiments, resulting in low H_2 production rates and significant CH_4 production. To improve the H_2 production in MECs, researchers have found that pre-acclimating the anode biofilm with butyrate enhances the treatment of corn stalk fermentation effluent [20].

Despite the progress made with MECs, an ongoing challenge in their operation is the increase in pH levels in the cathode [25]. This challenge was addressed by Yang et al. [39], who developed an automatic control system of periodic polarity reversal in MECs. This system offers an effective pH control strategy, thereby maximizing H_2 production.

7 Production of CH

While H_2 has its advantages, the practical application of CH (methane) surpasses that of H_2 in terms of transportation and storage [38]. Methanogens that use H_2 for growth, including Methanobacterium and Methanobrevibacter, dominate the cathode electrode in H_2-producing MECs and prevent H_2 generation.

Nonetheless, there are a number of benefits to directly producing CH in MECs as opposed to conventional anaerobic digestion methods. According to Calantonio and Kim, these benefits include high CH production, methanogen tolerance, and waste streams with decreased organic matter contents. Zhen et al. [45] have reported that MECs using hybrid graphite felt biocathodes may produce CH_4 at rates of 80.9 mL L^{-1} and a coulombic efficiency value of 194.4%.

An enhanced cathode/anode ratio has been found to be effective in improving energy recovery in MEC systems [10]. Furthermore, hydrogenotrophic methanogens benefit from low voltage (0.3 V) applied to anaerobic digestion systems, which also boosts CH_4 yields from glucose [9].

8 Production of Formic Acid (HCOOH)

The chemical and pharmaceutical industries rely heavily on formic acid (HCOOH). Due to its economic and technical potential for CO_2 compensation and valuable material production, the electrochemical reduction of CO_2 to HCOOH has attracted interest [43, 44].

Innovative methods have been devised by researchers to generate HCOOH in MECs. Wang et al. [36], for instance, created a MEC reactor with a two-layer rolling gas diffusion electrode loaded with Sn. This reactor produced 0.064 ± 0.006 mol m^{-2} of production yield and a maximum Faraday efficiency of $40.09 \pm 3.91\%$. Another study linked the electrochemical reduction of CO_2 into HCOOH to a series of microbial fuel cells (MFCs). With a Faraday efficiency of 64.8%, this configuration effectively produced HCOOH at a rate of 4.27 mg L^{-1} h^{-1} [43, 44].

The reduction of CO_2 into HCOOH using MECs offers a plausible and relatively environmentally friendly alternative to current fossil fuel-based production methods [33]. To evaluate the viability of MECs, researchers Sadhukhan et al. [33] and Shemfe et al. [34] evaluated the life cycle, techno-economic, and dynamic modeling aspects, utilizing the case of formic acid. The findings showcased significant environmental benefits associated with MECs compared to the conventional fossil fuel-based industrial methods. The MECs proved to be efficient in providing similar functionalities in terms of product synthesis and wastewater treatment.

The production of formic acid using electrochemical reduction techniques, particularly employing MECs and MFC stacks, shows promising results in terms of yield and environmental sustainability. As researchers continue to refine and optimize these processes, we can anticipate a future where formic acid production is more efficient, economically viable, and aligned with our environmental goals.

9 Production of Hydrogen Peroxide (H_2O_2)

Hydrogen peroxide (H_2O_2) is widely recognized as a chemical that is safe for the environment and merely breaks down into water and oxygen [19]. The production of H_2O_2 has proven to be achievable through the use of microbial electrolysis cells (MECs). Miran et al. [24] successfully employed an MEC reactor to biodegrade chlorinated phenol and produce H_2O_2. This investigation showed that an external voltage of 0.4 V produced the maximum concentration of H_2O_2 (13.3 g L^{-1} m^{-2} or 974 μM). Improving the rate of synthesis of H_2O_2 in the MEC system with a small external voltage has yielded encouraging results.

Rozendal et al. [32] presented a unique approach for producing H_2O_2 based on wastewater's bio-electrochemical oxidation in a distinct study. H_2O_2 was produced at an external voltage of 0.5 V through the use of an anode connected to the cathodic reduction of oxygen in this process. Based on acetate oxidation, the system produced 1.17 mmol L^{-1} h^{-1} in the aerated cathode with an average efficiency of 83%. Ki et al. [14] produced H_2O_2 at the cathode by using primary sludge as an electron donor at the anode. After six hours of batch cathode operation, they were able to reach a maximum H_2O_2 concentration of 230 mg L^{-1}.

10 Nitrogen Recovery

An important component for fertilizers, nitrogen, is now produced commercially using the Haber–Bosch process, which uses a lot of energy [30]. According to Cao et al. [3], the need for nitrogen fertilizer is expected to rise at an average annual growth rate of 1.5% throughout the next ten years. A global problem is producing

Table 1 Performance of MEC for total ammonia–nitrogen recovery

Substrate	Current density (mA m^{-2})	Total ammonia–nitrogen recovery (%)	Reference
Urine	1700 ± 200	31 ± 59	[42]
Pig slurry	3500	60	[4]
Urine	1600	46	[18]
Landfill leachate	760	63.7 ± 6.6	[29]
Sidestream centrate	760	99.7 ± 1.3	[47]

ammonia sustainably [21]. Microbial electrolysis cells (MECs) have been used to recover ammonia from wastewater in recent years. Nevertheless, recovered ammonia must be separated from MEC catholyte using high energy input processes that involve solution mixing or gas stripping.

Qin et al. [30] suggested a passive separation method that involves subjecting the cathode directly to the gas phase in order to overcome this difficulty. This technique turned out to be similar to the generation that is done now in MECs employing active aeration. The percentage of nitrogen in the MEC catholyte was 21.1 ± 4.8% after passive separation. Combining anaerobic digestion (AD) with bioelectrochemical systems (BES) is a promising approach because of the complexity and resistance to degradation of sewage [11]. Cerrillo et al. [5] established a series connection between the bioanode compartment of a two-chambered MEC and a thermophilic lab-scale AD system fed with pig slurry. In the anode compartment of the MEC, this integration produced an ammonium removal rate of 14.46 g N–NH_4^+ m^{-2} d^{-1}. A summary of other research on ammonia recovery can be found in Table 1.

11 Phosphate Recovery

Phosphate recovery in microbial electrolysis cells (MECs) occurs via a distinct method from that of ammonia. In their 2017 study, Almatouq et al. examined phosphate recovery in a dual-chamber MEC and produced hydrogen gas in addition to reaching a maximum precipitation efficiency of 95%. Furthermore, Cusick and Logan [7] discovered that the rate at which hydrogen gas is produced in struvite reactors was unaffected by crystal buildup. Nevertheless, MEC performance may be adversely affected by scaling on the cathode and exchange membrane in addition to pH variations between the anode and cathode [1].

In lab-scale MEC experiments, careful evaluation of different cathode designs and electric currents is necessary for efficient phosphorus recovery. Yuan and Kim [41] discovered that successful struvite production from dewatering centrates required the use of a foil-type cathode and a high electric current density (> 2 A m^{-2} for peak currents). About 92% of the phosphorous was removed and 96% of it was recovered using this procedure. For struvite precipitates to be more pure, more study on phosphorus recovery in MECs is required.

12 Metals Recovery

Metal extraction from wastewater poses significant challenges, as they are often resistant to natural degradation and can lead to environmental and health hazards. Traditional methods, such as reduction, oxidation, ion exchange, precipitation, solvent extraction, electro-driven separations, and adsorption, have limitations in terms of cost, safety, and efficiency [8].

In the past decade, microbial electrochemical cells (MECs) have shown promise in recovering metals, e.g., $Cr_2O_7^{2-}$, Ca^{2+}, Mg^{2+}, Cd^{2+}, and Cu^{2+} [22, 26, 27, 31]. MECs have also been utilized for recovery of uranium, silver, chromium, ferric ion, ferrous iron oxidation, removal of cadmium (II), removal of mercury, and recovery and removal of copper. The reduction of metal ions in MECs requires energy input at the cathode, but because the anode supplements the voltage, the total voltage required is still less than in conventional electrolysis. [26].

To enhance metal recovery efficiency, MEC configurations, the choice of anode and cathode materials, Conductivity and metal ion concentration are crucial operating factors that need more research [8].

13 Products from MECs

Various value-added products can be obtained through microbial electrochemical cells (Table 2), including hydrogen gas (H_2), methane, and hydrogen peroxide (H_2O_2). The cathodic chamber of MECs presents two pathways for hydrogen gas generation: H_2O reduction and proton reduction. However, the productivity of hydrogen gas can be compromised in membrane-less MECs due to methanogenesis. Methanogens in the anodic biofilm community oxidize hydrogen gas, reducing its availability for hydrogen production. The implementation of a membrane partition can help prevent this gas crossover [30].

14 Summary and Conclusion

While there have been some efforts to conduct large-scale microbial fuel cell (MFC) experiments, it is important to note that MFC technology is still far from being fully developed. Various critical challenges, particularly relating to techno-economic

Table 2 MECs with different bioproducts [46]

Product	Power supply	Recovery rate ($L\ L^{-1}\ d^{-1}$)	Energy efficiency (%)
H_2	1.06 V	0.3	–
	0.5 V	0.02	169
	0.6 V	0.53	204
	0.4 V	0.2	267
	Anode poised at + 0.5 V	0.018	57
CH_4	0.8 V	0.17–0.75	240–84
	0.9 V	0.12	67
	Anode poised at 0.0 V	0.53	70
H_2O_2	0.5 V	1.25 ± 0.13	83.1 ± 4.8

issues such as the cost of materials and hydrogen (H_2) management, remain unresolved. These challenges must be addressed before this technology can be practically applied.

It is worth mentioning that the hydrogen production achieved through current small-scale laboratory research is still insufficient. However, increasing the size of MFC reactors is crucial for their further application in treatment of wastewater and as a key technology for sustainable energy production.

Moving forward, there are several key points that require attention in subsequent work:

Alleviating Methanogen Consumption of H_2: Efforts should be made to find effective ways to reduce the consumption of hydrogen by methanogens, which will enhance the overall efficiency of MFC systems.

Enhancing the Bioanode Sensors' Accuracy: The development of in situ, self-sustaining, and real-time bioanode sensors must keep an eye on the quality of the water is essential. This will provide accurate data and ensure effective control over the treatment process.

Exploring Novel MFC Configurations at an Industrial Scale: It is important to investigate new configurations of MFC systems that are suitable for large-scale applications. This will help optimize the efficiency and effectiveness of the technology in real-world scenarios.

Integration of MFC with Other Disciplines: The incorporation of computer science, materials science, and sensory technologies with MFC holds great potential. Such interdisciplinary collaborations can lead to synergistic effects and further enhance the capabilities of MFC systems.

Despite the current challenges, MFC technology shows great promise in the fields of wastewater treatment, pollutant removal, and sustainable energy production. However, in order to fully capitalize on these possibilities, it is imperative to address the aforementioned key points and continue research and development efforts. With time and dedication, we can overcome these hurdles and unlock the full potential of MFC technology for a greener and more sustainable future.

References

1. Almatouq A, Babatunde AO (2017) Concurrent hydrogen production and phosphorus recovery in dual chamber microbial electrolysis cell. Biores Technol 237:193–203. https://doi.org/10.1016/j.biortech.2017.02.04
2. Bodirsky BL, Popp A, Lotze-Campen H, Dietrich JP, Rolinski S, Weindl I, Schmitz C, Müller C, Bonsch M, Humpenöder F, Biewald A, Stevanovic M (2014) Reactive nitrogen requirements to feed the world in 2050 and potential to mitigate nitrogen pollution. Nat Commun 5(1):3858. https://doi.org/10.1038/ncomms4858
3. Cao X, Huang X, Liang P, Xiao K, Zhou Y, Zhang X, Logan BE (2009) A new method for water desalination using microbial desalination cells. Environ Sci Technol 43(18):7148–7152. https://doi.org/10.1021/es901950j
4. Cerrillo M, Viñas M, Bonmatí A (2016) Removal of volatile fatty acids and ammonia recovery from unstable anaerobic digesters with a microbial electrolysis cell. Biores Technol 219:348–356. https://doi.org/10.1016/j.biortech.2016.07.103
5. Cerrillo M, Viñas M, Bonmatí A (2018) Anaerobic digestion and electromethanogenic microbial electrolysis cell integrated system: increased stability and recovery of ammonia and methane. Renew Energy 120:178–189. https://doi.org/10.1016/j.renene.2017.12.062
6. Chen X, Zhou H, Zuo K, Zhou Y, Wang Q, Sun D, Huang X (2017) Self-sustaining advanced wastewater purification and simultaneous in situ nutrient recovery in a novel bioelectrochemical system. Chem Eng J 330:692–697. https://doi.org/10.1016/j.cej.2017.07.130
7. Cusick RD, Logan BE (2012) Phosphate recovery as struvite within a single chamber microbial electrolysis cell. Biores Technol 107:110–115. https://doi.org/10.1016/j.biortech.2011.12.038
8. Dominguez-Benetton X, Varia JC, Pozo G, Modin O, Ter Heijne A, Fransaer J, Rabaey K (2018) Metal recovery by microbial electro-metallurgy. Prog Mater Sci 94:435–461. https://doi.org/10.1016/j.pmatsci.2018.01.007
9. Gajaraj S, Huang Y, Zheng P, Hu Z (2017) Methane production improvement and associated methanogenic assemblages in bioelectrochemically assisted anaerobic digestion. Biochem Eng J 117:105–112. https://doi.org/10.1016/j.bej.2016.11.003
10. Guo K, Prévoteau A, Rabaey K (2017) A novel tubular microbial electrolysis cell for high rate hydrogen production. J Power Sources 356:484–490. https://doi.org/10.1016/j.jpowsour.2017.03.029
11. Hassanein A, Witarsa F, Guo X, Yong L, Lansing S, Qiu L (2017) Next generation digestion: complementing anaerobic digestion (AD) with a novel microbial electrolysis cell (MEC) design. Int J Hydrogen Energy 42(48):28681–28689. https://doi.org/10.1016/j.ijhydene.2017.10.003
12. Heidrich ES, Edwards SR, Dolfing J, Cotterill SE, Curtis TP (2014) Performance of a pilot scale microbial electrolysis cell fed on domestic wastewater at ambient temperatures for a 12 month period. Biores Technol 173:87–95. https://doi.org/10.1016/j.biortech.2014.09.083
13. Kadier A, Simayi Y, Kalil MS, Abdeshahian P, Hamid AA (2014) A review of the substrates used in microbial electrolysis cells (MECs) for producing sustainable and clean hydrogen gas. Renewable Energy 71:466–472. https://doi.org/10.1016/j.renene.2014.05.052
14. Ki D, Popat SC, Rittmann BE, Torres CI (2017) H_2O_2 production in microbial electrochemical cells fed with primary sludge. Environ Sci Technol 51(11):6139–6145. https://doi.org/10.1021/acs.est.7b00174
15. Krithika T, Iswarya T, Sowndarya T (2022) Bioremediation of organic pollutants through microbial fuel cells. In: Microbial fuel cells for environmental remediation. Springer Nature Singapore, Singapore, pp 183–194. https://doi.org/10.1007/978-981-19-2681-5_10

16. Krithika T, Iswarya T, Sowndarya T (2023) Introduction of nanohybrid materials. In: Nanohybrid materials for treatment of textiles dyes. Springer Nature Singapore, Singapore, pp 1–17. https://doi.org/10.1007/978-981-99-3901-5_1
17. Krithika T, Kavitha R, Dinesh M, Angayarkanni J (2021) Assessment of ligninolytic bacterial consortium for the degradation of azo dye with electricity generation in a dual-chambered microbial fuel cell. Environ Challenges 4:100093. https://doi.org/10.1016/j.envc.2021.100093
18. Kuntke P, Zamora P, Saakes M, Buisman CJN, Hamelers HVM (2016) Gas-permeable hydrophobic tubular membranes for ammonia recovery in bio-electrochemical systems. Environ Sci: Water Res Technol 2(2):261–265. https://doi.org/10.1039/C5EW00299K
19. Li N, An J, Zhou L, Li T, Li J, Feng C, Wang X (2016) A novel carbon black graphite hybrid air-cathode for efficient hydrogen peroxide production in bioelectrochemical systems. J Power Sources 306:495–502. https://doi.org/10.1016/j.jpowsour.2015.12.078
20. Li X, Zhang R, Qian Y, Angelidaki I, Zhang Y (2017) The impact of anode acclimation strategy on microbial electrolysis cell treating hydrogen fermentation effluent. Biores Technol 236:37–43. https://doi.org/10.1016/j.biortech.2017.03.160
21. Liu Y, Qin M, Luo S, He Z, Qiao R (2016) Understanding ammonium transport in bioelectrochemical systems towards its recovery. Sci Rep 6(1):22547
22. Luo H, Xu P, Jenkins PE, Ren Z (2012) Ionic composition and transport mechanisms in microbial desalination cells. J Membr Sci 409:16–23. https://doi.org/10.1016/j.memsci.2012.02.059
23. Mary JE, Krithika T, Kavitha R (2020) Biodegradation of textile dye by ligninolytic bacteria isolated from Western Ghats. Int J Res Rev 7(4):22–29
24. Miran W, Nawaz M, Jang J, Lee DS (2017) Chlorinated phenol treatment and in situ hydrogen peroxide production in a sulfate-reducing bacteria enriched bioelectrochemical system. Water Res 117:198–206. https://doi.org/10.1016/j.watres.2017.04.008
25. Nam JY, Logan BE (2012) Optimization of catholyte concentration and anolyte pHs in two chamber microbial electrolysis cells. Int J Hydrogen Energy 37(24):18622–18628. https://doi.org/10.1016/j.ijhydene.2012.09.140
26. Nancharaiah YV, Mohan SV, Lens PNL (2015) Metals removal and recovery in bioelectrochemical systems: a review. Biores Technol 195:102–114. https://doi.org/10.1016/j.biortech.2015.06.058
27. Pandit S, Sengupta A, Kale S, Das D (2011) Performance of electron acceptors in catholyte of a two-chambered microbial fuel cell using anion exchange membrane. Biores Technol 102(3):2736–2744. https://doi.org/10.1016/j.biortech.2010.11.038
28. Paucar NE, Sato C (2021) Microbial fuel cell for energy production, nutrient removal and recovery from wastewater: a review. Processes 9(8):1318. https://doi.org/10.3390/pr9081318
29. Qin M, Molitor H, Brazil B, Novak JT, He Z (2016) Recovery of nitrogen and water from landfill leachate by a microbial electrolysis cell–forward osmosis system. Biores Technol 200:485–492. https://doi.org/10.1016/j.biortech.2015.10.066
30. Qin M, White C, Zou S, He Z (2018) Passive separation of recovered ammonia from catholyte for reduced energy consumption in microbial electrolysis cells. Chem Eng J 334:2303–2307. https://doi.org/10.1016/j.cej.2017.11.190
31. Rodenas Motos P, Molina G, Ter Heijne A, Sleutels T, Saakes M, Buisman C (2017) Prototype of a scaled-up microbial fuel cell for copper recovery. J Chem Technol Biotechnol 92(11):2817–2824. https://doi.org/10.1002/jctb.5353
32. Rozendal RA, Leone E, Keller J, Rabaey K (2009) Efficient hydrogen peroxide generation from organic matter in a bioelectrochemical system. Electrochem Commun 11(9):1752–1755. https://doi.org/10.1016/j.elecom.2009.07.008
33. Sadhukhan J, Martinez-Hernandez E, Murphy RJ, Ng DK, Hassim MH, Ng KS, Kin WY, Jaye IFMd, Leung Pah Hang MY, Andiappan V (2018) Role of bioenergy, biorefinery and bioeconomy in sustainable development: strategic pathways for Malaysia. Renew Sustain Energy Rev 81:1966–1987. https://doi.org/10.1016/j.rser.2017.06.007
34. Shemfe M, Gadkari S, Yu E, Rasul S, Scott K, Head IM, Sadhukhan J (2018) Life cycle, techno-economic and dynamic simulation assessment of bioelectrochemical systems: a case

of formic acid synthesis. Biores Technol 255:39–49. https://doi.org/10.1016/j.biortech.2018.01.071
35. Thiruppathi K, Rangasamy K, Ramasamy M, Muthu D (2021) Evaluation of textile dye degrading potential of ligninolytic bacterial consortia. Environ Challenges 4:100078. https://doi.org/10.1016/j.envc.2021.100078
36. Wang Q, Dong H, Yu H, Yu H, Liu M (2015) Enhanced electrochemical reduction of carbon dioxide to formic acid using a two-layer gas diffusion electrode in a microbial electrolysis cell. RSC Adv 5(14):10346–10351. https://doi.org/10.1039/C4RA14535F
37. Wang Y, Lin Z, Su X, Zhao P, Zhou J, He Q, Ai H (2019) Cost-effective domestic wastewater treatment and bioenergy recovery in an immobilized microalgal-based photoautotrophic microbial fuel cell (PMFC). Chem Eng J 372:956–965. https://doi.org/10.1016/j.cej.2019.05.004
38. Weger L, Abánades A, Butler T (2017) Methane cracking as a bridge technology to the hydrogen economy. Int J Hydrogen Energy 42(1):720–731. https://doi.org/10.1016/j.ijhydene.2016.11.029
39. Yang Y, Ren H, Ben-Tzvi P, Yang X, He Z (2017) Optimal interval of periodic polarity reversal under automated control for maximizing hydrogen production in microbial electrolysis cells. Int J Hydrogen Energy 42(31):20260–20268. https://doi.org/10.1016/j.ijhydene.2017.06.028
40. Yu Z, Leng X, Zhao S, Ji J, Zhou T, Khan A, Kakde A, Liu P, Li X (2018) A review on the applications of microbial electrolysis cells in anaerobic digestion. Biores Technol 255:340–348. https://doi.org/10.1016/j.biortech.2018.02.003
41. Yuan P, Kim Y (2017) Increasing phosphorus recovery from dewatering centrate in microbial electrolysis cells. Biotechnol Biofuels 10(1):1–8. https://doi.org/10.1186/s13068-017-0754-8
42. Zamora P, Georgieva T, Ter Heijne A, Sleutels TH, Jeremiasse AW, Saakes M, Buisman CJN, Kuntke P (2017) Ammonia recovery from urine in a scaled-up microbial electrolysis cell. J Power Sources 356:491–499. https://doi.org/10.1016/j.jpowsour.2017.02.089
43. Zhao HZ, Zhang Y, Chang YY, Li ZS (2012) Conversion of a substrate carbon source to formic acid for carbon dioxide emission reduction utilizing series-stacked microbial fuel cells. J Power Sources 217:59–64. https://doi.org/10.1016/j.jpowsour.2012.06.014
44. Zhao H, Zhang Y, Zhao B, Chang Y, Li Z (2012) Electrochemical reduction of carbon dioxide in an MFC–MEC system with a layer-by-layer self-assembly carbon nanotube/cobalt phthalocyanine modified electrode. Environ Sci Technol 46(9):5198–5204. https://doi.org/10.1021/es300186f
45. Zhen G, Lu X, Kobayashi T, Kumar G, Xu K (2016) Promoted electromethanosynthesis in a two-chamber microbial electrolysis cells (MECs) containing a hybrid biocathode covered with graphite felt (GF). Chem Eng J 284:1146–1155. https://doi.org/10.1016/j.cej.2015.09.071
46. Zhou M, Wang H, Hassett DJ, Gu T (2013) Recent advances in microbial fuel cells (MFCs) and microbial electrolysis cells (MECs) for wastewater treatment, bioenergy and bioproducts. J Chem Technol Biotechnol 88(4):508–518. https://doi.org/10.1002/jctb.4004
47. Zou S, Qin M, Moreau Y, He Z (2017) Nutrient-energy-water recovery from synthetic sidestream centrate using a microbial electrolysis cell-forward osmosis hybrid system. J Clean Prod 154:16–25. https://doi.org/10.1016/j.jclepro.2017.03.199

Applications of Microbial Electrolysis Cells in Anaerobic Digestion

Muhammad Fikri Zulkornain, Hassan Mohamed, and Mohd Nur Ikhmal Salehmin

Abstract This chapter explores the application of microbial electrolysis cells (MECs) in anaerobic digestion (AD). The overview of AD and MECs is defined, and the integration of these two technologies is discussed in terms of their advantages in treating biogas and producing high-value products. The fundamentals of this integration are explained in detail, along with the effect of the parameters such as feedstock type, temperature, hydraulic retention time, pH, voltage applied and electrode. Next, the existing configuration for the system integration is also discussed comprehensively. Lastly, the challenges and future prospects emphasise the technical inadequacy, economic feasibility and environmental aspects.

Keywords Anaerobic digestion · Microbial electrolysis cells · AD-MEC integration · Hydrogen · Methane · Bioelectrochemical process

1 Introduction

AD is a biological process degrading organic matter without or limited oxygen consumption. This process treats and manages organic waste-based material such as slurry or manure, crops or residues, residues and food waste by transforming it into biogas. The most prevalent model reactor is the fixed dome or hydraulic digester [1]. It just consists of a simple digester without any moving components. Among the dominant intermediate products of AD is acetate, which is directly turned into methane and carbon dioxide during the process. Furthermore, AD also produces digestate as a by-product. The digestate can produce ethanol, biochar and cultivator of insects and fungi [2–5]. Figure 1 shows the overall AD process.

AD consists of multiple processes to obtain the biogas as a product: hydrolysis, acidogenesis, acetogenesis and methanogenesis. Figure 2 illustrates the processes involved in the AD process. Hydrolysis is a process of breaking down the complex

M. F. Zulkornain · H. Mohamed · M. N. I. Salehmin (✉)
Institute of Sustainable Energy (ISE), Universiti Tenaga Nasional (UNITEN), Putrajaya Campus, Jalan Ikram-Uniten, 43000 Kajang, Selangor, Malaysia
e-mail: nur.ikhmal@uniten.edu.my

A. A. Yaqoob and A. Ahmad (eds.), *Microbial Electrolysis Cell Technology*, Sustainable Materials and Technology, https://doi.org/10.1007/978-981-97-3356-9_12

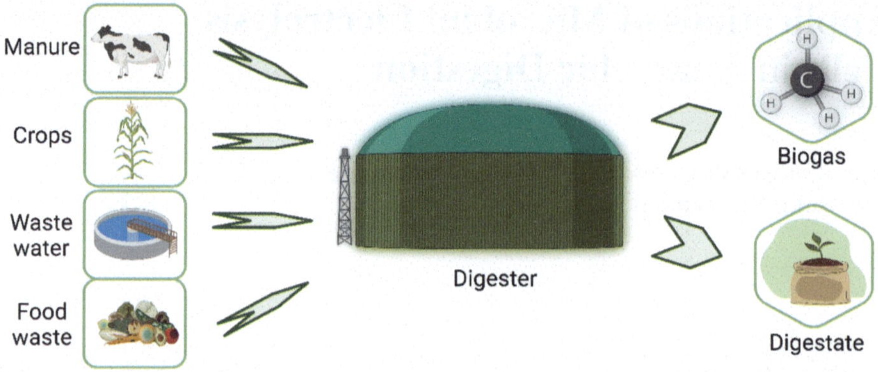

Fig. 1 Overall AD process

organic polymers that exist in the substrate. This step is essential for AD as it degrades the insoluble organic polymers such as carbohydrates, proteins and fats into simple monomers. Hydrolysis is carried out by numerous hydrolases and enzymes secreted from the microorganism, such as esterases, glycosidases and peptidases, to break the ester bonds, glycoside bonds and peptide bonds, respectively [6]. The microorganisms involved in the hydrolysis process are *Bacteroides sp, Clostridium sp, Micrococcus sp, Bacillus sp* and *Gammaproteobacteria sp* [7].

After hydrolysis, acidogenic bacteria further breaks down the monomers from the acidogenesis process. Through fermentation by acidogenic bacteria, the monomers are turned into organic acids, alcohols and other volatile fatty acids. Subsequently, the acetogenesis process takes place where the formed organic acid is converted into hydrogen, acetic acid and carbon dioxide via the acetogenic bacteria. Finally, the products are turned into biogas, mainly comprised of methane and carbon dioxide through methanogenesis. The methanogenic archaea such as *Methanobacterium* and *Methanosarcina* are the main characters of this process. The archaea generate biogas, or mainly methane, a valuable renewable energy resource, from the intermediate in the earlier process of AD [8].

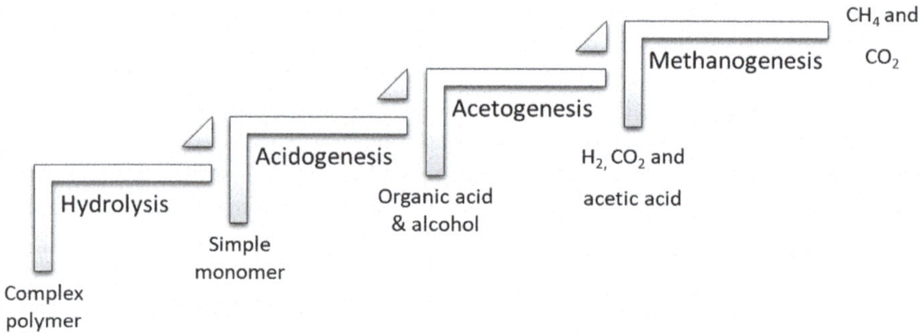

Fig. 2 Processes in anaerobic digestion (AD)

1.1 Microbial Electrolysis Cell (MEC)

MEC is a bioelectrochemical process recently studied for its potential in wastewater treatment, bioenergy generation and bioremediation [9]. MEC uses microorganisms in the electrolyte to facilitate the anodic and cathodic reactions leading to electron transfer and electrical energy from the organic matter [10]. This process is known for its sustainable energy generation and environmental remediation capability. Apart from wastewater, MEC can also use other organic matter, such as agricultural waste and lignocellulosic biomass, making it highly useful in waste management [11].

The MEC microorganism promotes hydrogen production from organic matter through anaerobic respiration. The microorganism decomposes the organic matter and releases electrons to the anode electrode in anaerobic conditions. The electron then moves to the cathode electrode, and the proton and electron reactions produce hydrogen [12]. Figure 3 shows an illustration of a simple MEC system. Previous studies have proven that MEC has various advantages, such as low energy requirements, high conversion efficiency and using microbial catalysts instead of metals [13].

Ongoing research and development on MEC are being done to enhance the performance, increase the efficiency and broaden its possible applications. Studies have also shown that it has potential in sustainable energy solutions without or limited effect on the environment [14]. In conclusion, MEC can convert any biodegradable waste into hydrogen, fuels and other valuable products, thus making it an eco-friendly and sustainable solution for managing and treating organic waste [15].

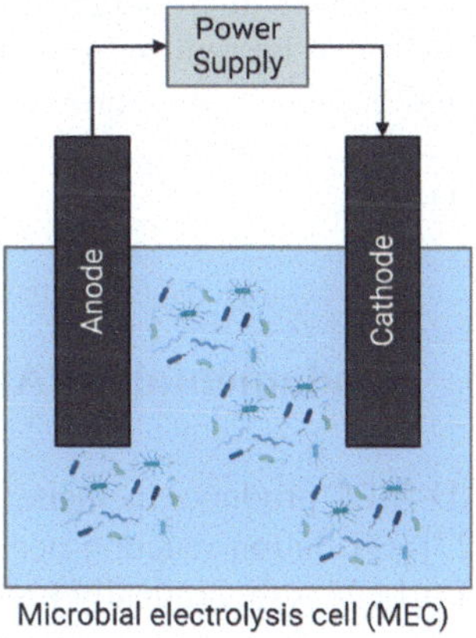

Fig. 3 Illustration of simple MEC system

2 Integrating AD with MEC

MEC and AD technologies have developed rapidly due to their benefits and advantages, especially in managing waste and renewable energy generation [16]. Compared to other electrochemical systems, such as microbial fuel cells (MFC) and microbial desalination cells (MDC), MEC is the most frequently paired system with AD. Integrating these technologies (AD-MEC) has proven to be one of the most promising approaches to solving wastewater issues while producing valuable products.

Among the benefits of MEC is a variant of outcomes from the process, such as electricity, hydrogen, minerals and chemicals throughout the process [17]. This shows that the process not only secures the removal of contaminants in the wastewater but also unfolds a new opportunity for sustainable energy production and recovery. The synergic relationship between MEC and AD can contribute to comprehensive and sustainable answers for the issues regarding wastewater treatment and management.

As further development occurred throughout the year, MEC has been proven to be a versatile technology as it can produce methane from organic compounds. Besides, the capability of MEC to generate energy from organic material via electron transfer or mediated pathways has tremendous potential for scalable renewable energy generation [18]. Much ongoing research has been studied to improve the scalability and replicability of MEC for large-scale organic waste into precious products.

AD performance can be improved with the integrated AD-MEC systems as VFA formation is mitigated, solid retention time (SRT) is decreased, and hydrolysis/acidogenesis rates are heightened [19]. Enhancement of methane generation can be triggered from direct interspecies electron transfer as MEC can cultivate the electroactive microorganism [20]. Furthermore, the oxidation–reduction potential (ORP) can be reduced due to enhanced electron transfer among the electroactive microorganism, thus promoting biofilm formation and facilitating a methanogenesis environment [21, 22]. Table 1 presents the selected investigation on integrated AD-MEC.

3 Fundamentals of AD-MEC

AD-MEC produces methane using the direct electron exchange within the electrodes [27]. Oxidation reaction occurs at the anode where the organ; an external voltage of 0.14 V is theoretically neededClick or tap here to enter text.. Figure 4 shows the bioelectrochemical electron pathway in AD-MEC. The H_2-mediated methanation is not a spontaneous reaction; an external voltage of 0.14 V is theoretically required.

Anode reaction

$$CH_3COO^- + 4H_2O \rightarrow 2HCO_{3-} + 9H^+ + 8e^- \tag{1}$$

$$CH_3CH_2COO^- + 7H_2O \rightarrow 3HCO_{3-} + 16H^+ + 14e^- \tag{2}$$

Table 1 Selected investigation of integrated AD-MEC system

Substrate	Operating conditions	Findings	References
Sewage sludge	Thermal-alkaline pretreated, 0.6–1.8 V	• CH_4 production increased from 20–79.3% for thermal-alkaline pretreated • The COD removal rate and sludge VSS increase as higher voltage is applied	[23]
Waste-activated sludge	Sequential homogenous protocatechuic acid (PCA) enhanced Fenton treatment	• Pretreatment increased the biogas production to 1.46-fold • Increased phosphorus recovery is observed in the form of struvite with a purity of 74.4%	[24]
Food waste	Carbon-modified copper foam electrode	• The CH_4 production rose to 338.1 mL CH_4/L than the control reactor • The methane produced from the electrode increases to 40.8% of total methane production	[20]
Quinoline wastewater	30 °C, 0.4–1.5 V, influent quinoline of 100–550 mg/L	• At optimal voltage (1.0 V), degradation and mineralisation of quinoline are improved • CH_4 production increases 1.5-fold at higher quinoline concentrations	[25]
Synthetic brewery wastewater	35 °C, 0.05–0.2 V, pH 7.2	• The best voltage of 0.1 V increased the COD removal by 16.72% and cumulative CH_4 production by 23.39%	[26]

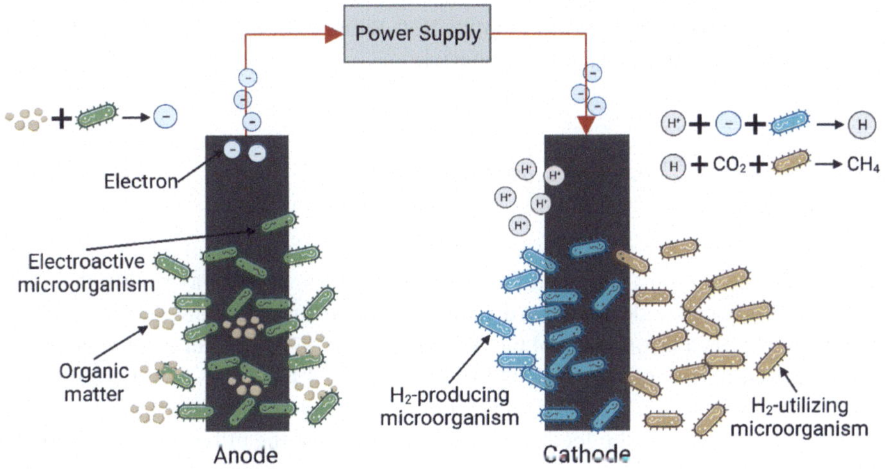

Fig. 4 Bioelectrochemical electron pathway in AD-MEC

$$CH_3CH_2CH_2COO^- + 10H_2O \rightarrow 4HCO_{3-} + 23H^+ + 20e^- \quad (3)$$

Cathode reaction (H_2-mediated methanation)

$$2H^+ + 2e^- \rightarrow H_2 \quad (4)$$

$$CO_2 + 4H_2 \rightarrow CH_4 + 2H_2O \quad (5)$$

In a conventional AD process, the total methane production consists of 2/3 of acetolastic methanogenesis and 1/3 of hydrogenotrophic methanogenesis [28]. The limited H_2 presence is the cause of the lower value of hydrogenotrophic methanogenesis. In AD-MEC, theoretically, hydrogenotrophic methanogenesis may increase theoretically as more H_2 production is obtained by reducing H^+ in the cathode. Some studies have concluded that 0.3–0.6 V of external voltage could improve hydrogenotrophic methanogenesis [29]. More hydrogenotrophic methanogenesis could decrease the pH drop in the reactor as it can quickly reduce the H^+ from the organic matter into H_2, thus improving AD performance [30]. Moreover, microbes in hydrogenotrophic methanogenesis are also more resistant to environmental conditions.

4 Effect of Parameters

The AD-MEC is one the most promising technologies in managing and treating wastewater and converting it into valuable products such as methane and hydrogen. Several parameters need to be studied to improve the AD-MEC system's performance further, such as feedstock type and composition, temperature, pH, hydraulic retention time, voltage applied and electrode material.

4.1 Feedstock Type and Composition

The AD-MEC system's performance and efficiency highly depend on the feedstock type and waste composition [31]. Different feedstocks, such as agricultural waste, food waste and sewage sludge, have different chemical compositions, and these variations can affect the performance of the AD-MEC system. Agricultural waste contains higher concentrations of cellulose and lignin, and food waste has a higher moisture content and organic acids [32]. The presence of organic matter, high nutrients and low moisture content in the feedstock used can improve the efficiency of AD-MEC [33].

Furthermore, higher energy content feedstock, such as organic waste with high carbohydrates and lipids, can also improve the overall energy efficiency and gas production in AD-MEC [34]. Lower energy content feedstock may apply additional pretreatment and processes for optimal performance. Microorganisms or bacteria degrade biodegradability, or the rate of organic matter is also crucial. Higher biodegradable feedstock can boost the conversion rate of organic matter into biogas and hydrogen, thus improving the overall system [31, 35].

Moreover, nutrients such as nitrogen and phosphorus are essential for microbial activity in the system. A balanced nutrient composition can facilitate microorganisms' growth and metabolic functions for biogas and hydrogen production [36]. On the other hand, the presence of inhibitory components such as heavy metal, ammonia and organic acid can impede microbial activity, thus decreasing gas production [37].

4.2 Temperature

Temperature is another vital parameter for the AD-MEC system [38, 39]. Different operating temperatures affect the anaerobic digestion process and microbial electrolysis cell process. In the AD process, the temperature can affect the growth and activity of the microorganism needed for organic matter decomposition [31]. Higher temperatures or thermophilic conditions (50–60 °C) can improve digestion by enhancing microbial activity and biogas production [40, 41]. On the other hand, lower temperatures or mesophilic conditions (30–40 °C) may decrease the digestion rates and reduce biogas production.

For the AD-MEC system, methane production also increases from 15 °C to 50 °C as the temperature increases. However, methane production declined as the temperature increased to 70 °C [42]. At higher temperatures, it is found that hydrogenotrophic methanogens are higher than acetoclastic methanogens. Furthermore, higher temperature also enhances the specific growth rate of methanogens, thus improving methane production [43].

It is essential to ensure stable temperature throughout the process as it can affect microbial activity and process stability. Fluctuations in temperature can potentially reduce the efficiency of AD-MEC as it can change the microbial community dynamics and enzymatic activities [44]. Besides, the energy required to maintain the temperature should also be balanced with energy recovery from biogas production to optimise the overall energy performance for the AD-MEC system [45].

4.3 Hydraulic Retention Time (HRT)

Hydraulic retention time (HRT), or the time the wastewater resides in the system, is also a crucial parameter. This parameter shows the duration of contact for substrate and microorganism, improving organic matter degradation and biogas production [46, 47]. The organic loading rate (OLR) will increase at lower HRT, resulting in higher biogas production [48]. However, a significantly shorter HRT may lead to incomplete digestion and reduce methane yield. Longer HRT can contribute to more time for microbial activity and improve the organic matter decomposition process [49].

Moreover, higher HRT can increase the organic removal rate, current density and coulombic recovery in AD-MEC [50, 51]. A study also shows that an increase in HRT can improve COD reduction, decrease effluent pH and enhance methane production [49]. Thus, the optimisation of HRT is crucial as it directly affects the organic loading rates, microbial activity and methane production while simultaneously assuring a stable process with minimal risk [52].

4.4 pH

The pH level is another parameter that influences the performance of the AD-MEC system [53]. Maintaining an optimal pH value is critical as it will affect the composition and function of the microbial community, thus affecting the efficiency of organic matter decomposition [54]. Besides that, fluctuations in pH can also mess with microbial activity and reduce the system's performance [55]. With proper control and monitoring of the pH value, the system can improve the efficiency of the organic substrate utilisation via microbial activity.

Methane production is obtained from electro-methanogenesis at a high pH value (7.8). In contrast, at neutral pH (7.0), methane is produced from the electro-methanogenesis and heterotrophic (hydrogenotrophic/acetolastic) methanogenesis pathways [56]. The proton motive force and cell membrane potential depend on the pH value, thus affecting microbial growth and metabolism. It is found that the preferred pH for methanogens and acidogens is 7.8–8.2 and 5.5–6.5, correspondingly thus formulating an optimised pH for a combined culture ranging from 6.8 to 7.4 [57].

4.5 Applied Voltage

The applied voltage is the parameter that mainly affects the electrochemical reactions for the production of hydrogen gas [58]. The applied voltage directly affects the electron transfer process occurring in the microbial electrolysis cells, which is crucial in producing hydrogen gas [59]. However, it is found that applied voltage has a minimal impact on microbial activity and its metabolism for biogas or methane production [60]. A balanced applied voltage should be considered for the reliable and consistent operation of integrated AD-MEC, especially for hydrogen production.

With the presence of voltage, it is found that no VFA accumulation is detected, even at an elevated OLR [61]. The optimum applied voltage of 1 V is established from previous studies for the best methane production condition [62, 63]. Generally, methanogenesis enhancement occurs at the applied voltage below 1.0 V, and an applied voltage higher than 1.0 V is preferred for hydrogen production while increasing the pH value simultaneously. A larger electrode spacing in an AD-MEC system needs a higher applied voltage to negate the increased transfer resistance [23].

4.6 Electrode

The electrode is also essential as it can affect the efficiency and performance of MEC components in the AD-MEC system [64]. Graphite felt, graphite brushes, carbon cloth and carbon paper are among the existing electrode materials. It can facilitate the transfer of electrons and enhance the efficiency of the AD-MEC system. In the AD-MEC system, the electrode material must be carefully considered regarding mechanical strength, chemical stability, biocompatibility, electrical conductivity and the presence of catalysts or dopants [65, 66]. Furthermore, the roughness of the electrode surface, porosity and surface charge should also be considered for the attachment and growth of microorganisms on the electrode surface.

Previous studies have proved that nickel cathode produces a higher hydrogen evolution rate (HER), methane production and current density compared to copper and stainless-steel cathode in AD-MEC [67]. Besides, carbon brush presented a better result for methane production than graphite electrodes [68]. Cathodes with superior microbial adsorption and better electrochemical properties, such as carbon-based cathodes, are also proven to produce higher phenanthrene decomposition and methane yield [69]. The electrode will act as a conductor for electron transfer and electroactive microorganisms.

Furthermore, the ratio of surface area at the cathode side and spatial volume at the anode or the cathode/anode (C/A) ratio also influence methane production and energy recovery efficiency. Methane production improves with a higher C/A ratio, improving bioelectrochemical performance [70]. The electrode position also plays a role in better system performance. A bottom-position electrode is preferred to a top-position electrode for the organic removal efficiency [71]. Besides, modification or pretreatment of electrodes also enhances the surface characteristics by promoting biofilm formation and lessening the overpotential of activation [72].

5 Types of Configurations

5.1 In-situ AD-MEC

The most common configuration used and researched to enhance the AD and MEC system is the in-situ AD-MEC, in which the AD and MEC coexist in a single reactor. In-situ anaerobic digestion and microbial electrolysis cells (AD-MEC) are among the rising technologies for sustainable waste management and generating renewable and clean energy. In-situ AD-MEC [55] refers to the process in which the waste is treated at the same place as it is generated. This technology can produce high-value products such as methane and clean electricity from biodegradable waste. This approach can dismiss transportation costs and emissions while generating clean energy using waste [73]. This technology can also reduce the CO_2 produced from AD on the anode and generate CH_4 from hydrogenotrophic methanogens.

The improvement in AD with in-situ MEC can be attributed to several factors, such as the enrichment of electroactive microbials (EAB) and electrotrophic methanogenic archaea (EMA). Organic acids are oxidised via EAB, and EMA uses the electron transfer to the electrode and the transferred electrons for methane production through the electro-methanogenesis process [27]. During in-situ AD-MEC, the enhancement of hydrogenotrophic methanogens also occurred as MEC facilitates access to more H_2 for the electrode or bulk solution. Hydrogenotrophic methanogens are preferred over acetoclastic methanogens as they have higher growth rates, improved CH4 production rates and greater resistance to environmental changes [74]. Furthermore, in-situ AD-MEC can overcome the thermodynamic limitations by applying an external voltage to enhance organic acid degradation [75].

5.2 *Ex-situ AD-MEC*

Another configuration standard for AD-MEC is ex-situ AD-MEC, which refers to the AD and MEC processes operating in separate reactors. One of the approaches is to operate AD as a pre-fermentation phase before the MEC process. This particular approach maximises overall energy recovery by improving the biodegradability of organic matter for the overall process [76]. A researcher has concluded that the MEC process has a higher COD removal rate, current density and coulombic efficiency as the feedstock is pre-fermented compared to without pre-fermented feedstock.

Furthermore, Krishnan et al. have studied integrating the dual-stage thermophilic dark fermentation (DF) and mesophilic MECs using palm oil mill effluent (POME). The result shows that the COD removal rate and total energy recovery have increased rapidly. Besides, the post-treatment process of AD using MEC to improve the AD process is also a viable option. It is reported that an AD process of pig slurry is more stable with MEC as its post-treatment process with NH^{4+} removal at the anode side and production of CH_4 at the cathode side [77]. Integrating an anaerobic osmotic membrane bioreactor (AnOMBR) and MEC can also improve methane yield and content [78].

5.3 *Other Novel Integration Configurations*

Apart from in-situ and ex-situ AD-MEC, other novel configurations are also being explored. Among the configurations is integrating MEC with two-phase AD, isolating the methanogenesis stage from the acidogenesis and hydrolysis stage in two different reactors. It is found that electrode polarisation increased the microbial richness and modified the microbial communities due to applied voltage. During the first stage, the gas production rate is reduced. However, it improved in the second phase, thus improving the overall methane content [79]. Besides, a three-stage integrated configuration (HAF-MAD-HMEC) has also proven to be better compared to a two-stage or single-stage process in terms of biohythane yield and total organic carbon (TOC) removal [80]. MEC's advantages in exoelectrogen enrichment have improved the degradation of volatile fatty acids (VFAs) and other organic matter. Another configuration of dark fermentation (DF), MEC and AD (DF-MEC-AD) is also reported, and each stage's arrangement affects the overall process's performance, thus increasing the potential for further studies to improve the overall integration process [81].

Another novel configuration is by adding filler to reactors. The addition of Fe–C filler to reactors subjected to magnetic field and external voltage has improved the AD performance [82]. Applying electric and magnetic fields (EMF) has improved the degradation rates of lignocellulosic compounds, increased the diversity of microbes and enriched the functional microorganism, resulting in increased methane content. Furthermore, a novel approach of supplying side-stream voltage to a main AD reactor appears to have a similar effect as an auxiliary AD-MEC (a-ADMEC) via sludge recirculation from the a-ADMEC to the main AD [83]. The result demonstrated a similar performance for side-stream AD and mainstream AD voltage applications as the recirculation rate is raised at high OLR. The current density is enhanced at a higher recirculation rate as it alters the microbials from acetolastic methanogens into hydrogenotrophic ones. However, it is essential to note that a-ADMEC is only 1/7 of the size of the main AD reactor, and the sludge recirculation rate is only around 1/10 of the mechanical mixing optimal energy, indicating lower additional energy consumption. Table 2 presents the summary of findings of different AD-MEC configurations.

6 Challenges and Future Outlook

This technology has some challenges and constraints despite all the challenges and benefits of integrated AD-MEC, such as the co-generation of electricity and valuable products, primarily hydrogen and methane. Among the significant challenges is the replicability and efficiency of AD-MEC for large-scale applications. During wastewater treatment, the complex relationship and interaction between microbial communities and electrochemical reactions is also a crucial challenge to maintain a stable performance while optimising energy conversion [84]. Besides, the low external voltage requirement adds more to the operational cost and energy required, profoundly affecting the overall economic feasibility of the system.

As electro-methanogenesis studies continue to develop, concern has arisen regarding the reliability and stability of AD-MEC for a long time. The complexity of electron uptake pathways, such as via direct or mediated pathways, must be further

Table 2 Summary of findings of different AD-MEC configurations

In-situ AD-MEC	
• The most common configuration for AD-MEC • AD and MEC system in a single reactor • Can dismiss transportation cost CO_2 emission • Enhancement of EAV and EMA for CH_4 production improvement • Higher ratio of hydrogenotrophic methanogen compared to acetoclastic methanogen, thus limiting the pH drop after the process	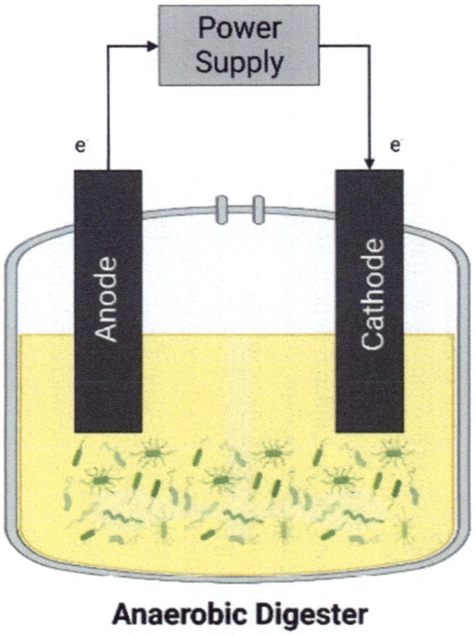

(continued)

Table 2 (continued)

Ex-situ AD-MEC	
• Second most common configuration of AD-MEC • Two reactors to separate the AD and MEC system • Improvement in the biodegradability of organic matter, thus maximising the overall energy recovery • Higher COD removal rate, current density and coulombic efficiency with ex-situ AD-MEC • AD with MEC as post-treatment can increase NH^{4+} reduction and CH_4 production	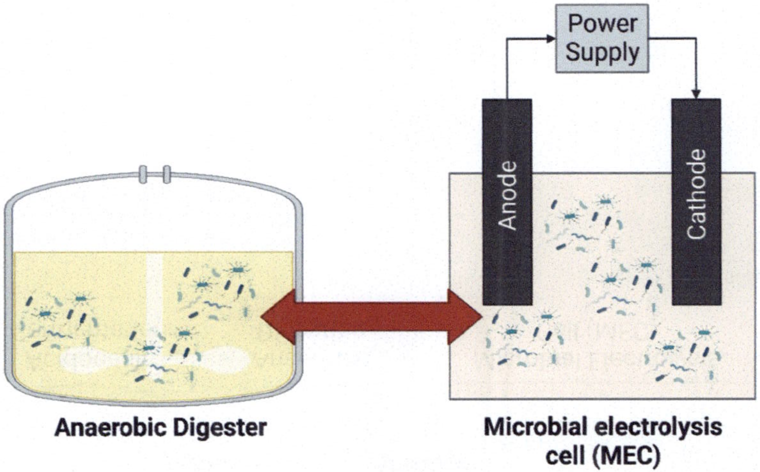

(continued)

Table 2 (continued)

Multiple stage AD-MEC	
• Electrode polarisation can improve microbial richness and modify the microbial communities • The overall methane content can be increased with multiple-stage AD-MEC • Three-stage integrated configuration is established to be superior to the single- and two-stage process to produce biohythane and reduce total organic carbon (TOC) • The degradation of volatile fatty acid and other organic matter is also improved with multiple-stage AD-MEC	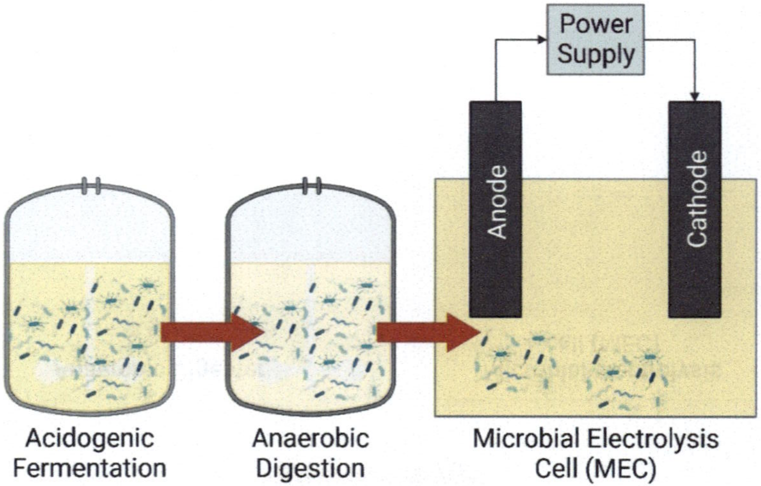

(continued)

Table 2 (continued)

Side-stream sludge recirculation	
• The side-stream voltage supplied by sludge recirculation has a similar effect as the main stream voltage supply to the AD reactor • High sludge recirculation rate can improve the operation stability at high OLR and enhance H_2-producing bacterial community • The aBEAD having 1/7 capacity of main AD shows potential for BEAD technology in new and existing AD facility	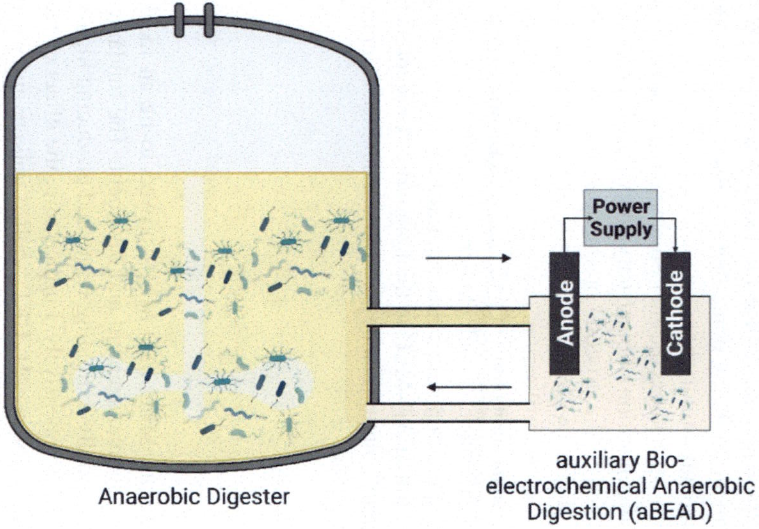

investigated to optimise the efficiency and reduce the likely drawbacks [85]. As ongoing research and development are being done, there is also a need to address these technological and fundamental limitations.

Furthermore, there is also a need to address the environmental impact of AD-MEC. Despite the advantage of the co-generation of electricity and valuable products, the ecological footprint report related to the process, production and utilisation must be assessed comprehensively [86]. Assessment of the life cycle analysis, such as greenhouse gas emissions and resource utilisation, is essential to enhance the overall sustainability of AD-MEC for waste management and energy generation [87].

In a nutshell, a balanced consideration of each challenge, namely technical inadequacy, economic feasibility and environmental aspects, must be addressed for the sustainable and reliable integration of AD-MEC in the long term. A comprehensive and sustainable waste management solution can be possible as a detailed examination of the limitation is done and resolved.

7 Conclusion

In conclusion, integrating AD with MEC (AD-MEC) proved to be an ideal and potential solution for organic waste management and treatment. The fundamentals and advantages of AD-MEC, especially in treating waste and producing high-value products, have been discussed comprehensively. Furthermore, the effect of parameters such as feedstock type, temperature, pH, HRT, applied voltage and electrode is also explained. The different kinds of existing configurations are also mentioned in detail, and lastly, the challenges and future prospects are discussed in terms of technical inadequacy, economic feasibility and environmental aspects.

References

1. Aworanti OA, Agbede OO, Agarry SE et al (2023) Decoding anaerobic digestion: a holistic analysis of biomass waste technology, process kinetics, and operational variables. Energies (Basel) 16. https://doi.org/10.3390/EN16083378
2. Monlau F, Francavilla M, Sambusiti C et al (2016) Toward a functional integration of anaerobic digestion and pyrolysis for a sustainable resource management. In: Comparison between solid-digestate and its derived pyrochar as soil amendment. https://doi.org/10.1016/j.apenergy.2016.02.084
3. O'Brien BJ, Milligan E, Carver J, Roy ED (2019) Integrating anaerobic co-digestion of dairy manure and food waste with cultivation of edible mushrooms for nutrient recovery. Bioresour Technol 285:121312. https://doi.org/10.1016/J.BIORTECH.2019.121312
4. Teater C, Yue Z, Maclellan J et al (2010) Assessing solid digestate from anaerobic digestion as feedstock for ethanol production. https://doi.org/10.1016/j.biortech.2010.09.099
5. Wee CY, Su JJ (2019) Biofuel produced from solid-state anaerobic digestion of dairy cattle manure in coordination with black soldier fly larvae decomposition. Energies 12:911. https://doi.org/10.3390/EN12050911

6. Wainaina S, Lukitawesa, Kumar Awasthi M, Taherzadeh MJ (2019) Bioengineering of anaerobic digestion for volatile fatty acids, hydrogen or methane production: a critical review. Bioengineered 10:437–458. https://doi.org/10.1080/21655979.2019.1673937
7. Wang P, Wang H, Qiu Y et al (2018) Microbial characteristics in anaerobic digestion process of food waste for methane production—a review. Bioresour Technol 248:29–36. https://doi.org/10.1016/J.BIORTECH.2017.06.152
8. Anukam A, Mohammadi A, Naqvi M, Granström K (2019a) A review of the chemistry of anaerobic digestion: methods of accelerating and optimizing process efficiency. Processes 7:504. https://doi.org/10.3390/PR7080504
9. Ivase TJP, Nyakuma BB, Oladokun O et al (2020) Review of the principal mechanisms, prospects, and challenges of bioelectrochemical systems. Environ Prog Sustain Energy 39:13298. https://doi.org/10.1002/EP.13298
10. Chakraborty I, Sathe SM, Khuman CN, Ghangrekar MM (2020) Bioelectrochemically powered remediation of xenobiotic compounds and heavy metal toxicity using microbial fuel cell and microbial electrolysis cell. Mater Sci Energy Technol 3:104–115. https://doi.org/10.1016/J.MSET.2019.09.011
11. Kaushik A, Singh A (2020) Metal removal and recovery using bioelectrochemical technology: the major determinants and opportunities for synchronic wastewater treatment and energy production. J Environ Manage 270. https://doi.org/10.1016/J.JENVMAN.2020.110826
12. Xiang LJ, Dai L, Guo KX et al (2020) Microbial electrolysis cells for hydrogen production. Chin J Chem Phys 33:263–284. https://doi.org/10.1063/1674-0068/CJCP2005075
13. Yu Z, Leng X, Zhao S et al (2018) A review on the applications of microbial electrolysis cells in anaerobic digestion. Bioresour Technol 255:340–348. https://doi.org/10.1016/J.BIORTECH.2018.02.003
14. Waqas M, Rehan M, Aburiazaiza AS, Nizami AS (2018) Wastewater biorefinery based on the microbial electrolysis cell: opportunities and challenges. Progress Recent Trends Microbial Fuel Cells 347–374. https://doi.org/10.1016/B978-0-444-64017-8.00017-8
15. Tao L, Song M, Jiang G (2023) Enhanced depolluting capabilities of microbial bioelectrochemical systems by synthetic biology. Synth Syst Biotechnol 8:341–348. https://doi.org/10.1016/J.SYNBIO.2023.05.005
16. Zhang Y, Liu M, Zhou M et al (2019) Microbial fuel cell hybrid systems for wastewater treatment and bioenergy production: synergistic effects, mechanisms and challenges. Renew Sustain Energy Rev 103:13–29. https://doi.org/10.1016/J.RSER.2018.12.027
17. Klein EM, Knoll MT, Gescher J (2023) Microbe-anode interactions: comparing the impact of genetic and material engineering approaches to improve the performance of microbial electrochemical systems (MES). Microb Biotechnol 16:1179–1202. https://doi.org/10.1111/1751-7915.14236
18. Wang XT, Zhang YF, Wang B, et al (2022b) Enhancement of methane production from waste activated sludge using hybrid microbial electrolysis cells-anaerobic digestion (MEC-AD) process—a review. Bioresour Technol 346. https://doi.org/10.1016/J.BIORTECH.2021.126641
19. Wang W, Chang JS, Lee DJ (2022) Integrating anaerobic digestion with bioelectrochemical system for performance enhancement: a mini review. Bioresour Technol 345:126519. https://doi.org/10.1016/J.BIORTECH.2021.126519
20. An Z, Feng Q, Zhao R, Wang X (2020) Bioelectrochemical methane production from food waste in anaerobic digestion using a carbon-modified copper foam electrode. Processes 8:416. https://doi.org/10.3390/PR8040416
21. Feng Q, Song YC, Ahn Y (2018) Electroactive microorganisms in bulk solution contribute significantly to methane production in bioelectrochemical anaerobic reactor. Bioresour Technol 259:119–127. https://doi.org/10.1016/J.BIORTECH.2018.03.039
22. Wang D, Han H, Han Y et al (2017) Enhanced treatment of Fischer Tropsch (F-T) wastewater using the up-flow anaerobic sludge blanket coupled with bioelectrochemical system: effect of electric field. Bioresour Technol 232:18–26. https://doi.org/10.1016/J.BIORTECH.2017.02.010

23. Xiao B, Chen X, Han Y et al (2018) Bioelectrochemical enhancement of the anaerobic digestion of thermal-alkaline pretreated sludge in microbial electrolysis cells. Renew Energy 115:1177–1183. https://doi.org/10.1016/J.RENENE.2017.06.043
24. Hou H, Li Z, Liu B et al (2020) Biogas and phosphorus recovery from waste activated sludge with protocatechuic acid enhanced Fenton pretreatment, anaerobic digestion and microbial electrolysis cell. Sci Total Environ 704:135274. https://doi.org/10.1016/J.SCITOTENV.2019.135274
25. Gao Y, Kong X, Zhou A et al (2020) Enhanced degradation of quinoline by coupling microbial electrolysis cell with anaerobic digestion simultaneous. Bioresour Technol 306:123077. https://doi.org/10.1016/J.BIORTECH.2020.123077
26. Sun M, Zhang Z, Lv M et al (2020) Enhancing anaerobic digestion performance of synthetic brewery wastewater with direct voltage. Bioresour Technol 315:123764. https://doi.org/10.1016/J.BIORTECH.2020.123764
27. Zakaria BS, Dhar BR (2019) Progress towards catalyzing electro-methanogenesis in anaerobic digestion process: fundamentals, process optimization, design and scale-up considerations. Bioresour Technol 289:121738. https://doi.org/10.1016/J.BIORTECH.2019.121738
28. Kotsyurbenko OR, Chin KJ, Glagolev MV et al (2004) Acetoclastic and hydrogenotrophic methane production and methanogenic populations in an acidic West-Siberian peat bog. Environ Microbiol 6:1159–1173. https://doi.org/10.1111/J.1462-2920.2004.00634.X
29. Park JG, Lee B, Kwon HJ, Jun HB (2019) Contribution analysis of methane production from food waste in bulk solution and on bio-electrode in a bio-electrochemical anaerobic digestion reactor. Sci Total Environ 670:741–751. https://doi.org/10.1016/J.SCITOTENV.2019.02.112
30. Park JG, Heo TY, Kwon HJ et al (2020) Effects of voltage supply on the methane production rates and pathways in an anaerobic digestion reactor using different electron donors. Int J Hydrogen Energy 45:9459–9468. https://doi.org/10.1016/J.IJHYDENE.2020.01.237
31. Gong H, Liu M, Li K et al (2020) Optimizing dry anaerobic digestion at pilot scale for start-up strategy and long-term operation: organic loading rate, temperature and co-digestion. Bioresour Technol 316. https://doi.org/10.1016/J.BIORTECH.2020.123828
32. Li Y, Han Y, Zhang Y et al (2020) Anaerobic digestion of different agricultural wastes: a techno-economic assessment. Bioresour Technol 315:123836. https://doi.org/10.1016/J.BIORTECH.2020.123836
33. Mao C, Xi J, Feng Y et al (2019) Biogas production and synergistic correlations of systematic parameters during batch anaerobic digestion of corn straw. Renew Energy 132:1271–1279. https://doi.org/10.1016/J.RENENE.2018.09.009
34. Hassanein A, Witarsa F, Lansing S et al (2020) Bio-electrochemical enhancement of hydrogen and methane production in a combined Anaerobic Digester (AD) and Microbial Electrolysis Cell (MEC) from dairy manure. Sustainability 12:8491. https://doi.org/10.3390/SU12208491
35. Xing BS, Han Y, Cao S, Wang XC (2020) Effects of long-term acclimatization on the optimum substrate mixture ratio and substrate to inoculum ratio in anaerobic codigestion of food waste and cow manure. Bioresour Technol 317:123994. https://doi.org/10.1016/J.BIORTECH.2020.123994
36. Li W, Khalid H, Amin FR et al (2020) Biomethane production characteristics, kinetic analysis, and energy potential of different paper wastes in anaerobic digestion. Renew Energy 157:1081–1088. https://doi.org/10.1016/J.RENENE.2020.04.035
37. Atelge MR, Atabani AE, Banu JR et al (2020) A critical review of pretreatment technologies to enhance anaerobic digestion and energy recovery. Fuel 270:117494. https://doi.org/10.1016/J.FUEL.2020.117494
38. Hossain MS, Karim T ul, Onik MH et al (2022) Impact of temperature, inoculum flow pattern, inoculum type, and their ratio on dry anaerobic digestion for biogas production. Sci Rep 12:1–13. https://doi.org/10.1038/s41598-022-10025-1
39. Rocamora I, Wagland ST, Villa R et al (2020) Dry anaerobic digestion of organic waste: a review of operational parameters and their impact on process performance. Bioresour Technol 299:122681. https://doi.org/10.1016/J.BIORTECH.2019.122681

40. Anukam A, Mohammadi A, Naqvi M, Granström K (2019b) A review of the chemistry of anaerobic digestion: methods of accelerating and optimizing process efficiency. Processes 7:504. https://doi.org/10.3390/PR7080504
41. Jiang J, He S, Kang X et al (2020) Effect of organic loading rate and temperature on the anaerobic digestion of municipal solid waste: process performance and energy recovery. Front Energy Res 8:522334. https://doi.org/10.3389/FENRG.2020.00089/BIBTEX
42. Yang HY, Bao BL, Liu J et al (2018) Temperature dependence of bioelectrochemical CO_2 conversion and methane production with a mixed-culture biocathode. Bioelectrochemistry 119:180–188. https://doi.org/10.1016/J.BIOELECHEM.2017.10.002
43. Yu J, Kim S, Kwon OS (2019) Effect of applied voltage and temperature on methane production and microbial community in microbial electrochemical anaerobic digestion systems treating swine manure. J Ind Microbiol Biotechnol 46:911–923. https://doi.org/10.1007/S10295-019-02182-6
44. Mei X, Xing D, Yang Y et al (2017) Adaptation of microbial community of the anode biofilm in microbial fuel cells to temperature. Bioelectrochemistry 117:29–33. https://doi.org/10.1016/J.BIOELECHEM.2017.04.005
45. Lin R, Cheng J, Ding L, Murphy JD (2018) Improved efficiency of anaerobic digestion through direct interspecies electron transfer at mesophilic and thermophilic temperature ranges. Chem Eng J 350:681–691. https://doi.org/10.1016/J.CEJ.2018.05.173
46. Liu X, Khalid H, Amin FR et al (2018) Effects of hydraulic retention time on anaerobic digestion performance of food waste to produce methane as a biofuel. Environ Technol Innov 11:348–357. https://doi.org/10.1016/J.ETI.2018.06.004
47. Yun JH, Cho KS (2017) Effect of hydraulic retention time on suppression of methanogens during a continuous biohydrogen production process using molasses wastewater. J Environ Sci Health Part A 52:37–44. https://doi.org/10.1080/10934529.2016.1221221
48. Abbagoni BM, Yeung H, Lao L (2022) Non-invasive measurement of oil-water two-phase flow in a vertical pipe using ultrasonic Doppler sensor and gamma-ray densitometer. Chem Eng Sci 248:117218. https://doi.org/10.1016/J.CES.2021.117218
49. Nelabhotla ABT, Khoshbakhtian M, Chopra N, Dinamarca C (2020) Effect of hydraulic retention time on MES operation for biomethane production. Front Energy Res 8:536886. https://doi.org/10.3389/FENRG.2020.00087/BIBTEX
50. Ki D, Parameswaran P, Popat SC et al (2017) Maximizing Coulombic recovery and solids reduction from primary sludge by controlling retention time and pH in a flat-plate microbial electrolysis cell. Environ Sci (Camb) 3:333–339. https://doi.org/10.1039/C6EW00305B
51. Luo S, Jain A, Aguilera A, He Z (2017) Effective control of biohythane composition through operational strategies in an innovative microbial electrolysis cell. Appl Energy 206:879–886. https://doi.org/10.1016/J.APENERGY.2017.08.241
52. Majid MSA, Faisal F, Fahlevi H et al (2022) Decomposing total factor productivity of SMEs in agricultural sector in Aceh, Indonesia. IOP Conf Ser Earth Environ Sci 951:012051. https://doi.org/10.1088/1755-1315/951/1/012051
53. Zwain HM, Naje AS, Vakili M, Dahlan I (2021) Temperature analysis of a novel MAIB reactor during the treatment of wastewater from recycled paper mill. Water Pract Technol 16:592–604. https://doi.org/10.2166/WPT.2021.023
54. Westerholm M, Schnürer A, Westerholm M, Schnürer A (2019) Microbial responses to different operating practices for biogas production systems. In: Anaerobic digestion. https://doi.org/10.5772/INTECHOPEN.82815
55. Van DP, Fujiwara T, Tho BL et al (2020) A review of anaerobic digestion systems for biodegradable waste: configurations, operating parameters, and current trends. Environ Eng Res 25:1–17. https://doi.org/10.4491/EER.2018.334
56. Nelabhotla ABT, Dinamarca C (2019) Bioelectrochemical CO_2 reduction to methane: MES integration in biogas production processes. Applied Sciences (Switzerland) 9. https://doi.org/10.3390/APP9061056
57. Lee M, Nagendranatha Reddy C, Min B (2019) In situ integration of microbial electrochemical systems into anaerobic digestion to improve methane fermentation at different substrate

concentrations. Int J Hydrogen Energy 2380–2389. https://doi.org/10.1016/J.IJHYDENE.2018.08.051

58. Shabib A, Abdallah M, Shanableh A, Sartaj M (2021) Design configurations and operating parameters of bio-electrochemically enhanced anaerobic digestion. Int J Environ Sci Dev 12:63–68. https://doi.org/10.18178/IJESD.2021.12.2.1319
59. Zhang X, Li X, Zhao X, Li Y (2019) Factors affecting the efficiency of a bioelectrochemical system: a review. RSC Adv 9:19748–19761. https://doi.org/10.1039/C9RA03605A
60. Luo L, Xu S, Jin Y et al (2018) Evaluation of methanogenic microbial electrolysis cells under closed/open circuit operations. Environ Technol 39:739–748. https://doi.org/10.1080/09593330.2017.1310934
61. Park J, Lee B, Shin W et al (2018) Application of a rotating impeller anode in a bioelectrochemical anaerobic digestion reactor for methane production from high-strength food waste. Bioresour Technol 259:423–432. https://doi.org/10.1016/J.BIORTECH.2018.02.091
62. Choi KS, Kondaveeti S, Min B (2017) Bioelectrochemical methane (CH_4) production in anaerobic digestion at different supplemental voltages. Bioresour Technol 245:826–832. https://doi.org/10.1016/J.BIORTECH.2017.09.057
63. Flores-Rodriguez C, Nagendranatha Reddy C, Min B (2019) Enhanced methane production from acetate intermediate by bioelectrochemical anaerobic digestion at optimal applied voltages. Biomass Bioenergy 127:105261. https://doi.org/10.1016/J.BIOMBIOE.2019.105261
64. Kumari U, Shankar R, Mondal P (2018) Electrodes for microbial fuel cells. In: Progress and recent trends in microbial fuel cells, pp 125–141. https://doi.org/10.1016/B978-0-444-64017-8.00008-7
65. Yaqoob AA, Ibrahim MNM, Rafatullah M et al (2020a) Recent advances in anodes for microbial fuel cells: an overview. Materials 13:2078. https://doi.org/10.3390/MA13092078
66. Yaqoob AA, Ibrahim MNM, Rodríguez-Couto S (2020) Development and modification of materials to build cost-effective anodes for microbial fuel cells (MFCs): an overview. Biochem Eng J 164:107779. https://doi.org/10.1016/J.BEJ.2020.107779
67. Wang L, He Z, Guo Z et al (2019) Microbial community development on different cathode metals in a bioelectrolysis enhanced methane production system. J Power Sources 444:227306. https://doi.org/10.1016/J.JPOWSOUR.2019.227306
68. Liu C, Yuan X, Gu Y et al (2020) Enhancement of bioelectrochemical CO_2 reduction with a carbon brush electrode via direct electron transfer. ACS Sustain Chem Eng 8:11368–11375. https://doi.org/10.1021/ACSSUSCHEMENG.0C03623/SUPPL_FILE/SC0C03623_SI_001.PDF
69. Lin C, Wu P, Liu Y et al (2019) Enhanced biogas production and biodegradation of phenanthrene in wastewater sludge treated anaerobic digestion reactors fitted with a bioelectrode system. Chem Eng J 365:1–9. https://doi.org/10.1016/J.CEJ.2019.02.027
70. Guo Z, Liu W, Yang C et al (2017) Computational and experimental analysis of organic degradation positively regulated by bioelectrochemistry in an anaerobic bioreactor system. Water Res 125:170–179. https://doi.org/10.1016/J.WATRES.2017.08.039
71. Sangeetha T, Guo Z, Liu W et al (2017) Energy recovery evaluation in an up flow microbial electrolysis coupled anaerobic digestion (ME-AD) reactor: Role of electrode positions and hydraulic retention times. Appl Energy 206:1214–1224. https://doi.org/10.1016/J.APENERGY.2017.10.026
72. Noori MT, Vu MT, Ali RB, Min B (2020) Recent advances in cathode materials and configurations for upgrading methane in bioelectrochemical systems integrated with anaerobic digestion. Chem Eng J 392:123689. https://doi.org/10.1016/J.CEJ.2019.123689
73. Long X, Wang H, Yang S et al (2021) Characteristics of gas-production, electricity-generation, and anode biofilm microorganisms based on electrochemical anaerobic digestion of different substrates. Environ Prog Sustain Energy 40:e13612. https://doi.org/10.1002/EP.13612
74. Park JG, Jiang D, Lee B, Jun HB (2020b) Towards the practical application of bioelectrochemical anaerobic digestion (BEAD): insights into electrode materials, reactor configurations, and process designs. Water Res 184. https://doi.org/10.1016/J.WATRES.2020.116214

75. Baek G, Kim J, Kim J, Lee C (2018) Role and potential of direct interspecies electron transfer in anaerobic digestion. Energies 11:107. https://doi.org/10.3390/EN11010107
76. Barbosa SG, Rodrigues T, Peixoto L et al (2019) Anaerobic biological fermentation of urine as a strategy to enhance the performance of a microbial electrolysis cell (MEC). Renew Energy 139:936–943. https://doi.org/10.1016/J.RENENE.2019.02.120
77. Cerrillo M, Viñas M, Bonmatí A (2018) Anaerobic digestion and electromethanogenic microbial electrolysis cell integrated system: increased stability and recovery of ammonia and methane. Renew Energy 120:178–189. https://doi.org/10.1016/j.renene.2017.12.062
78. Gao T, Zhang H, Xu X, Teng J (2021) Integrating microbial electrolysis cell based on electrochemical carbon dioxide reduction into anaerobic osmosis membrane reactor for biogas upgrading. Water Res 190. https://doi.org/10.1016/J.WATRES.2020.116679
79. Sasaki K, Sasaki D, Tsuge Y et al (2018) Changes in the microbial consortium during dark hydrogen fermentation in a bioelectrochemical system increases methane production during a two-stage process. Biotechnol Biofuels 11:1–12. https://doi.org/10.1186/S13068-018-1175-Z/FIGURES/5
80. Kaur M, Srikanth S, Kumar M et al (2019) An integrated approach for efficient conversion of Lemna minor to biogas. Energy Convers Manag 180:25–35. https://doi.org/10.1016/J.ENCONMAN.2018.10.106
81. Yahya M, Herrmann C, Ismaili S et al (2021) Development and optimization of an innovative three-stage bioprocess for converting food wastes to hydrogen and methane. Biochem Eng J 170:107992. https://doi.org/10.1016/J.BEJ.2021.107992
82. Qu G, Lv P, Cai Y et al (2020) Enhanced anaerobic fermentation of dairy manure by microelectrolysis in electric and magnetic fields. Renew Energy 146:2758–2765. https://doi.org/10.1016/j.renene.2019.06.050
83. Park JG, Kwon HJ, Sposob M, Jun HB (2020c) Effect of a side-stream voltage supplied by sludge recirculation to an anaerobic digestion reactor. Bioresour Technol 300. https://doi.org/10.1016/J.BIORTECH.2019.122643
84. Kong F, Ren HY, Pavlostathis SG et al (2020) Overview of value-added products bioelectrosynthesized from waste materials in microbial electrosynthesis systems. Renew Sustain Energy Rev 125:109816. https://doi.org/10.1016/J.RSER.2020.109816
85. Garcia FT, Ten Caten CS, de Campos EAR et al (2022) Mortality risk factors in micro and small businesses: systematic literature review and research agenda. Sustainability 14:2725. https://doi.org/10.3390/SU14052725
86. Ardakani MN, Badalians Gholikandi G (2020) Microbial fuel cells (MFCs) in integration with anaerobic treatment processes (AnTPs) and membrane bioreactors (MBRs) for simultaneous efficient wastewater/sludge treatment and energy recovery—a state-of-the-art review. Biomass Bioenergy 141:105726. https://doi.org/10.1016/J.BIOMBIOE.2020.105726
87. Rodrigues ICB, Leão VA (2020) Producing electrical energy in microbial fuel cells based on sulphate reduction: a review. Environ Sci Pollut Res Int 27:36075–36084. https://doi.org/10.1007/S11356-020-09728-7

Microbial Electrolysis Cells for the Treatment of Dye Effluent and Coupled with Generation of Hydrogen

Mahdi Esmaelzadeh, Mahsa Taheri, Mahdi Masoudi, and Narges Fallah

Abstract Dyes find extensive use in industries such as textiles, leather, and cosmetics, but their discharge poses a significant environmental risk. Bioelectrochemical methods for treating dye wastewater offer an environmentally friendly and economically viable solution, embraced by numerous scientists. Microbial electrolysis cells (MECs) have emerged as a promising technology for azo dye treatment and simultaneous hydrogen gas production. MECs employ an anaerobic biocathode where exoelectrogens catalyze the breakdown of azo bonds, generating hydrogen gas. These cells exhibit rapid decolorization and mineralization of a wide range of azo dyes. Importantly, the bioelectrochemical reduction process prevents the accumulation of harmful aromatic amine intermediates, a common issue in anaerobic digestion. MECs operate efficiently under neutral pH and ambient temperatures, and their simultaneous wastewater treatment and hydrogen production improve overall economics. However, MECs face challenges such as substrate competition with organic materials, exoelectrogen sensitivity to dye toxicity, and the potential for methane formation. Successful deployment requires biocathode acclimation, reactor configuration optimization, electrode material selection, voltage adjustment, and co-substrate supply. While MECs have demonstrated high decolorization efficiency, overcoming obstacles related to intermediate mineralization and scalability is crucial for practical application. Integration with photocatalysis or aerobic polishing can effectively address intermediate byproducts. Techno-economic analyses highlight the promising potential of MECs for sustainable azo dye wastewater treatment

M. Esmaelzadeh · M. Masoudi · N. Fallah
Department of Chemical Engineering, Amirkabir University of Technology, Tehran, Iran
e-mail: Mahdiesmaelzadeh7706@gmail.com

M. Masoudi
e-mail: mahdi.masoudi@aut.ac.ir

N. Fallah
e-mail: nfallah2001@aut.ac.ir

M. Taheri (✉)
Department of Civil and Environmental Engineering, Amirkabir University of Technology, Tehran, Iran
e-mail: mahsa_taheri88@yahoo.com

A. A. Yaqoob and A. Ahmad (eds.), *Microbial Electrolysis Cell Technology*, Sustainable Materials and Technology, https://doi.org/10.1007/978-981-97-3356-9_13

and bioenergy generation. Realizing this potential necessitates further research on exoelectrogen acclimation, reactor design, and large-scale trials.

Keywords Dye removal · Textile wastewater · Bioelectrochemical · Hydrogen production · Microbial electrolysis cell

1 Introduction

The textile industry, a key player in the global economy, offers employment and contributes to the gross domestic product (GDP) of many nations [1]. The rapid global population growth fuels extensive expansion across industries, notably impacting textiles and contributing significantly to water pollution. Manufacturing processes within the textile sector result in a substantial environmental challenge known as dye effluent. Reports highlight the staggering use of 6.1 million liters of water daily to produce 800 kg of fabric, releasing effluents laden with diverse pollutants into the environment [2]. These effluents pose severe threats, adversely affecting ecosystems, disrupting aquatic life, and jeopardizing human and environmental health. Paint effluents from the industry contain harmful pollutants, emphasizing the importance of proper treatment to prevent water contamination.

Efficient removal of dyes and impurities from wastewater is crucial. Common methods like coagulation, flocculation, sedimentation, and chemical oxidation, though effective for some contaminants, often demand large chemical quantities, generate sludge, and exhibit limited efficacy against dyes and resistant organic compounds. Moreover, these methods are energy-intensive and can produce secondary pollutants [3, 4]. In contrast, biological treatment offers sustained effectiveness with minimal sludge production and environmental friendliness, requiring fewer reagents and less energy for complete dye mineralization [5, 6].

In recent years, there has been a growing interest in sustainable treatment technologies for wastewater, with microbial electrolysis cells (MECs) emerging as a promising solution. MECs utilize microorganisms to catalyze electrochemical reactions, effectively treating dye wastewater and producing hydrogen gas simultaneously. Comprising an anode and cathode separated by a proton exchange membrane, MECs leverage electroactive bacteria at the anode to oxidize organic compounds in dye effluent, generating electrons and protons. These electrons then combine with protons at the cathode, producing hydrogen gas. MECs offer advantages over conventional methods by efficiently removing dyes and resistant organic compounds, providing a sustainable and environmentally friendly alternative to chemical oxidation [7, 8]. Additionally, hydrogen gas produced as a byproduct serves as a clean, renewable energy source, reducing the environmental impact of dye wastewater treatment and offering an eco-friendly alternative to traditional thermochemical processes that emit greenhouse gases [9, 10].

This chapter aims to introduce colored wastewater and its treatment methods, with a specific focus on exploring the potential of MECs for effectively treating dye wastewater while simultaneously generating hydrogen gas. The objective is to assess MEC performance in terms of color removal efficiency, hydrogen production efficiency, and energy consumption. Additionally, this chapter aims to introduce the optimal conditions of the design and performance parameters of MECs, with the aim of maximum efficiency by understanding the basic principles and mechanisms governing MECs.

2 Dye Effluents

The discharge of dye effluents, which originates from various industrial sectors such as textile production, paper manufacturing, leather processing, and dye creation itself, encompasses a diverse range of dyes and contaminants. These waste streams differ in composition depending on the specific type of dye used, the techniques employed during manufacturing processes, as well as the respective industry they stem from. Gaining knowledge about distinct types of dye effluents holds great importance when it comes to implementing efficient treatment methods utilizing MECs. Among the types of dyes in wastewater are [11–13] the following:

- Reactive dyes, commonly utilized in the textile industry, give rise to effluents that possess an intense coloration and contain high levels of COD as well as reactive dye intermediates.
- Acid dyes, extensively employed across various sectors, produce effluents characterized by heightened color intensity, and acidic compounds.
- Basic dyes, being cationic and water-soluble in nature, find their usage primarily in acrylic and polyester fibers. Consequently, they generate effluents exhibiting enhanced color saturation along with the presence of cationic compounds.
- Direct dyes are water-soluble substances directly applied onto fibers during production processes. These particular dyes yield moderately intense colored effluents containing COD quantities alongside direct dye intermediates.
- Disperse dyes, with their hydrophobic and insoluble nature, find application in synthetic fibers. Consequently, the resultant effluents exhibit diminished color intensity, chemical oxygen demand (COD), and the presence of hydrophobic compounds.
- Vat dyes are insoluble but require the assistance of a reducing agent for effective coloring of cellulosic fibers. This process yields effluents that display low color intensity, COD levels as well as evidence of reducing agents.
- Sulfur dyes necessitate a reducing agent alongside an alkaline environment for water insolubility. As a consequence, these dyes contribute to effluents with reduced color intensity and COD along with residual traces of reducing agents.
- Azo dyes represent widely employed synthetic variants which can be either soluble or insoluble in water. Their extensive use generates effluents that demonstrate heightened color intensity due to their inherent properties. Moreover, such effluents typically reveal significant levels of COD as well as aromatic amines.

- Lastly, encompassing various types like disperse-reactive, metal-complexed solvent-based and natural varieties; other types of dyes exhibit unique traits leading to distinct environmental impacts when utilized in industrial processes.

The effluents from dye manufacturing processes harbor a host of pollutants, encompassing both organic compounds that possess the quality of being aromatic and resistant to degradation, as well as inorganic compounds like salts, acids, and alkalis. Moreover, heavy metals have found their way into these deleterious mixtures [14]. In order to conceive successful strategies for the treatment of such polluted wastewater using MECs, it is imperative to grasp a comprehensive understanding of their composition and behavior.

3 Conventional Methods of Dye Wastewater Treatment

The treatment of effluents containing dye presents formidable challenges due to the intricate and intricate nature of these pollutants. The composition of dyes varies greatly, and the presence of multiple pollutants further complicates finding a universal approach for their treatment. Organic dyes and their byproducts often prove resistant to conventional methods, persisting in the environment with long-lasting consequences. In addition, the existence of diverse pollutants, both organic and inorganic in nature, adds an additional layer of complexity that necessitates a comprehensive strategy for successful removal. Dye effluents are characterized by their high intensity coloration and elevated chemical oxygen demand (COD), rendering them particularly burdensome for various industries dealing with different types of dyes. The primary hurdle lies in achieving effective color removal given the remarkable stability exhibited by these dyes. Elevated concentrations of COD serve as a cogent evidence of the existence of substantial organic contaminants, necessitating their expeditious elimination to forestall any deleterious repercussions upon the delicate ecosystem teeming with aquatic fauna and the overarching environment. Furthermore, confronting us with formidable obstacles are both the noxious properties inherent in synthetic dyes found pervasively within industrial effluents and their penurious capacity for biodegradation. Despite perseverance in employing time-honored methodologies over prolonged periods, they persistently encounter impediments ranging from effectiveness and financial restrain to environmental ramifications, which are succinctly forthwith [3–6]:

I. Physical Methods

   - Sedimentation: Gravity-induced settling removes large particles but may not efficiently handle dissolved dye compounds.
   - Filtration: Various filters, like sand and activated carbon, effectively remove suspended solids and some dyes but may not cover all types.

- Adsorption: Activated carbon's high surface area efficiently removes a wide range of dye compounds but may require frequent replacement.

II. Membrane Separation: Uses semi-permeable membranes like reverse osmosis for high dye removal efficiencies. However, it may demand high operating pressures and energy consumption.
III. Chemical Methods

- Coagulation: Involves adding coagulants like alum to destabilize and aggregate dye compounds, effective for some but not all dyes.
- Oxidation: Introduces oxidizing agents to break down dye compounds into less harmful molecules. Different methods offer varied advantages and limitations.
- Reduction: Uses reducing agents to convert dye compounds into colorless or less toxic substances. Methods differ in efficiency, cost, and environmental impact.

IV. Biological Methods

- Aerobic Processes: Oxygen-consuming microorganisms break down dye compounds, with methods like activated sludge process offering varying outcomes. Of course, azo rusts are resistant to aerobic degradation and are mostly biodegradated anaerobically.
- Anaerobic Processes: Oxygen-independent microorganisms convert dye compounds into simpler substances, with methods like anaerobic baffled reactor having distinct pros and cons.

In conclusion, while conventional methods have been widely employed, they exhibit limitations.

4 Working Principles of Microbial Electrolysis Cells

Microbial electrolysis cells provide a hopeful resolution for the purification of dye effluent, while also yielding hydrogen as a byproduct. These cells function as bioelectrochemical systems, employing microorganisms to drive electrochemical reactions in an environmentally conscious and enduring fashion.

The fundamental components that comprise the MEC system are of utmost importance and consist of the anode, cathode, and proton exchange membrane (PEM) which serve to partition their respective compartments. The anode, fortified with materials such as carbon cloth or graphite, serves as a platform for microbial adhesion. On the other hand, the cathode is usually composed of a noble metal catalyst like platinum or nickel, thereby enabling seamless facilitation of hydrogen evolution reaction. Lastly, the PEM plays a crucial role by ensuring unobstructed proton transportation while effectively preventing any undesired mixing between anolyte and catholyte solutions [15].

MECs function on the principles of microbial fuel cells (MFCs) and electrolysis. Bacteria form a biofilm on the anode, breaking down organic matter in dye effluent to generate electrons and protons. These particles transfer to the cathode, where a small voltage is applied to combine them, producing hydrogen gas. Simultaneously, water is reduced at the cathode through electrolysis, generating additional hydrogen gas. Microorganisms catalyze oxidation and reduction reactions, while an external power source enhances cathode reactions [16].

5 Mechanisms of Dye Effluent Treatment in MECs

The treatment of dye effluent in MECs involves diverse mechanisms for pollutant removal. However, information on decolorization, azo dye degradation, and simultaneous hydrogen production by MECs is limited based on literature research.

- Microbial Oxidation: Microorganisms, especially anaerobic bacteria, efficiently degrade azo dyes using enzymes like azoreductases, laccases, and peroxidases. Electrically active biofilms on MEC electrodes enhance transformation pathways [17, 18].
- Electrolytic Splitting of Water: Water electrolysis at the cathode produces H_2 gas, acting as an electron donor. Hydrogenotrophic microbes like Clostridium spp. then use H_2 to enhance decoloration, albeit slower than direct electrochemical reactions [19].
- Electrochemical Processes: MECs employ electrochemical processes between the anode and cathode, promoting degradation of organic compounds. The generated electrical current produces reactive oxygen species, enhancing the treatment efficiency of MECs [20].
- Direct Electrochemical Reduction: A key mechanism is the direct electrochemical reduction of dye molecules at the cathode, significantly increasing decolorization with higher cathodic potential [16, 21].
- Adsorption and Precipitation: Dyes adsorb onto negatively charged electrodes and cellular biofilms, leading to decolorization. Precipitation of reduced insoluble aromatic amines due to applied voltage further reduces color [5, 22].
- Hydrogen Gas Production: Hydrogen gas production at the cathode enhances effluent mixing, promoting dye degradation and improving overall treatment efficiency [23].

In summary, MECs utilize microbial, electrochemical, and physical processes for efficient dye effluent treatment, although comprehensive information on certain aspects remains limited.

Despite the said content, it should be noted that although bioelectrochemical systems (BESs) can decolorize azo dyes through reductive cleavage of azo bonds, this process generates toxic aromatic amine intermediates that accumulate due to incomplete anaerobic reduction. Most studies on using BESs to treat azo dye wastewater have focused on dye decolorization rates and mechanisms rather than degradation of the aromatic amine intermediates. The accumulation of these toxic intermediates can inhibit the exoelectrogens in BESs. Also, discharge of these intermediates is environmentally harmful. Therefore, some researches have focused more on degradation of this toxic intermediates like aromatic amine, phenolic compounds, etc. [17].

Photocatalysis has been demonstrated as an efficient approach for degradation of recalcitrant organics, including dyes. In short, it can be said in an photocatalytic MEC reactor, the active species of ·OH generated in the photocatalysis process can simultaneously oxidize the toxic intermediates and produce H_2 with a simplified configuration and operation [17].

Another method for degradation of toxic intermediates is to use the fermentation process. Fermentation is an important metabolic process that occurs in bioanodes, in addition to exoelectrogenesis, when fermentable substrates are used. Fermentation first breaks down relatively complex substrates into lower molecular weight molecules. These fermentation products then serve as the electron donors for exoelectrogenesis in the bioanodes. The two-step process of fermentation followed by exoelectrogenesis is key when using fermentable substrates in bioelectrochemical system anodes. And in general, it can be said that the fermentation process plays an effective role in eliminating toxic intermediates and especially about phenol and catechol [24].

6 Electrolysis Reactions in MECs

Electrolysis is pivotal in MECs for dye effluent treatment, converting chemical energy in the effluent into electrical energy and generating hydrogen gas. These reactions transpire at the anode and cathode, guided by the microbial community within the system [25, 26].

I. Anode Reactions

- Organic Matter Oxidation: Exoelectrogens oxidize organic matter in the dye effluent, releasing electrons and protons. This process, facilitated by extracellular enzymes, targets various pollutants.
- Electron Transfer to the Anode: Electrons from oxidation transfer to the anode electrode via a conductive biofilm on the anode surface, serving as a pathway for electron transport.
- Proton Generation: Oxidation produces protons, contributing to a pH decrease in the anode compartment.

II. Cathode Reactions

- Proton Reduction: Protons from the anode migrate to the cathode compartment, where they are reduced to form hydrogen gas.
- Hydrogen Gas Production: Proton reduction at the cathode yields hydrogen gas, released and collectible. The efficiency of this production gauges MEC performance.

7 Microbial Community in MECs

The microbial community within MECs is integral to both the degradation of organic pollutants and the production of hydrogen gas. Employing mixed cultures from wastewater or anaerobic sludge as a source of electrochemically active microbes is a common practice, eliminating the need for sterilization and pure cultures. Through the acclimation of mixed cultures on electrodes, a selection process occurs over time, favoring microbes proficient in efficient electrochemical interactions. The anode chamber of MECs provides an optimal environment for the growth of electroactive microorganisms that can transfer electrons directly to the electrode surface. Electrogenic bacteria, responsible for the electron transfer process, form biofilms on the anode surface, comprising a diverse array of microorganisms like Geobacter, Shewanella, and other electroactive bacteria, known as exoelectrogens. These microorganisms utilize organic matter in dye effluent as a carbon source, releasing electrons during metabolic processes. The MEC microbial community, consisting of bacteria, archaea, and fungi, engages in complex interactions and synergistic relationships, ensuring efficient dye pollutant degradation and hydrogen gas production. The diversity of the microbial community is crucial for the stability and resilience of the MEC system. Research indicates that the microbial community composition in MECs is influenced by various factors, including dye effluent type, operational conditions, and electrode materials. Specific microbial species, such as Geobacter and Shewanella, have been associated with enhanced electron transfer and hydrogen production [27, 28]. Factors like pH, temperature, hydraulic retention time, nutrient availability, and biofilm formation dynamics influence the dynamics of the microbial community within MECs. Adjusting operational parameters can favor the growth of specific microorganisms and promote the targeted degradation of dye compounds. For instance, maintaining a slightly acidic pH enhances the activity of bacteria involved in dye degradation, while optimizing the temperature can encourage the growth of thermophilic microorganisms with higher dye degradation capabilities [29]. The choice of electrode materials, including conductive materials like carbon-based electrodes, significantly influences microbial community composition, attachment, and biofilm formation [30]. A comprehensive understanding and manipulation of microbial community dynamics are crucial for enhancing MEC performance and stability in dye effluent treatment.

8 Hydrogen Production Pathways in MECs

Generating hydrogen via MECs includes the oxidation of organic substances, leading to the chemical transformation of hydrogen in the cathode compartment [31]. Multiple techniques in the MEC framework have been utilized for hydrogen generation, with direct cathode reduction being the most uncomplicated. In this approach, hydrogen development takes place directly on the cathode surface when protons and electrons unite. Electrohydrogenesis is a procedure employing exoelectrogenic microorganisms to release electrons that amalgamate with protons, showcasing this process [31].

Another approach is in situ biocatalysis, incorporating fermentative bacteria into the MEC system. Fermentation metabolites migrate to the anode chamber, where they undergo oxidation, releasing protons and electrons. As these bacteria degrade organics at the anode, hydrogen is produced, crossing over to the cathode for recovery. A key advantage of this route is the direct utilization of low-value organics without external pathways for substrate conversion. However, poor conversion efficiencies have hindered large-scale applications [1, 32, 33]. Conversely, ex situ biocatalysis separates fermentation from the MEC, feeding effluent from the MEC anode to a separate bioreactor containing fermentative bacteria. While allowing independent optimization, this approach incurs higher capital costs due to dual reactor systems [1, 32, 33].

Certain photosynthetic bacteria can generate hydrogen and organic acids via photofermentation. Integrating these light-driven reactions into an MEC is an emerging tactic to boost yields and offer synergies in substrate conversion capacities through combined microbial electrochemical and photobiological systems. However, the need for photobioreactors increases process complexity, and stability issues arise under continuous operation. Providing optimal illumination intensity is critical to balancing H_2 generation against biomass growth kinetics [34, 35].

Combination of MECs with other technologies like microbial fuel cells, anaerobic digestion, or dark fermentation can offer synergies for enhanced hydrogen production. Effluents from microbial fuel cell (MFC) anodes, for instance, can provide substrates for MECs, enabling higher overall energy and H_2 recoveries from organics. Though lab-scale studies display encouraging results, techno-economic studies are vital for assessing commercial feasibility. Cost factors, component compatibility, and process control require detailed assessment prior to implementation [36, 37]. The choice of the method depends on factors such as desired hydrogen purity and yield, complexity, cost, and ease of operation. The multiplicity of options suggests that a hybridized approach, combining favorable aspects of different methods, maybe the most effective route for maximizing hydrogen yields from organic feedstocks through MECs.

9 Optimizing MECs for Dye Effluent Treatment

9.1 Microbial Community

One of the key elements impacting the performance of MEC is the meticulous selection of microbial varieties. Bacteria with electrochemical activity, particularly the Geobacter species known for their strong redox capabilities, are frequently utilized in MEC anodes. Research suggests that these bacteria have a significant influence on hydrogen production by directly affecting the formation of biofilms [38–40].

Moreover, electrochemically active bacteria (EAB) play a crucial role in the energy recovery mechanism of microbial electrolysis cells. They facilitate the transfer of electrons from the substrate to the surface of the anode, a critical step in capturing energy. These microorganisms, capable of transporting electrons from feedstocks to external electron acceptors, are widespread in various natural settings and wastewater. Despite the limited characterization of EAB species so far, further investigation is vital to explore the diversity and physiology of electrochemically active bacteria [38–40].

9.2 Electrode

In microbial electrolysis cells, the selection of electrode materials and configurations significantly impacts system performance. Electrodes serve as a connection between the microbial community and electrochemical reactions within the cell. Anode materials are crucial, attracting electrochemically active bacteria (EAB) to form biofilms that transfer electrons. Effective anode materials require a large surface area for microbial attachment and growth. Materials like carbon cloth, carbon felt, and graphite granules, commonly used due to ample surface area and good electrical conductivity, meet these criteria. Anode materials should also be chemically stable and resistant to fouling, considering potential organic compounds in dye effluents. Choosing a chemically resistant anode material is vital for long-term stability. Low internal resistance in anode materials minimizes energy losses, with high electrical conductivity, like carbon-based materials, being preferred for efficient electron transfer. Heat-treated carbon/graphite and carbon nanotubes enhance electron transfer and surface area [25, 39, 41, 42]. Cathode materials, like the anode, require specific features for optimal MEC performance. Catalysts for proton reduction in hydrogen production, such as precious metals like platinum, face cost limitations, leading researchers to explore alternatives like non-precious metal catalysts (e.g., nickel, cobalt, stainless steel, molybdenum sulfide, polyaniline, palladium, and iron-based) [25, 39, 41, 42]. Nickel-based materials demonstrate higher hydrogen production and coulombic efficiency. Nanomaterials, like palladium nanoparticles, improve hydrogen recovery. The cathode material should facilitate efficient proton and electron transport, often achieved with porous structures like carbon cloth or gas diffusion electrodes, enhancing mass transfer and overall performance. Optimizing electrode spacing in single-chamber MECs can balance the trade-off between internal resistance and membrane fouling and affect the overall performance of the system [39, 41].

9.3 Operating Parameters

Operating parameters are critical for the performance and efficiency of MECs in treating dye effluent. These parameters, encompassing applied voltage, hydraulic retention time (HRT), pH, temperature, substrate type, substrate concentration, and electrolyte conductivity, play pivotal roles in optimizing MEC functionality [39, 40, 42].

I. Applied Voltage: Crucial for electrochemical reactions, higher applied voltages enhance hydrogen production, but excessive levels lead to wastage and electrode degradation. Optimal voltage balance is around 0.11–0.23 V.
II. Hydraulic Retention Time (HRT): Duration of dye effluent in MECs impacts microorganism–dye interactions. Longer HRTs improve dye removal efficiency, but excessive times may reduce system throughput and increase costs.
III. pH: Maintaining optimal pH (around 9) is vital for efficient MEC functioning, influencing microbial activity. Continuous monitoring and pH control are challenging yet crucial for hydrogen production.
IV. Temperature: Influencing microbial activity, higher temperatures enhance MEC performance. Operating at 35–40 °C optimizes anaerobic conditions, promoting COD removal and coulombic efficiency.
V. Substrate Type: Different substrates (sugars, acids, acetate, etc.) impact hydrogen production. Acetate is widely used for high hydrogen recovery, while various substrates require careful selection based on production goals.
VI. Substrate Concentration: Critical for treatment efficiency, optimizing substrate concentration is essential. Higher concentrations enhance dye removal but must be carefully controlled to avoid inhibition and instability.
VII. Electrolyte Conductivity: Affecting hydrogen production rate, higher conductivity improves electron/ion transfer. Increasing conductivity from 7.5 to 20 mS/cm enhances yield while reducing Ohmic losses.

In summary, meticulous consideration and optimization of these operating parameters are crucial for maximizing MEC efficiency in dye effluent treatment.

We have presented the optimal values for the specified parameters in Table 1, derived from a limited yet comparable body of research in the field of hydrogen production, concomitant with the treatment of colored wastewater in MEC.

10 Future Perspectives and Research Directions

Microbial electrolysis cells are promising for treating dye effluent while producing valuable hydrogen. Future research can enhance their efficiency and applicability:

a. Enhanced Performance and Efficiency

Table 1 Research has been conducted in the field of hydrogen production simultaneously with color wastewater treatment in MEC

Reactor type*	Pollutant	Pollutant concentration	Applied voltage	Co-substrate	Co-substrate concentration	HRT	pH
PMEC	Methyl orange	50–300 mg/L	0.8 v	Acetate	20–50 mg/L	12 h	–
MREC	Orange G	400 mg/L	-	Acetate sodium	20 mM	–	2–7
CW-MEC	Methyl orange	1000, 1500, 2000 mg/L	–	–	–	–	3
SCMEC	Ponceau S	10–40 mg/L	0.7 v	Sodium acetate	50 mM	–	7.5
MEC	Furanic and phenolic compounds	1200 mg/L	0.6 v	Acetate	326 mg/L	–	–
SCMEC	Alizarin yellow R	30–150 mg/L	-	NaAc, glucose, NaAc + glucose	–	10 h	
MFMEC	Congo red	200 mg/L	0.9 v	Glucose	500 mg/L	–	7

Reactor type*	Microorganism	Anode material	Cathode material	Pollutant removal	COD removal	H_2 produced	References
PMEC	Anaerobic bacteria	Graphite brush	Ni foam coated with TiO_2 catalyst	98–76%	80–65%	113–68 ml	[17]
MREC	Anaerobic bacteria	Carbon fiber brush	Graphite plate	–	82%	–	[43]
CW-MEC	-	Graphite	Graphite	95.2, 78.6, 65.3%	85%	–	[44]
SCMEC	Mixed microbial culture	Graphite	Graphite	85–100%	94.78%	18 mL	[16]
MEC	Exoelectrogenesis bacteria	Porous carbon felt	Carbon	–	57%	26.5 mL/d	[24]
SCMEC	Electrochemically active bacteria	Graphite	MoS_2-GO nickel foam	90%	–	28.7 mL	[45]
MFMEC	Mixture of anaerobic sludge and aerobic sludge	Graphite felts	Graphite felts	87.9%	–	–	[46]

*PMEC (photocatalytic microbial electrolysis cell), MREC (microbial reverse electrodialysis electrolysis cell), CW-MEC (constructed wetland coupled with a microbial electrolysis cell), SCMEC (single-chamber microbial electrolysis cell), MEC (microbial electrolysis cell), MFMEC (membrane-free microbial electrolysis cells)

- Exploring new electrode materials like carbon-based materials can improve catalytic activity. Modifying electrode structures enhances microbial attachment and electron transfer efficiency.
- Manipulating microbial communities, through bioaugmentation or environmental conditions, improves performance.
- Fine-tuning operational parameters, such as pH and temperature, optimizes MEC performance.

b. Integration with Other Treatment Technologies

- Combining MECs with advanced oxidation processes (AOPs) enhances the breakdown of complex dye molecules before MEC treatment, reducing energy consumption.
- Integrating MECs with membrane filtration enables the separation and recovery of valuable products from treated effluent.

c. Scale-up and Commercialization

- Developing scalable and cost-effective designs for large-scale MECs involves optimizing reactor configurations and implementing automated monitoring systems.
- Techno-economic analyses evaluate the economic feasibility of MEC-based dye effluent treatment systems, attracting investment for commercialization.

d. Environmental and Regulatory Considerations

- Conducting environmental impact assessments helps identify and mitigate negative impacts.
- Developing regulatory frameworks ensures compliance and facilitates MEC technology adoption.

In conclusion, ongoing advancements in performance, integration, scale-up, and regulatory considerations will revolutionize MECs, providing a sustainable solution for dye effluent treatment across industries.

References

1. Chen X et al (2021) Circular economy and sustainability of the clothing and textile industry. Mater Circular Econ 3:1–9
2. Khandare RV, Govindwar SP (2015) Phytoremediation of textile dyes and effluents: current scenario and future prospects. Biotechnol Adv 33(8):1697–1714
3. Hai FI, Yamamoto K, Fukushi K (2007) Hybrid treatment systems for dye wastewater. Crit Rev Environ Sci Technol 37(4):315–377
4. Bhatia D et al (2017) Biological methods for textile dye removal from wastewater: a review. Crit Rev Environ Sci Technol 47(19):1836–1876
5. Ahmad A et al (2015) Recent advances in new generation dye removal technologies: novel search for approaches to reprocess wastewater. RSC Adv 5(39):30801–30818

6. Popli S, Patel UD (2015) Destruction of azo dyes by anaerobic–aerobic sequential biological treatment: a review. Int J Environ Sci Technol 12:405–420
7. Wang W, Lee D-J, Lei Z (2022) Integrating anaerobic digestion with microbial electrolysis cell for performance enhancement: a review. Biores Technol 344:126321
8. Asrul MAM et al (2021) Mathematical model of biohydrogen production in microbial electrolysis cell: a review. Int J Hydrogen Energy 46(75):37174–37191
9. Shayan Nasr M, Esmaeilnezhad E, Choi HJ (2022) An overview on the enhanced gas condensate recovery with novel and green methods. Environ Sci Pollution Res 29(18):26160–26181
10. Zhu G et al (2014) Review of bio-hydrogen production and new application in the pollution control via microbial electrolysis cell. Desalin Water Treat 52(28–30):5413–5421
11. Ismail GA, Sakai H (2022) Review on effect of different type of dyes on advanced oxidation processes (AOPs) for textile color removal. Chemosphere 291:132906
12. Adeyemo AA, Adeoye IO, Bello OS (2017) Adsorption of dyes using different types of clay: a review. Appl Water Sci 7:543–568
13. Manzoor J, Sharma M (2020) Impact of textile dyes on human health and environment. In: Impact of textile dyes on public health and the environment. IGI Global, pp 162–169
14. Yaseen D, Scholz M (2019) Textile dye wastewater characteristics and constituents of synthetic effluents: a critical review. Int J Environ Sci Technol 16:1193–1226
15. Gautam R et al (2023) Bio-hydrogen production through microbial electrolysis cell: structural components and influencing factors. Chem Eng J 455:140535
16. Cebecioglu, R., D. Akagunduz, and T. Catal, *Hydrogen production in single-chamber microbial electrolysis cells using Ponceau S dye.* 3 Biotech, 2021. **11**(1): p. 27.
17. Hou Y et al (2017) Accelerated azo dye degradation and concurrent hydrogen production in the single-chamber photocatalytic microbial electrolysis cell. Biores Technol 224:63–68
18. Sudha M et al (2014) Microbial degradation of azo dyes: a review. Int J Curr Microbiol App Sci 3(2):670–690
19. Oliveira JMS et al (2023) Microbial communities and metabolic pathways involved in reductive decolorization of an azo dye in a two-stage AD system. Chemosphere 310:136731
20. Li Y, Xie S, Yao J (2023) Singlet oxygen generation for selective oxidation of emerging pollutants in a flow-by electrochemical system based on natural air diffusion cathode. Environ Sci Pollut Res 30(7):17854–17864
21. Méndez-Martínez AJ et al (2012) Electrochemical reduction and oxidation pathways for reactive black 5 dye using nickel electrodes in divided and undivided cells. Electrochim Acta 59:140–149
22. Verma AK (2017) Treatment of textile wastewaters by electrocoagulation employing Fe-Al composite electrode. J Water Process Eng 20:168–172
23. Jayabalan T et al (2021) Influence of Nickel molybdate nanocatalyst for enhancing biohydrogen production in microbial electrolysis cell utilizing sugar industrial effluent. Biores Technol 320:124284
24. Zeng X, Borole AP, Pavlostathis SG (2015) Biotransformation of furanic and phenolic compounds with hydrogen gas production in a microbial electrolysis cell. Environ Sci Technol 49(22):13667–13675
25. Verea L et al (2014) Performance of a microbial electrolysis cell (MEC) for hydrogen production with a new process for the biofilm formation. Int J Hydrogen Energy 39(17):8938–8946
26. Peng H et al (2019) A strategy for enhancing anaerobic digestion of waste activated sludge: driving anodic oxidation by adding nitrate into microbial electrolysis cell. J Environ Sci 81:34–42
27. Hu Y et al (2021) Biofilm biology and engineering of Geobacter and Shewanella spp. for energy applications. Front Bioeng Biotechnol 9:786416
28. Liu D-F, Li W-W (2020) Potential-dependent extracellular electron transfer pathways of exoelectrogens. Curr Opin Chem Biol 59:140–146
29. Tang J et al (2018) Complete genome sequence of the dissimilatory azo reducing thermophilic bacterium Novibacillus thermophiles SG 1. J Biotechnol 284:6–10

30. Hemdan BA et al (2023) Bacterial community structure of electrogenic biofilm developed on modified graphite anode in microbial fuel cell. Sci Rep 13(1):1255
31. Bora A et al (2022) Microbial electrolysis cell (MEC): reactor configurations, recent advances and strategies in biohydrogen production. Fuel 328:125269
32. Zhao J, Li Y, Dong R (2021) Recent progress towards in-situ biogas upgrading technologies. Sci Total Environ 800:149667
33. Jafary T et al (2017) A comprehensive study on development of a biocathode for cleaner production of hydrogen in a microbial electrolysis cell. J Clean Prod 164:1135–1144
34. Then M et al (2021) Dynamic modeling of hydrogen production from photo-fermentation in microbial electrolysis cell using sago waste. In: IOP conference series: materials science and engineering. IOP Publishing
35. Ding C, Yang K-L, He J (2016) Biological and fermentative production of hydrogen. Handbook of biofuels production. Elsevier, pp 303–333
36. Bundhoo ZM (2017) Coupling dark fermentation with biochemical or bioelectrochemical systems for enhanced bio-energy production: a review. Int J Hydrogen Energy 42(43):26667–26686
37. Sun M et al (2008) An MEC-MFC-coupled system for biohydrogen production from acetate. Environ Sci Technol 42(21):8095–8100
38. Commault AS et al (2013) Influence of anode potentials on selection of geobacter strains in microbial electrolysis cells. Biores Technol 139:226–234
39. Kadier A et al (2020) Essential factors for performance improvement and the implementation of microbial electrolysis cells (MECs). In: Bioelectrochemical systems: vol 1 principles and processes, pp 139–168
40. Shao Q et al (2019) Effects of different substrates on microbial electrolysis cell (MEC) anodic membrane: biodiversity and hydrogen production performance. Water Sci Technol 79(6):1123–1133
41. Posadas-Hernández M et al (2023) Enhanced biohydrogen production in a membraneless single-chamber microbial electrolysis cell during high-strength wastewater treatment: effect of electrode materials and configurations. Int J Hydrogen Energy 48(2):495–513
42. Park S-G et al (2022) Addressing scale-up challenges and enhancement in performance of hydrogen-producing microbial electrolysis cell through electrode modifications. Energy Rep 8:2726–2746
43. Li X et al (2017) Novel bio-electro-Fenton technology for azo dye wastewater treatment using microbial reverse-electrodialysis electrolysis cell. Biores Technol 228:322–329
44. Srivastava P et al (2018) Degradation of azo dye in constructed wetland coupled with a microbial electrolysis cell. In: IWA international conference on wetland systems for water pollution control (16th: 2018).
45. Hou Y et al (2021) Dye wastewater treatment and hydrogen production in microbial electrolysis cells using MoS2-graphene oxide cathode: effects of dye concentration, co-substrate and buffer solution. Process Biochem 102:51–58
46. Huang W et al (2018) Enhancement of Congo red decolorization by membrane-free structure and bio-cathode in a microbial electrolysis cell. Electrochim Acta 260:196–203

Microbial Electrolysis Cell for Heavy Metal Removal from Wastewater

Growther Lali, V. Mahalakshmi, K. Sukirtha, and S. Sangeetha

Abstract Heavy metals find their applications in various industries like medical, agriculture, textile, cosmetic, etching and microfabrication, electroplating, tanneries, pharmaceutical and other technological industries. This wide application had resulted in presence or accumulation of these heavy metals in the environment. Though they are naturally occurring elements with a greater density and higher atomic weight, their toxicity to human is a major concern. These heavy metals are carcinogens and causes systemic damage to human. Release of these heavy metals in the wastewaters of such industries lead to water pollution. Wastewaters are now a major source of water due to water scarcity and population growth. In India, water demand from domestic and industry will be 85%. Demand of water will increase from 22 to 32% in 2025 and 2050, respectively. Sustainable clean water and sanitation is a major challenge across the globe. Recycling and reuse of wastewater can meet 50% of the water requirements for domestic and industrial applications. Without the reuse of treated wastewater, it is impossible to meet the demands of the present and future generations. Untreated wastewater leads to environmental pollution and human health hazard. Thus, heavy metal removal from industrial wastewaters plays a major role in achieving sustainability. The conventional treatment methodologies are insufficient in removing trace amounts of heavy metals in wastewaters, where they are present in trace amounts. Thus, innovative wastewater treatment methods are required to remove and recover these elements. Microbial Electrolysis Cells (MECs) are a promising technology to remove and recover heavy metals. These MECs use microorganisms either as single culture or a consortium of microorganisms. Microbial metabolism with electrodeposition and precipitation is the mechanisms used in MECs. Several factors influence the removal and recovery of heavy metals in such systems. This review summarizes the need for heavy metal removal, the types of MECs used for removal, their significance, efficacy and the challenges in implementing this technology at the large scale.

G. Lali (✉) · K. Sukirtha · S. Sangeetha
Department of Microbiology, Hindusthan College of Arts & Science, Coimbatore 641028, India
e-mail: laligrowther@hicas.ac.in

V. Mahalakshmi
Department of Microbiology, Madras Christian College, Chennai, India

A. A. Yaqoob and A. Ahmad (eds.), *Microbial Electrolysis Cell Technology*, Sustainable Materials and Technology, https://doi.org/10.1007/978-981-97-3356-9_14

Keywords Microbial electrolysis cell · Heavy metals · Removal · Biocathode · Electrogenic bacteria

1 Introduction

An adaptable and efficient bioelectrochemical technology for the treatment of wastewater is the microbial electrolysis cell (MEC). MECs have been applied in amalgamation of removal and recovery of heavy metals industrial water treatment [35]. In this process, microorganisms are the biocatalysts that convert the chemical energy in organic compounds to hydrogen [77]. Several microorganisms are reported to as electrogenic bacteria. *Aeromonas hydrophila, Escherichia coli, Thermincola* sp., *Saccharomyces, Geothrix fermentans, K. Pneumoniae, Gluconobacter oxydans, S. Putrefaciens, S. oneidensis* [14], *Rhodoferax ferrireducens, Xanthomonas, Pseudomonas, Prevotella* [42] and *E. cloacae* [37]. Microorganisms used in the MEC and their biocatalysis are based on substrates available in the industrial wastewater. This emerging technology plays a major role in recovery of products from waste using microorganisms and is the future of biorefineries. This technology is self-sustainable and eco-friendly.

2 Heavy Metals

Heavy metals are characterized by their high density and relatively high atomic mass. While they are typically found in relatively small concentrations, they are distributed throughout the Earth's crust. The definition of a heavy metal generally includes having at least 5 g/cm^3 density and this distinguishes them from lighter metals. Other definitions of heavy metals include metals with a greater atomic number and mass. However, these definitions are often indefinite and prone to inaccuracy. These varying definitions can also include non-metals, potentially encompassing up to 99 elements out of the 118 total elements in the periodic table. When strictly considering the density criterion, the majority of elements classified as "heavy metals" are found in columns 3–16 of periods 4–6, that include lanthanides, post-transition metals and transition metals [41].

2.1 Sources of Heavy Metal Pollution

The primary sources of these hazardous metals, such as HMs, are often attributed to human activities, particularly in industries like non-ferrous metallurgy, mining, electroplating, mineral processing, chemical industries and leather tanning industries. These activities lead to the release of these harmful elements into the environment.

Prolonged exposure to heavy metals can result in various health issues in both humans and other organisms [1]. The sources of heavy metals are depicted in Fig. 1.

The usage of pesticides and fertilizers, as well as weathering of rocks, were the main causes of metal pollution in Africa, accounting for a combined contribution of more than 56.7%. With rock weathering and mining accounting for 97.1% of the total, these were the primary sources for Asia. The primary metal sources in Europe and Asia contributed a combined 56.2% of the total metal content. The primary sources in North America were industry and mining, together with the use of pesticides and fertilizers, which together accounted for more than 90.4% of the total [83].

This historical awareness of heavy metal toxicity dates back to publications such as those by Moore and Hawkes in 1908, Lomholt in 1924, and Mudd and Anderson in 1942 [26]. Additionally, small watersheds are more susceptible to human activities and natural factors and have a greater impact on the downstream aquatic environment compared to larger watersheds. Heavy metal accumulation in the sediment of water reservoirs has posed threats to the aquatic ecosystem. Furthermore, through the food chain, heavy metals can enter the human body, posing risks to human. Consequently, there is a critical need for controlling and addressing heavy metal pollution in reservoirs, particularly in small basins that serve as essential water supply sources. To effectively protect aquatic ecosystems, it is essential to comprehend the pollution levels and sources of heavy metals in sediments [83].

Christophoridis et al. [20] investigated the impact of both human activities and natural processes on heavy metal input. They observed the concentration of heavy metal pollution in sediments was closely linked to the density of the human population in the area. The more densely populated an area, the higher the concentration of

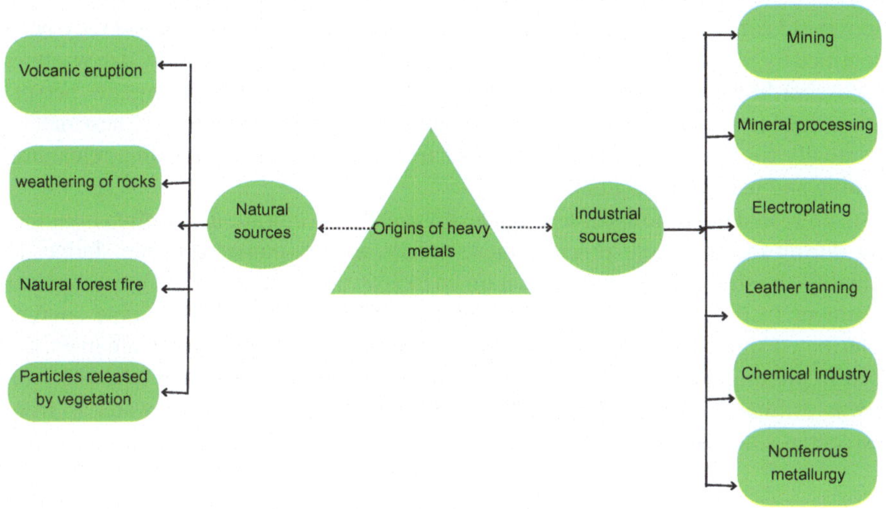

Fig. 1 Sources of heavy metals

heavy metal pollution in the sediments, highlighting the influence of human activities on heavy metal contamination. These studies shed light on the sources and factors influencing heavy metal pollution in sediment, emphasizing the role of both anthropogenic activities and natural processes [83]. The issue of heavy metal soil contamination due to wastewater irrigation is a widespread and serious problem in various countries, including India, China, Germany and France.

Heavy metal concentrations in river and lake waters exhibited variations based on their continent of origin. In general, Africa, Asia and South America showed higher concentrations, while Europe and North America displayed lower levels. For example, the average chromium (Cr) concentration followed this order from highest to lowest: South America > Africa > Asia > Europe > North America. Out of the 12 heavy metals analysed, North America exceeded the threshold levels set by WHO and USEPA standards for lead (Pb) and aluminium (Al), and in Europe, cadmium (Cd), zinc (Zn), nickel (Ni), aluminium (Al), manganese (Mn) and arsenic (As) surpassed these standards. However, in South America, Asia, and Africa, seven, nine and ten heavy metals, respectively, had concentrations that exceeded the thresholds defined by both standards.

2.2 *Heavy Metals in Household Wastewater*

Trace elements in domestic wastewater can come from a variety of sources, such as detergents, cosmetics, medications, toilet paper and other personal hygiene items. Food scraps should also not be disregarded, particularly in locations like Longyearbyen where they are ground up in sinks and disposed some of into the sewer system. Household emissions—especially those from food and decorative paints—have been identified by Sörme and Laderkvist [70] as the second-leading cause of Cd in wastewater. According to Moriyama et al. [51], tap water, food, cosmetics, detergents, perspiration and dust are just a few of the ways that metals can enter our homes. These metals eventually find their way into wastewater from tasks including cooking, cleaning, bathing and using the lavatory.

In untreated household wastewater, heavy metal concentrations are often minimal. But during the treatment process, these metals are accumulated in the sludge. In actuality, the sludge is thought to contain up to 96% of all metals added to wastewater treatment facilities. In a UK study, Comber and Gunn found that home sources contributed significantly to zinc (46.0%) and copper (64.2%), mostly from materials used in plumbing as well as nickel (25.7%), chromium (15.3%) and lead (16.9%). It was estimated that 16–49% of the total cadmium dose in wastewater is from residential sources.

2.3 Heavy Metals in Industrial Wastewater

Various industrial sectors produce wastewater streams containing heavy metals. Heavy metals are generated by industries involved in fermentation, electrical, electronics, textile, rayon, electrolysis, coating, mining, metal processing, tanning, petrochemicals, etching, paper manufacturing, electrolysis applications anodizing-cleaning and milling industries. Electrical manufacturing appliances, for instance, contribute to a significant amount of heavy metal waste, such as cadmium, lead and nickel. Arsenic-containing wastes are generated from wood processing industries. Inorganic pigments containing chromium compounds and cadmium sulphide are another source of heavy metal contamination. Industries that discharge substantial volumes of metal-contaminated wastewater, particularly those containing heavy metals like cadmium, chromium, copper, nickel, arsenic, lead, titanium, vanadium, platinum and zinc pose the highest level of hazard among chemical-intensive industries.

Additionally, heavy metals like nickel, vanadium and chromium can be found in conversion catalysts produced by the petroleum refining sector. Even photographic operations, particularly those involved in film production, result in wastewater with high concentrations of silver and ferrocyanide [31]. The metals listed above as well as others that are frequently found are cobalt, silver, iron, manganese, molybdenum, boron and antimony.

2.4 Heavy Metals-Environment and Health Hazards

Among the pollutants, in wastewaters, heavy metals are of particular concern due to the significant increase in the number of industrial activities and the release of these heavy metals into wastewater [7, 69]. Furthermore, due to their toxic nature, heavy metals can lead to various health hazards and systemic impacts in lungs, kidneys, urinary bladder, brain, reproductive organs, heart and the immune system [3, 5]. Therefore, it is of paramount importance to implement treatments that eliminate heavy metal concentrations in drinking water to ensure compliance with permissible levels [28, 44, 60, 74].

Heavy metals have the potential to be absorbed by aquatic fauna and flora due to their high solubility in aquatic environments. Once these metals are in the food chain, they can accumulate in significant concentrations within the human body. In cases where these metals are ingested in concentrations exceeding permissible limits, they can lead to severe health issues [8]. It is important to note that there is a correlation between their weight and toxicity. HMs also include metalloids like arsenic (As), which can be toxic even at low exposure levels. Cadmium (Cd), lead (Pb), chromium (Cr), arsenic (As), copper (Cu), mercury (Hg), zinc (Zn) and nickel (Ni), among other heavy metals, are recognized for their potential harm to the aquatic ecosystem.

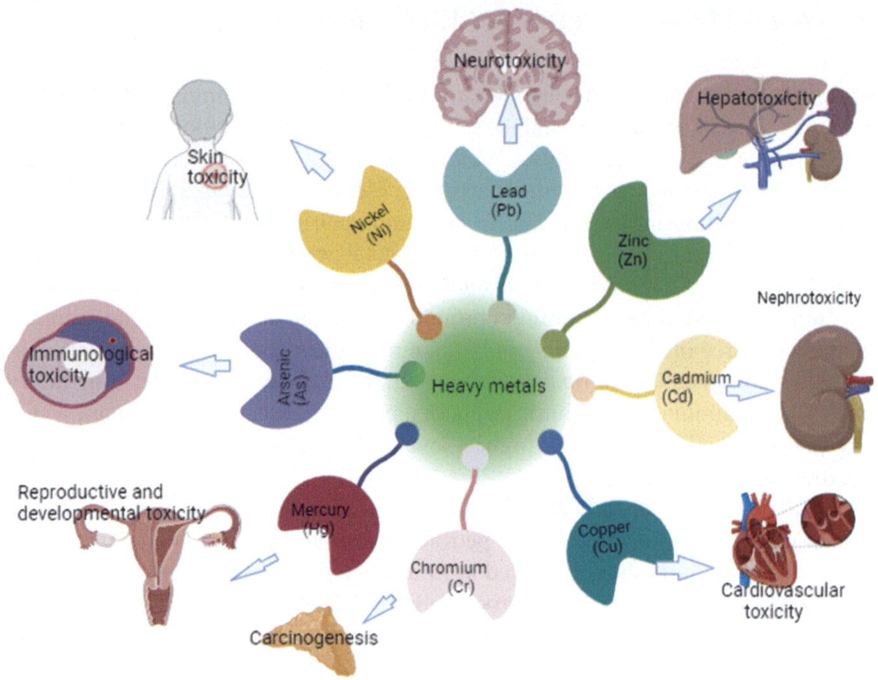

Fig. 2 Health hazards of heavy metals

They are considered both biologically non-essential and harmful. The health hazards of heavy metals are shown in Fig. 2.

Environmental contamination, particularly water pollution, has consistently been a matter of public concern. Wastewater contaminated with metals, originating from various human activities such as agriculture, industries (including mining, battery production, nuclear processes, textile dyeing, tanneries and more), domestic sewage and other sources, poses a continuous threat to both human health and the environment.

Degradation of the soil, which lowers agriculture yields, and destruction of marine habitats are two of the most direct effects on the environment. According to Sankaran et al. [66], plants absorb heavy metals from contaminated water, which can then enter the food chain and have detrimental effects on both people and animals. In human, because the body cannot metabolize certain heavy metals, they tend to accumulate in soft tissues and can be harmful to people's health [66].

Metal ions, such as lead, nickel, cadmium, and chromium, are naturally occurring elements found in the Earth's crust and are often associated with high levels of toxicity. Despite their toxic properties, these metals serve various purposes in consumer products and fundamental engineering processes. They are used in industries such as paper and pulp, leather tanning, plastic stabilization, photography, fertilizer production, pigment manufacturing, batteries, metal plating, mining, smelting, alloy production and more. The problem arises when wastewater from these industrial activities is released into the environment, causing pollution [54].

These heavy metal ions have the potential to combine with proteins to produce complex toxic compounds that can malfunction the kidneys, liver, spleen and other organs. Numerous diseases that are considered public nuisances and have an international impact include blood lead poisoning, Itai disease, Minamata disease and other heavy metal poisoning incidents [77].

3 Methods of Heavy Metal Removal

Ultrafiltration, coagulation, flocculation, adsorption, membrane filtration and ion exchange are examples of physical techniques. Chemical techniques include, among others, neutralization, solvent extraction, chemical precipitation and electrochemical treatment. Several researchers also try combining these approaches.

3.1 Conventional Methods-Physical Methods

3.1.1 Ultrafiltration (UF)

In ultrafiltration, suspended particles, heavy metals, and macromolecules are eliminated from an inorganic solution using a permeable membrane. The two primary variables in this process are the molecular weight of the molecule (1000–100,000 Da) and the pore size of the membrane (5–20 nm) [9]. There are two types of ultrafiltration techniques: Micellar-enhanced ultrafiltration (MEUF) and polymer-enhanced ultrafiltration (PEUF) [45].

3.1.2 Coagulation/Flocculation

Coagulation or flocculation are two methods for removing heavy metals from solutions. Coagulation, to put it briefly, is the process of reducing a particle's net surface charge in order to neutralize its charge and enable colloidal particles to stabilize through electrostatic repulsion. Coagulation needs larger-form dissolved organic materials and/or coagulants. Flocculation, which occurs when inorganic and dissolved organic polymers collide and interact with one another more, gradually combines the destabilized particles to expand their size. Filtration, flotation or straining are simple methods for separating the smaller particles that have aggregated into bigger ones. The two main types of metal coagulants that are frequently used are those that are based on iron and those that are based on aluminium [55]. A biopolymer with easy biodegradability, chitosan functions as a flocculant and coagulant [79]. The main disadvantages of this method include sludge production, chemical application and the transfer of hazardous substances into the solid phase.

3.1.3 Adsorption

Adsorption forms a coating of adsorbate, or metal ions, on the surface of adsorbents. Adsorption is a reversible process under some conditions [13]. Charged pollutants typically adsorb on differently charged adsorbents due to electrostatic attraction [15]. Physical adsorption and chemisorption are the two main types of adsorptions. Adhesion is known as physical adsorption and is caused by the non-specific (i.e. unaffected by the material's properties) Vander Waals force, in contrast, chemisorption occurs chemicals combine to form a strong, attractive connection [72].

3.1.4 Membrane Filtration

A range of porous and thin-film composite membranes are used in wastewater treatment. Membrane processes are influenced by various driving factors: low pressure, high pressure and osmotic pressure. Reverse osmosis, ultrafiltration and nanofiltration are three membrane filtering techniques that can be used to remove heavy metals from wastewater, depending on the size of the particle that can be retained [9].

3.1.5 Ion Exchange

Ion exchange transfers soluble ions from the liquid phase into the solid phase economically, using simple, low-cost materials and an easy-to-use operation. It has been demonstrated to be a successful technique for removing heavy metals from aqueous solutions, particularly when the metal concentration is low. The aqueous phase's pH affects the procedure [32].

4 Chemical Methods for Heavy Metal Removal

4.1 Neutralization

Precipitation-neutralization has been utilized to extract metals from sediments. All heavy metals from the acid mine wastewater have been eliminated by neutralizing it with apatite, and microelements, and so forth have been collected in the phosphorus-rich precipitate. To remove heavy metals by precipitation, alkaline reagents such as lime, ferrous salts, limestone and other compounds like Mg $(OH)_2$, $MgCO_3$, $BaCl_2$ and $CaCl_2$ are frequently used to raise the pH of acidic water to basic. Several reagents have been tested in the acid effluent neutralization step, including kiln dust, soluble silicate, fly ash and dolomite, lime, limestone, FMBO (1:1)-diatomite1, iron and manganese solutions, Portland cement, phosphate and magnesium oxide [57]. When calcium oxide (CaO) is compared to other limes, it shows that it is the most cost-effective, versatile and efficient way to neutralize cations that are in the form of hydroxides or complex anions. Furthermore, it is not affected by changes in temperature.

4.2 Solvent Extraction

In hydrometallurgy, solvent extraction is one of the most commonly used and widely accepted techniques. The hydrometallurgical boron processing, nuclear fuel reprocessing, decontamination of wet-process phosphoric acid, tungsten, niobium, hafnium, indium, uranium, vanadium, cobalt, zirconium, molybdenum and platinum group metals are all found effective applications of it. Currently, there are a number of stable extractants with improved metal ion selectivity that come from ecologically friendly production paths [27, 76].

4.3 Chemical Precipitation

Chemical precipitation is most commonly used method due to its efficiency, low cost, process control and adaptability at a wide range of temperatures. The conversion of heavy metal ions to hydroxides, carbonates and sulphides occurs in a pH-dependent manner [18]. These compounds can be physically eliminated by filtration, flotation or sedimentation.

It can also be used to soften water, remove phosphate from wastewaters and other effluent, remove oil and grease. The treated water is then reused after the precipitate that was formed is separated from the solution in the technical system by sedimentation or filtration [30]. Sulphide and hydroxide precipitation are two common types of chemical precipitation [10, 46].

4.4 Electrochemical Treatment

Electrolysis: One technique to remove metals from wastewater streams is electrolytic recovery. In this procedure, a current is run through an aqueous metal-bearing solution that has an insoluble anode and a cathode plate using electricity.

5 Biological Methods

Biological methods for the removal of contaminants from wastewater are used in the biological removal of heavy metals from wastewater. Solids in the solution settle in part because of the action of microorganisms in this process. Wastewater is often treated using stabilization ponds, trickling filters and activated sludge. The most popular alternative is activated sludge, which breaks down organic material using microorganisms during the treatment process with agitation and aeration before allowing solids to settle out. Trickling through microorganisms can develop in filters that have beds of coarse media, like stones or plastic, 3–10 ft deep. Wastewater is collected and subjected to sedimentation after trickling filters empty at the bottom. For a variety of wastewater types, stabilization ponds or lagoons are an inexpensive, sluggish and rather ineffective biological approach. Algae, bacteria, sunshine and oxygen interact to support them [8].

5.1 Biomaterials as Metal Biosorbents

Algae are efficient heavy metal adsorbents; examples include the bioadsorption of Cu^{2+} and Zn^{2+} utilizing dried sea green macroalgae, Chaetomorphalinum [4]; and the biosorption of Cu^{2+}, Zn^{2+}, Cd^{2+}, and Pb^{2+} by *Caulerpa lentillifera* [58]. These days, it is thought that microbes are quite effective at removing metal ions. *Bacillus cereus, Escherichia coli, Pseudomonas aeruginosa* and other microorganisms to bind heavy metals have all been investigated [6, 17, 62]. Fungi are another commonly used biosorbent because they are simple to grow, yield a high amount of biomass quickly, and are easily manipulable genetically and morphologically [39, 40]. The most often utilized fungal species as biosorbents include *Saccharomyces cerevisiae* [49], *Aspergillus niger* [2], *Lentinus edodes*, and others.

6 Microbial Electrolysis Cell and Heavy Metal Removal

Microbial electrolysis cells (MECs) ability to alter the wastewater treatment process's energy balance have been highlighted [23, 29]. By decomposing organic material in the wastewater and releasing protons and electrons to the anode in the process, electrochemically active bacteria form a biofilm on an electrode in a MEC. According to Ditzig et al. [22], an electrical current can be produced in a circuit by releasing protons and applying an extra potential at the cathode. This current can then be used to produce energy or products like hydrogen gas.

Wastewater is becoming more widely recognized as a "misplaced resource" that may be used to produce energy and high-quality goods. Energy treatment is typically required for proper wastewater management, which has a major impact on climate change. Biological anaerobic treatment technologies have been the preferred choice for recovering the majority of this energy due to its stability and diversity. MECs are preferable to Microbial Fuel Cells (MFCs) in a number of ways. However, the final product produced by a MEC—hydrogen, methane, ethanol, hydrogen peroxide, etc.—has a big influence on its efficiency when it works, and because of its enormous energy output, there has been conjecture that it will be the power source of the future.

6.1 Mechanism of Heavy Metal Removal by MEC

MEC removal of Pb^{2+} from industrial effluent has been thoroughly studied because Pb^{2+}, even at very low concentrations, is extremely harmful to the nervous and cardiovascular systems. Research indicates that nearly 40 mg/L of Pb^{2+} could be removed from wastewater by a MEC operating in an acidic environment and with the negative electrode flooded with oxygen to act as an electron acceptor [12]. The process was determined to be the electro deposition of reduced Pb^{2+} at the cathode at a rate of 0.53 mg/L/h using the XRD technique. Additionally, an attempt was made to couple MFC with MEC's simultaneous metal shifting in order to generate electricity. They saw a quicker removal of Pb^{2+}, which went from a starting concentration of 100–2 mg/L in just three days. On the other hand, regardless of the voltage potential applied, shifting of 77–95% of Pb^{2+} was accomplished with an open circuit in two days.

Zhang et al. [82] noted that precipitation at the negative electrode is responsible for the Cd^{2+} shifting from the industrial effluent, and that the process is dependent on the concentration of hydrogen ions and feedstock. Therefore, it is always important to ensure that there is a higher concentration of hydrogen ions than the cathode in order to maximize the removal of Cd^{2+} as precipitates. If not, the precipitates would dissolve and raise the levels of cadmium. The volume of the feedstock allowed for the high concentration of hydrogen ions at the cathode to be reached. The team also observed that Cd^{2+} take off from the wastewater would not be impacted by the feedstock content if the voltage potential at the negative electrode was favourable.

The denitrifying bacteria affixed as a mat on the cathode's exterior showed promise in removing copper in MEC. Hydrogen was released when current flowed through the reactor's negative electrode, reducing copper. Denitrification is favoured by this hydrogen's subsequent reaction with acetate Watanabe et al. [75].

$$CH_3COO^- + 4H_2O + 4Cu^{2+}2HCO^-{}_3 + 9H^+ + 8e \rightarrow 4Cu$$
$$H^2 + Cu^{2+}2H^+ \rightarrow Cu$$

At the beginning of the system's operation, acetate served as an electron donor; towards the end, hydrogen was used, which led to copper reduction and hydrogen oxidation.

In order to remove zinc, a MEC was built that contains acidophilic sulphate-reducing bacteria as a stimulant in the negative electrode, which catalyses the subsequent reaction.

$$SO^{2-}{}_4 + 8e^- + 8H \rightarrow S^{2-} + 4H_2O$$

$$Zn^{2+} + S^{2-} \rightarrow ZnS$$

Teng et al. [71] reported that these reactors using acidic effluents were successful in removing nearly 99% of Zn along with sulphates. The sulphate-reducing bacteria produce hydrogen sulphide, which in turn reduces Zn as sulphide.

6.2 Chromium

There are two oxidation states of chromium (Cr) that are present in the environment: hexavalent Cr (VI) and trivalent Cr (III) [56]. According to research, Cr (VI) is highly carcinogenic, mutagenic and toxic [33, 63]. Cr (VI) is one of the most concerning environmental contaminants due to its wide range of applications, including tanneries and metal plating; therefore, remediation of this contamination is of utmost importance [59, 63].

6.3 Silver

Numerous industries' effluents contain silver, a naturally occurring precious metal [11]. Due to its many advantageous qualities, including its strong electrical and heat conductivity, malleability, ductility and light-reflectiveness, silver is used in the jewellery, electronics and photography industries [53]. As natural resources become scarcer, recovering silver from industrial water is important for the environment as well as the economy.

Initial concentrations ranging from 50 to 200 ppm were used in a cost-effective MFC to obtain silver recovery efficiencies as high as 99.91–98.26%, and after 8 h of operation [19]. In a different investigation, Tao et al. reported using acetate as an electron donor to cathodically reduce Ag^+ ions as well as $[AgS_2O_3]^-$ to Ag^0. Ag(I) was removed to 95% with power densities of 109 and 35 mW/m^{-3} for Ag^+ and $[AgS_2O_3]^-$, respectively.

6.4 Nickel

Qin et al. [61] investigated the effects of a number of factors on the performance of MEC, including pH, applied voltage and starting concentration. They came to the conclusion that MEC removed nickel several times more effectively than other techniques. A steady removal efficacy was noted as the concentration increased. They had demonstrated the potential of MEC. The removal efficacy increased from 51 ± 4.6% to 67 ± 5.3% with applied voltages between 0.5 and 1.1 V. However, because Ni^{2+} ions cannot be detected in the anodic chamber, the removal efficacy is only 9 ± 0.1% at 0.9 V and 1000 mg/L (pH 5.0) concentration. In microbial electrolysis cells, the bacterium *Curvibacter* has been found to be an effective organism for removing arsenic.

7 Construction of MEC

The selection of appropriate materials that can fulfil the requirements of the active elements (i.e. the anode and cathode) and function in the reactor's unique environmental conditions will be crucial to the success of future MEC designs. These materials should, at the very least, meet the following specifications [47]:

(i) The simple circulation of charges (ionic and electronic) to increase reaction rates and maximize the energy.
(ii) Biological, chemical and physical stability.
(iii) Economically viable: this is a crucial prerequisite for large-scale applications and commercial prototypes, but it is not a requirement for laboratory work. Reactor layouts, manufacturing procedures and the use of less expensive materials must all be optimized to reduce capital and operating costs.
(iv) High specific surface area: It is important to maximize the electrodes' surface area to maximize reaction rates.

7.1 Anode Material

Carbon-based materials are the most common electrode material because they fulfil many of the requirements [43]. In comparison with metals, they too have comparatively low conductivity, a shortcoming that can be made up for by using metallic current collectors. Even though they have a higher conductivity than carbon-based materials, however, their efficacy tends to be poor, most likely due to their lesser surface area and other qualities that make them less conducive to the formation of biofilms [25].

7.2 Cathode Material

The majority of carbon-based anode materials can also be used as cathode materials, however they usually need to be modified in some way. This reaction is the same as what happens in traditional electrolysis systems, which typically employ noble metals (such palladium and platinum) because of their exceptional catalytic activity and stability. These are relatively expensive metals, though, and their usage in low productivity, low added value tasks like wastewater treatment might not be economically justified. Because they are inexpensive, stable, easy to synthesis and have low over-potentials, materials based on nickel, molybdenum and other transition metals have shown promise as viable substitutes for Pt-based cathodes [21, 36, 64, 67, 80, 81].

8 Prospects of Heavy Metal Removal from Wastewater Using MEC

The presence of recalcitrant chemicals including heavy metals in industrial wastewaters and the treatment methodologies is unmanageable [34]. Removal of heavy metals through conventional treatment methodologies results in hazardous chemicals and environmental pollution. MEC is a green technology, which is used for removal of recalcitrant pollutants, heavy metal removal, recovery and as biosensors. MECs are used for production of hydrogen, methane, organic chemicals, to reduce greenhouse emissions and to recycle organic wastes and protects the environment [73]. High hydrogen yield, overcoming thermodynamic limitations, production from wide range of organic matters at relatively mild conditions are the advantages in using MEC's. This MECs will be the sustainable technology using electrogenic microbes. As the electro-catalyst, biocatalyst microorganisms develop an electroactive biofilm on the anode surface by themselves [65]. These electroactive microorganisms have the ability to remove harmful substances from a variety of wastewaters.

Energy generation is an advantage of using Microbial Electrolysis Cell technology, when compared to the conventional methods. Studies on heavy metal removal or recovery is still in its initial stage and Qin et al. [61] had reported 99% removal of Nickel with an initial concentration of 50 mg/L, while Luo et al. [48] had reported 97%, when the initial concentration was 5–9 mM [16]. Potentially toxic heavy metals like cadmium, lead and nickel removal have been reported with different substrates, though the reduction potential of these metals is higher than the voltage generated.

9 Challenges in Using MEC for Heavy Metal Removal

The incapacity of MFCs to treat all metal ions must be their biggest problem [52]. Heavy metals in the aquatic ecosystem lead to bioaccumulation and biomagnification. Studies have shown removal of heavy metals by cathodic reduction, biosorption or precipitation. Since heavy metals are present in very lower concentrations in industrial wastewaters, studies using such lower concentration are warranted. The complexity of the wastewater to be treated before effluent release is a challenge.

Use of a single microorganism in the MECs may not be feasible for industrial wastewaters, as the substrates vary and sequential bioremediation of such chemicals becomes impossible. When a consortium of microorganisms is used, the interaction of the microorganisms in the wastewaters needs to be studied. The inhibition of one organism by another organism or the product formed by one organism may inhibit the other and product formation may be inhibited. Therefore, for an effective MEC, it is essential to recognize microorganisms and their interactions with one another to ensure that competition between microbial species is not affecting substrate utilization or product synthesis. Consequently, it is essential to have a deep understanding of the microorganisms and the behaviour that is related to them [38]. The methodologies to analyse microbial community are expensive.

Xu et al. [78] examined the effects of a number of heavy metals, are toxic to the microorganism *S. oneidens* is MR-1, that has been used. The construction of the reactor set up, the substrates and the materials used affects the process. Thus, defining the process with selected cathode, anode materials to the appropriate industrial waste materials and treating all metal ions from wastewaters is another major challenge.

The architecture of the chamber, the design of the anode, cathode varies with the organisms inoculated, their survival and the ability to donate and accept electrons affect the process parameters. Studies on the process control in large-scale operations are very fewer. This is due to the difficulty in controlling the process parameters like temperature, pH and other inhibitors [24]. Refining the process, scaled up field trials and maximizing the efficacy is a challenge.

Recovery of heavy metals has been studied only at the lab scale and field level studies are required [68]. This scaling up process raises economic challenges like the components of the electrode and design of the reactor [50]. To accomplish this goal, a multidisciplinary combined effort of microbiologists, biochemists, engineers, electrochemists is required [65].

10 Conclusion and Future Perspectives

Thus, it is evident that heavy metals in wastewater cause a serious hazard to the environment and hence need to be removed during treatment and before disposal. The traditional methods of treating industrial effluents proved to be insufficient. Though MFC is being deployed in the process of heavy metals removal, yet they have their own limitations. Heavy metals such as zinc, copper, chromium, cobalt, silver, boron, vanadium, mercury and selenium with high redox potential traverse electrons from anode to cathode thus reducing them at the cathodic electrode surface of MFC. However, metals like Pb^{2+}, Cd^{2+}, Zn^{2+} and Ni^{2+} with negative reduction potential cannot be reduced at the negative electrode, since they won't accept electrons. A hopeful outlook is MEC to remove these metals with negative reduction potential and its necessity for the external energy curtails its efficacy. Integrated MFC-MEC design is a self-sustainable option, where needed energy is provided from MFC; thereby multiple heavy metals are removed during the operation of the system. Even though heavy metal removal is high in integrated systems, the energy generation is low, if modification is done for more energy production, the reactors are not economical. Hence, studies need to be done in scaling up the reactor design that is economical feasible. The biggest drawback in implementing this system for industrial operations is that so far researches have been conducted with simulator wastewater and not as real time monitoring with industrial effluents. Therefore, pilot studies with factual industrial wastewater need to be conducted, for heavy metals removal from these effluents is tougher than with synthetic wastewater. Future researches on improved reactor design with potential electrode material operating with factual industrial wastewater, may add value to MEC technology to be state-of-the-art technology for sustainable industrial applications.

References

1. Abd Elnabi MK, Elkaliny NE, Elyazied MM, Azab SH, Elkhalifa SA, Elmasry S, Mahmoud YAG et al (2023) Toxicity of heavy metals and recent advances in their removal: a review. Toxics 11(7):580
2. Acosta-Rodríguez I, Cárdenas-González JF, Rodríguez Pérez AS, Oviedo JT, Martínez-Juárez VM (2018) Bioremoval of different heavy metals by the resistant fungal strain *Aspergillus niger*. In *Bioinorganic chemistry and applications*
3. Ahmad SZN, Salleh WNW, Ismail AF, Yusof N, Yusop MZM, Aziz F (2020) Adsorptive removal of heavy metal ions using graphene-based nanomaterials: toxicity, roles of functional groups and mechanisms. Chemosphere 248:126008
4. Ajjabi LC, Chouba L (2009) Biosorption of Cu^{2+} and Zn^{2+} from aqueous solutions by dried marine green macroalga *Chaetomorpha linum*. J Environ Manage 90(11):3485–3489
5. Ali H, Khan E, Ilahi I (2019) Environmental chemistry and ecotoxicology of hazardous heavy metals: environmental persistence, toxicity, and bioaccumulation. J Chem
6. Arivalagan P, Singaraj D, Haridass V, Kaliannan T (2018) Removal of cadmium from aqueous solution by batch studies using Bacillus cereus. Ecol Eng 71:728–735
7. Awad M, El-Desoky MA, Ghallab A, Kubes J, Abdel-Mawly SE, Danish S, El Sabagh A et al (2021) Ornamental plant efficiency for heavy metals phytoextraction from contaminated soils amended with organic materials. Molecules 26(11):3360
8. Babel S, Kurniawan TA (2003) Low-cost adsorbents for heavy metals uptake from contaminated water: a review. J Hazard Mater 97(1–3):219–243
9. Barakat MA (2011) New trends in removing heavy metals from industrial wastewater. Arab J Chem 4(4):361–377
10. Bejan D, Bunce NJ (2015) Acid mine drainage: electrochemical approaches to prevention and remediation of acidity and toxic metals. J Appl Electrochem 45:1239–1254

11. Birloaga I, Vegliò F (2018) Overview on hydrometallurgical procedures for silver recovery from various wastes. J Environ Chem Eng 6(2):2932–2938
12. Bo T, Zhang L, Zhu X, He X, Tao Y, Zhang J (2014) Lead ions removal from aqueous solution in a novel bioelectrochemical system with a stainless steel cathode. RSC Adv 4(77):41135–41140
13. Burakov AE, Galunin EV, Burakova IV, Kucherova AE, Agarwal S, Tkachev AG, Gupta VK (2018) Adsorption of heavy metals on conventional and nanostructured materials for wastewater treatment purposes: a review. Ecotoxicol Environ Saf 148:702–712
14. Chabert N, Ali OA, Achouak W (2015) All ecosystems potentially host electrogenic bacteria. Bioelectrochemistry 106:88–96
15. Chai WS, Cheun JY, Kumar PS, Mubashir M, Majeed Z, Banat F, Ho S-H, Show PL (2021) A review on conventional and novel materials towards heavy metal adsorption in wastewater treatment application. J Clean Prod 296:126589
16. Chakraborty I, Sathe SM, Dubey BK, Ghangrekar MM (2020) Waste-derived biochar: applications and future perspective in microbial fuel cells. Biores Technol 312:123587
17. Chellaiah ER (2018) Cadmium (heavy metals) bioremediation by Pseudomonas aeruginosa: a minireview. Appl Water Sci 8(6):154
18. Chen T, Liu X, Li X, Zhao K, Zhang J, Xu J, Dahlgren RA et al (2009) Heavy metal sources identification and sampling uncertainty analysis in a field-scale vegetable soil of Hangzhou, China. Environ Pollut 157(3):1003–1010
19. Choi C, Cui Y (2012) Recovery of silver from wastewater coupled with power generation using a microbial fuel cell. Biores Technol 107:522–525
20. Christophoridis C, Bourliva A, Evgenakis E, Papadopoulou L, Fytianos K (2019) Effects of anthropogenic activities on the levels of heavy metals in marine surface sediments of the Thessaloniki Bay, Northern Greece: spatial distribution, sources and contamination assessment. Microchem J 149:104001
21. De Silva ML, Bergel A, Féron D, Basséguy R (2010) Hydrogen production by electrolysis of a phosphate solution on a stainless steel cathode. Int J Hydrog Energy 35:8561–8568
22. Ditzig J, Liu H, Logan BE (2007) Production of hydrogen from domestic wastewater using a bioelectrochemically assisted microbial reactor (BEAMR). Int J Hydrogen Energy 32(13):2296–2304
23. Escapa A, Mateos R, Martínez EJ, Blanes J (2016) Microbial electrolysis cells: an emerging technology for wastewater treatment and energy recovery. From laboratory to pilot plant and beyond. Renew Sustain Energy Rev 55:942–956
24. Ezziat L, Elabed A, Ibnsouda S, El Abed S (2019) Challenges of microbial fuel cell architecture on heavy metal recovery and removal from wastewater. Front Energy Res 7:1
25. Fan C, Yang C (2014) Yellow/orange emissive heavy-metal complexes as phosphors in monochromatic and white organic light-emitting devices. Chem Soc Rev 43(17):6439–6469
26. Fei Y, Hu YH (2023) Recent progress in removal of heavy metals from wastewater: a comprehensive review. Chemosphere 139077
27. Flett DS (2005) Solvent extraction in hydrometallurgy: the role of organophosphorus extractants. J Organomet Chem 690(10):2426–2438
28. Flora SJS, Mittal M, Mehta A (2008) Heavy metal induced oxidative stress & its possible reversal by chelation therapy. Indian J Med Res 128(4):501–523
29. Foley J, De Haas D, Hartley K, Lant P (2010) Comprehensive life cycle inventories of alternative wastewater treatment systems. Water Res 44(5):1654–1666
30. Fu F, Wang Q (2011) Removal of heavy metal ions from wastewaters: a review. J Environ Manage 92(3):407–418
31. Gunatilake SK (2015) Methods of removing heavy metals from industrial wastewater. Methods 1(1):14
32. Hamdaoui O (2009) Removal of copper (II) from aqueous phase by Purolite C100-MB cation exchange resin in fixed bed columns: modeling. J Hazard Mater 161(2–3):737–746
33. Hamilton JA, Miller JM (2018) From transects to transcripts: teasing apart the architecture of reproductive isolation

34. Haris M, Shakeel A, Hussain T, Ahmad G, Ansari MS, Khan AA (2021) New trends in removing heavy metals from industrial wastewater through microbes. In: Removal of emerging contaminants through microbial processes, pp 183–205
35. Janani R, Gurunathan B, Sivakumar K, Varjani S, Ngo HH, Gnansounou E (2022) Advancements in heavy metals removal from effluents employing nano-adsorbents: way towards cleaner production. Environ Res 203:111815
36. Jeremiasse AW, Bergsma J, Kleijn JM, Saakes M, Buisman CJN, Cohen Stuart M et al (2011) Performance of metal alloys as hydrogen evolution reaction catalysts in a microbial electrolysis cell. Int J Hydrog Energy 36:10482–10489
37. Kadier A, Jain P, Lai B, Kalil MS, Kondaveeti S, Fahad Saud Alabbosh K, Mohanakrishna G et al (2020) Biorefinery perspectives of microbial electrolysis cells (MECs) for hydrogen and valuable chemicals production through wastewater treatment
38. Kadir AA, Hassan MIH, Salim NSA, Sarani NA, Ahmad S, Rahmat NAI (2018) Stabilization of heavy metals in fired clay brick incorporated with wastewater treatment plant sludge: leaching analysis. J Phys Conf Ser 995(1):012071
39. Karim A, Gerliani N, Aïder M (2020) Kluyveromyces marxianus: An emerging yeast cell factory for applications in food and biotechnology. Int J Food Microbiol 333:108818
40. Karim A, Islam MA, Khalid ZB, Yousuf A, Khan MMR, Faizal CKM (2021) Microbial lipid accumulation through bioremediation of palm oil mill effluent using a yeast-bacteria co-culture. Renew Energy 176:106–114
41. Koller M, Saleh HM (2018) Introductory chapter: Introducing heavy metals. Heavy Metals 1:3–11
42. Krishna KV, Mohan SV (2016) Selective enrichment of electrogenic bacteria for fuel cell application: enumerating microbial dynamics using MiSeq platform. Biores Technol 213:146–154
43. Kumar CS, Jaikumar M, Robin RS, Karthikeyan P, Kumar CS (2013) Heavy metal concentration of sea water and marine organisms in Ennore Creek, Southeast Coast of India. J Toxicol Health 103:192–201
44. Kumar M, Puri A (2012) A review of permissible limits of drinking water. Indian J Occup Environ Med 16(1):40
45. Landaburu-Aguirre J, García V, Pongrácz E, Keiski RL (2009) The removal of zinc from synthetic wastewaters by micellar-enhanced ultrafiltration: statistical design of experiments. Desalination 240(1–3):262–269
46. Lewis AE (2010) Review of metal sulphide precipitation. Hydrometallurgy 104(2):222–234
47. Logan B, Call D, Merrill M, Cheng S (2010) U.S. patent application no. 12/649,945
48. Luo ZB, Wu C, Zhang C, Li H, Lipka U, Polle A (2014) The role of ectomycorrhizas in heavy metal stress tolerance of host plants. Environ Exp Bot 108:47–62
49. Massoud R, Hadiani MR, Hamzehlou P, Khosravi-Darani K (2019) Bioremediation of heavy metals in food industry: application of Saccharomyces cerevisiae. Electron J Biotechnol 37:56–60
50. Mishra S, Bharagava RN, More N, Yadav A, Zainith S, Mani S, Chowdhary P (2019) Heavy metal contamination: an alarming threat to environment and human health. In: Environmental biotechnology: for sustainable future, pp 103–125
51. Moriyama K, Mori T, Arayashiki H, Saitot H, Chino M (1988) The amount of heavy metals derived from domestic wastewater. In: Water pollution research and control Brighton. Pergamon, pp 1913–1916
52. Nancharaiah YV, Mohan SV, Lens PNL (2015) Metals removal and recovery in bioelectrochemical systems: a review. Biores Technol 195:102–114
53. Nawaz T, Sengupta S (2017) Silver recovery from greywater: role of competing cations and regeneration. Sep Purif Technol 176:145–158
54. Odumbe E, Murunga S, Ndiiri J (2023) Heavy metals in wastewater effluent: causes, effects and removal technologies
55. Olafisoye OB, Adefioye T, Osibote OA (2013) Heavy metals contamination of water, soil, and plants around an electronic waste dumpsite. Pol J Environ Stud 22(5)

56. Oliveira, H. (2012). Chromium as an environmental pollutant: insights on induced plant toxicity. J Bot
57. Papassiopi N, Virčiková E, Nenov V, Kontopoulos A, Molnár L (1996) Removal and fixation of arsenic in the form of ferric arsenates. Three parallel experimental studies. Hydrometallurgy 41(2–3):243–253
58. Pavasant P, Apiratikul R, Sungkhum V, Suthiparinyanont P, Wattanachira S, Marhaba TF (2006) Biosorption of Cu^{2+}, Cd^{2+}, Pb^{2+}, and Zn^{2+} using dried marine green macroalga Caulerpa lentillifera. Biores Technol 97(18):2321–2329
59. Peng W, Li X, Xiao S, Fan W (2018) Review of remediation technologies for sediments contaminated by heavy metals. J Soils Sediments 18:1701–1719
60. Qasem NA, Mohammed RH, Lawal DU (2021) Removal of heavy metal ions from wastewater: a comprehensive and critical review. Npj Clean Water 4(1):36
61. Qin B, Luo H, Liu G, Zhang R, Chen S, Hou Y, Luo Y (2012) Nickel ion removal from wastewater using the microbial electrolysis cell. Biores Technol 121:458–461
62. Quintela FM, Lima GP, Silveira ML, Costa PG, Bianchini A, Loebmann D, Martins SE (2019) High arsenic and low lead concentrations in fish and reptiles from Taim wetlands, a Ramsar site in southern Brazil. Sci Total Environ 660:1004–1014
63. Renitta J, Pamela J, Anoop Kumar Y, Desai N (2018) Biosorption and biotransformation of hexavalent chromium [Cr(VI)]: a comprehensive review. Chemosphere 207:255–266. https://doi.org/10.1016/j.chemosphere.2018.05.050
64. Ribot-Llobet E, Nam J, Tokash JC, Guisasola A, Logan BE (2013) Assessment of four different cathode materials at different initial pHs using unbuffered catholytes in microbial electrolysis cells. Int J Hydrog Energy 38:2951–2956
65. Rousseau R, Etcheverry L, Roubaud E, Basséguy R, Délia ML, Bergel A (2020) Microbial electrolysis cell (MEC): strengths, weaknesses and research needs from electrochemical engineering standpoint. Appl Energy 257:113938
66. Sankaran R, Show PL, Ooi CW, Ling TC, Shu-Jen C, Chen SY, Chang YK (2020) Feasibility assessment of removal of heavy metals and soluble microbial products from aqueous solutions using eggshell wastes. Clean Technol Environ Policy 22:773–786
67. Selembo PA, Merrill MD, Logan BE (2009) The use of stainless steel and nickel alloys as low-cost cathodes in microbial electrolysis cells. J Power Sources 190:271–278
68. Shah V, Daverey A (2021) Effects of sophorolipids augmentation on the plant growth and phytoremediation of heavy metal contaminated soil. J Clean Prod 280:124406
69. Singh M, Thind PS, John S (2018) Health risk assessment of the workers exposed to the heavy metals in e-waste recycling sites of Chandigarh and Ludhiana, Punjab, India. Chemosphere 203:426–433
70. Sörme L, Lagerkvist RJSOTTE (2002) Sources of heavy metals in urban wastewater in Stockholm. Sci Total Environ 298(1–3):131–145
71. Teng W, Liu G, Luo H, Zhang R, Xiang Y (2016) Simultaneous sulfate and zinc removal from acid wastewater using an acidophilicand autotrophic biocathode. J Hazard Mater 304:159–165
72. Tripathi A, Ranjan MR (2015) Heavy metal removal from wastewater using low cost adsorbents. J Bioremed Biodeg 6(6):315
73. Varanasi JL, Veerubhotla R, Pandit S, Das D (2019) Biohydrogen production using microbial electrolysis cell: recent advances and future prospects. Microb Electrochem Technol 843–869
74. Velarde L, Nabavi MS, Escalera E, Antti ML, Akhtar F (2023) Adsorption of heavy metals on natural zeolites: a review. Chemosphere 138508
75. Watanabe T, Motoyama H, Kuroda M (2001) Denitrification and neutralization treatment by direct feeding of an acidic wastewater containing copper ion and high-strength nitrate to a bio-electrochemical reactor process. Water Res 35(17):4102–4110. https://doi.org/10.1016/S0043-1354(01)00158-0
76. Wilson AM, Bailey PJ, Tasker PA, Turkington JR, Grant RA, Love JB (2014) Solvent extraction: the coordination chemistry behind extractive metallurgy. Chem Soc Rev 43(1):123–134
77. Wu X, Xie W, Ye J, Sun D, Ohnuki T, Li M, Li D et al (2023) Progress in heavy metals-containing wastewater treatment via microbial electrolysis cell: a review. J Water Process Eng 55:104228

78. Xu P, Sun CX, Ye XZ, Xiao WD, Zhang Q, Wang Q (2016) The effect of biochar and crop straws on heavy metal bioavailability and plant accumulation in a Cd and Pb polluted soil. Ecotoxicol Environ Saf 132:94–100
79. Yang TC, Zall RR (1984) Absorption of metals by natural polymers generated from seafood processing wastes. Ind Eng Chem Prod Res Dev 23(1):168–172
80. Yuan X, Zhang L, Li J, Wang C, Ji J (2014) Sediment properties and heavy metal pollution assessment in the river, estuary and lake environments of a fluvial plain, China. CATENA 119:52–60
81. Zhang Y, Merrill MD, Logan BE (2010) The use and optimization of stainless steel mesh cathodes in microbial electrolysis cells. Int J Hydrog Energy 35:12020–12028
82. Zhang Y, Yu L, Wu D, Huang L, Zhou P, Quan X, Chen G (2015) Dependency of simultaneous Cr(VI), Cu(II) and Cd(II) reduction on the cathodes of microbial electrolysis cells self-driven by microbial fuel cells. J Power Sources 273:1103–1113
83. Zhou Q, Yang N, Li Y, Ren B, Ding X, Bian H, Yao X (2020) Total concentrations and sources of heavy metal pollution in global river and lake water bodies from 1972 to 2017. Glob Ecol Conserv 22:e00925

Engineering Challenges of the Microbial Electrolysis Cells for Stable Performance

Muhammad Alamzeb and Ihsan Ullah

Abstract Keeping in mind the global warming and acute shortage of energy resources, microbial electrolysis cell (MEC) looks to be a very important sources of green energy. MECs work on the principle of conversion of organic matter into hydrogen, hydrogen peroxide, acetate methane, ethanol, formic acid and other sources and forms of energy by the electrochemical bacteria. MECs do not cause much environmental pollution. To date, MECs are the most efficient and cleanest source of H_2 production. However, MECs are still in their embryonic stages, and it faces many technical and engineering challenges in their scaling-up and wide range applicability. This book chapter provides a brief account about the basic principle, composition, kinds, and nanomaterials used in the manufacturing of MECs. This book chapter also through light on some of the engineering challenges and hurdles faced by MECs for stable and reliable performance. The data collected here will help the researchers and engineers in the better understanding of MECs because it not only contains data about the challenges faced by MECs but also provides suggestions about overcoming those obstacles. At the end of the chapter, a brief conclusion is provided about MEC and its future prospects.

Keywords MEC · H_2 generation · Methanogenesis · Electrolysis · Renewable energy

---

M. Alamzeb (✉)
Department of Chemistry, University of Kotli, Kotli 11100, Azad Jammu & Kashmir, Pakistan
e-mail: zebchemist@gmail.com

I. Ullah
Institute of Chemical Sciences, University of Swat, Swat, Pakistan

A. A. Yaqoob and A. Ahmad (eds.), *Microbial Electrolysis Cell Technology*, Sustainable Materials and Technology, https://doi.org/10.1007/978-981-97-3356-9_15

1 Introduction and Background

Microbial electrolytic cell (MEC) produces hydrogen through bio-electrochemical processes. This technique for hydrogen production is very environmentally friendly. MECs produce hydrogen by using bio-catalytic anodic oxidation and cathodic reduction methodology. MECs operation is less sustainable because if almost always require an external input of energy, however, it can made self-sustainable if it is integrated with technologies like photoelectrochemical cells [1]. The generation of H_2 from MECs using the process of water electrolysis is of significant importance because it this way electrical energy can be converted into chemical energy. This chemical energy can be stored, transported, and eventually convert back to electrical energy whenever the need arises. Various kinds of electrolysis techniques are in vogue but still MECs need to be upgraded and updated for economical and large-scale production of green energy. The green energy obtained from MECs is very costly too, hence, the cost of MECs needs to be reduced for its large-scale applicability [2]. Water electrolysis is a promising method for producing hydrogen, which can be used to convert electricity into storable, transportable, and re-convertible chemical energy. Even though several kinds of water electrolysis technologies are advanced, more work must be done they can claim to be claimed as stable, cost-effective electricity networks. Improved and more affordable electrocatalysts are especially needed [3, 4]. It has been believed that microbial electrolysis provides a valuable, novel approach. In 2005, the notion of MECs was originally put forth. An anode contaminated with microbes and an ordinary hydrogen cathode make up the core of a MEC. The anode compartment must contain the microbe culture medium. As the electro-catalyst, certain microorganisms grow on the anode surface and produce an electroactive biofilm. The anode may oxidize a wide range of inexpensive carbon compounds because of its biofilm [2, 5].

2 Structure and Operating Principle

MECs and MFCs are closely related hydrogen generating technologies. The basic mechanism of MFCs is well explained in the literature [6–8]. MECs produce methane or hydrogen by utilizing electric current, whereas MFCs generate electricity from the breakdown of organic matter [9, 10]. An MEC's anode and cathode processes are identical to those of an MFC, and a water electrolyze, respectively. MEC can be broken down into three main parts in theory: the separator, cathode chamber, and anode chamber [11]. In a MEC, organic matter is broken down into CO_2, electrons, and protons by electrochemically active microorganisms that are growing on the anode's surface. Electron and protons pass through the electrolyte and reaches to the cathode after passing through electrical circuit and at cathode it produces hydrogen. The microbial biofilm present on the electrode could promote oxidation at anode. In order to develop MEC anodes, MEC suggests designing certain microbial biofilms.

To do this, it will find new inoculum sources, modify the microbial population, and determine the structural traits of the biofilm that would best enhance its electroactive qualities [12]. The single-celled cyanobacterium, Archaea, dechlorinating bacteria, methanogens, and homoacetogens are the most studied microbial cultures for use in MECs [13, 14]. Wastewater and biowaste offer the highest chance of MEC technological success as well as instant income. It may be possible to generate enough hydrogen from abundant and renewable cellulosic biomass for industrial and transportation usage [15]. The products and MEC efficiency are determined by the identification of the particular microorganisms. When electrogenic bacteria (like acetic acid) consume an energy source, they release protons and electrons, which can result in an electrical potential of up to 0.3 V. This voltage is used to produce electricity in a traditional MFC. An external power source provides an extra voltage to the cell in a MEC. Protons are reduced by the combined voltage to produce hydrogen gas. The type of organic materials utilized plays a very important role in the efficient production of H_2 gas. The H_2 production efficiency of lactic acid is more than 80%, although untreated cellulose or glucose have values that are almost 63%. Normal water electrolysis has a 60–70% efficiency rate. MECs can generate 144% more useful energy than they use in electrical energy by converting useless biomass into usable hydrogen. Through a comparable mechanism, MECs can also create methane, depending on the organisms present at the anode [16].

3 Types of MECs

3.1 Double-Chamber MECs

In double beam MECs (as shown in Fig. 1), a membrane acts as separator and reduces the crossing over of organic matter and bacteria from the anodic chamber into cathodic chamber. The membrane also plays a role in the reduction of contaminants like H_2S, CO_2, and CH_4 [17]. This membrane in the two-chambered MEC also prevents the chances of short circuit. It is very important because the internal strength can be reduced by reducing the distance between the electrodes. The double-chambered type was initially used as supportive evidence for the microbial electrolysis by two groups at almost the same time. In neutral state, the transport rate of hydroxyl ions (OH^-) and protons (H^+) over the whole membranes is restricted which then results in the fall of the pH all over the membrane [18]. Various kinds and types of double-chambered MECs have been created by using porous membrane of different types, including barrel-shaped, concentric tube-shaped, circular like a disc, and cruse type. Recently, a flat-type MEC was developed to lessen input resistance [19].

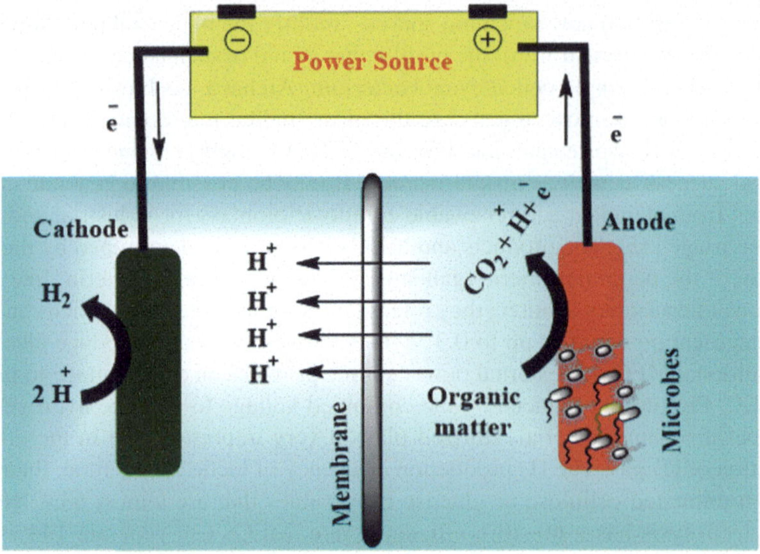

Fig. 1 Schematic representation of two chamber MEC

3.2 Single-Chamber MECs

When membrane is eliminated from a two chamber MEC, and both the electrodes are immersed in the same solution in a single chamber, then it is known as single-chamber MEC (as shown in Fig. 2). Single-chambered MEC can be used to reduce the membrane-related losses and problems [20]. It removes all of the issues with membranes, including fouling, degradation, and exorbitant cost. It is also easy to make and autoclave. Different designs of one-chamber microbial electrolysis cells have been developed; the primary structure is made of a range of materials, such as a plastic block and a glass cruse. For higher throughput bio-electrochemical research, glass serum cruses can be utilized to cultivate limited microbial electrolysis cells that can be utilized to choose potential cathode materials or Electrochemically Active Bacteria (EAB). In one-chamber microbial electrolysis cells, the most challenging tasks are methanogenic bacteria using generated H_2 and electrochemically active bacteria re-oxidizing H_2. Methanogenesis was prevented by using UV irradiation [21]. Using a vacuum and a gas-permeable hydrophobic membrane to swiftly extract H_2, the active gas harvesting technique has recently been employed to eliminate H_2 usage [22].

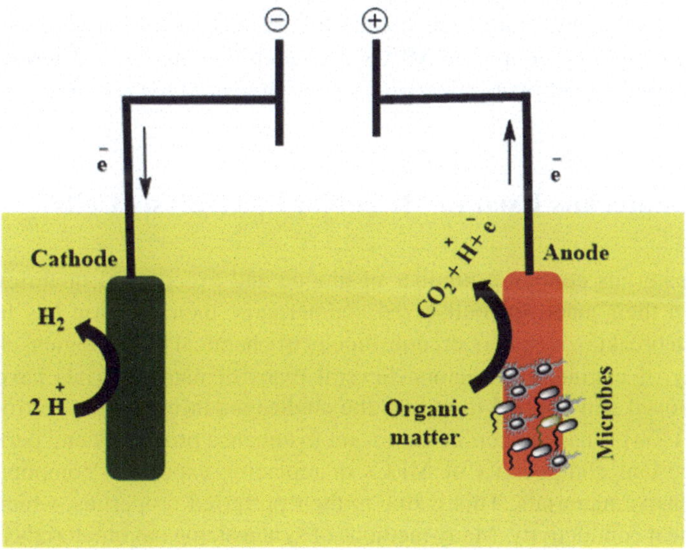

Fig. 2 Single-chamber MEC

4 Electrode Catalyst Materials

The electrode of MECs can be made up of different material including but not limited to graphite granules, carbon, graphite felt, carbon paper or graphite brushes, etc., much like in an MFC [23–25]. To increase the current production of microbial electrochemical cells, highly efficient anode materials must be developed. A newly constructed multi-anode microbial electrolysis cell (MEC) was used to develop and assess graphite anodes decorated with Au and Pd nanoparticles. The current density and the particle density, as determined by area fraction and particle counts, did not, however, show any discernible association. These findings showed that, while the chemical makeup, size, and shape of the nanoparticles affected the degree of the improvement, nano-decoration can significantly improve the performance of microbial anodes [26]. When compared to other materials, the use of graphite felt or carbon cloth as anode could have significant advantages for scale-up. Although platinum is the greatest cathode catalyst material available for use in MEC, its high cost makes it unfeasible [27]. Research into bio-cathodes as an alternative is being fueled by the high cost of platinum. Additionally, a great deal of research has been done on catalysts for MEC without precious metals. In MEC, inexpensive materials like carbon-based NiMo- and NiW-nanocomposites and stainless steel demonstrated good cathode performance [28]. Researchers at Penn State have discovered a technique to use stainless steel brushes in place of platinum catalyst in MECs which produce hydrogen without sacrificing any efficiency. Additionally, the trapped hydrogen stays in the reactor for a longer period of time, making it available to bacteria that eat hydrogen.

Since these results have only been studied in extremely acidic or alkaline environments, they cannot be applied to MECs. Research is required to determine which cathode materials function as effective catalysts at neutral pH [29, 30].

5 Nanomaterials Used in Microbial Electrolysis Cells

The nanomaterials contain a number of unique and exceptional qualities that are absent from their macro-or bulk-sized counterparts, their creation has frequently resulted in a breakthrough to overcome bioelectrochemical technological challenges in a variety of engineering sectors. Several types of nanomaterials have synthesized, improved and applied to reduce the challenges faced by bioelectrochemical systems [31–34]. These materials are meant to enhance properties and performance of the important components of MECs or replacing expensive components with lesser expensive materials. This is due to their preferred properties, which include high electrical conductivity. Many methods of synthesizing nanomaterials for MECs have been documented (Table 1), including thermal annealing, electrodeposition, and electrochemical anodization.

6 Engineering Challenges of MECs in Its Stable Performance

6.1 High Cost (Expensiveness)

The primary obstacle to the widespread implementation of MEC is unquestionably its reduced capital cost, mostly stemming from the high cost of construction materials. Using highly effective, scalable, and less priced anode, cathode, and membrane materials can lower the capital cost. A graphite fiber brush, which is created by combining graphite fibers into a non-corrosive metal core (such as titanium or certain stainless steels), is one of the most promising electrode designs for the anode. Brushes made of metals like stainless steel and tungsten can also be utilized. The distribution of microorganisms and the processes governing the transfer of protons and electrons within the anode chamber, however, is still largely unknown.

6.2 Cost of Cathode

A significant barrier to the widespread use of cathode materials is still their cost. In experiments with MECs, platinum (Pt) has proven to be the most effective material due to its strong catalytic activity. When a consistent cathode activity is needed, Pt is

Table 1 Nanomaterials used in microbial electrolysis cells (MECs)

Nanomaterial type	Size (nm)	Structure	Efficiency	References
Au nanoparticles	0.3 μm^2	0D	Current density: 74 μA/cm^2	[26]
Tio$_2$ nanotube	4.35 μm^2	1D	H_2 production rate: 1434 mmol/m^3/h	[35]
Pd nanoparticle	0.36 μm^2	0D	Current density: 74 μA/cm^2	[26]
Ni nanoparticle	7–50	0D	Enhanced CH_4 production (52%)	[36, 37]
CeO_2–rGO nanocomposite	–	2D	H_2 recovery efficiency of 95%	[38]
Pt–Ni nanoparticle	8–21	0D	77% hydrogen conversion efficiency	[37]
Ni_2P nanoparticle	7	0D	65% H_2 recovery	[39]
Fe_3O_4 nanoparticle	12–28	0D	15 mA/m current density	[40]
$N_{i2}P$ nanoparticle	07	0D	65% H_2 recovery efficiency	[41]
NiO nanoparticle	–	0D	27% recovery efficiency at cathode	[42]
Ni–Co–P nanoparticle	33–35	0D	90% hydrogen conversion efficiency	[43]
Co_3O_4 nanoparticle	–	0D	26% H_2 recovery efficiency	[42]
$NiMoO_4$ nanoparticle	< 50	0D	12% hydrogen conversion efficiency	[44]
$NiFe_2O_4$ nanoparticle	> 17	0D	H_2 production rate of 0.74 A/m^2	[45]
Ag nanoparticle	–	3D	62% hydrogen conversion efficiency	[46]
NiO–rGO nanocomposite	–	2D	H_2 production rate of 4.38 mmol/L/d	[47]
MoS_2 nanosheet	149–251	2D	H_2 production rate = 0.133 m^3/m^3/d	[48]
Activated carbon + Ni powder	0.5–1 μm	3D	H_2 produced at a rate of 0.27 m^3/m^3/d	[49]
Y Zeolites–NiO nanocomposite	–	3D	H_2 production rate: 0.81 m^3/m^3/d	[50]
MoS_2–Cu–rGO nanocomposite	–	2D	H_2 production rate: 0.449 m^3/m^3/d	[50]
$NiO/MoO_2/MoO_3/C$	–	3D	Current density of 37.5 A/m^2	[51]
$CoNi/CoFe_2O_4$ composite	–	3D	H_2 production rate: 1.25 m^3/m^3/d	[51]

(continued)

Table 1 (continued)

Nanomaterial type	Size (nm)	Structure	Efficiency	References
$Mg(OH)_2$/graphene nanocomposite	–	2D	84% H_2 recovery efficiency	[52]
Carbon nanoparticle	51	0D	47% H_2 conversion efficiency	[37]
Polyaniline nanofibers	50	1D	80% H_2 conversion efficiency	[53]
Graphene	–	2D	H_2 production rate: 2.3 m^3/d	[54]
SWCNT	–	1D	39% H_2 conversion efficiency	[37]

typically a better fit for foundation investigations. All agree, however, that Pt's high cost and detrimental effects on the environment make it impractical for an application to be scaled up [55–57]. The second problem with Pt is that it can become poisoned by substances like sulphurides, which will cause it to lose its catalytic abilities. For MEC technology to be used in real applications, it is essential to develop cathodes that are stable, high-performing, and inexpensive. Catalyst costs are a well-known method to enhance the overall performance of MECs by further lowering the over-potential and total internal resistance. Due to its straightforward design, inexpensive operating costs, lack of need for a metal catalyst or artificial electron mediator, strong stability, and ecologically benign characteristics, biocathode is thought to be a potential cathode catalyst substitute for noble metals. It is yet uncertain, therefore, how well biocathode works in pilot operations. It is necessary to look into more resilient and efficient biofilms [58].

6.3 Mode of Operation

The mode of operation of the MEC system is another factor that restricts the application of MEC for hydrogen production. Recently, there has been a report on the scaling-up of MECs for hydrogen production from bench experiments to pilot scale. It was observed that despite having electric energy recovery of more than 70 %, the Coulomb efficiency and hydrogen production rate was lower than the maximum values noted for these parameters in laboratory; hence, there is need for more optimization. If MECs are operated in continuous flow mode, then it may facilitate the scaling-up of MECs in future [59].

6.4 Power Sources

To make the entire process more economical and sustainable, renewable and sustainable power sources are required. MECs could provide a different way to store excess renewable energy in exchange (such as electricity from windmills).

6.5 Scalabality of MECs

Another crucial element in the upscaling of MECs is the reactor design. In laboratory studies, membrane-less single-chamber MECs are frequently employed. MECs are inefficient for producing hydrogen or for other uses, even though the removal of the membrane significantly lowers construction costs. This is because methane production is always detected as a result of the growth of methanogens. The pilot scale test makes this issue more serious. The first pilot scale single-chamber MEC based on the principle of continuous flow was reported by Cusick et al. [60] for the simultaneous production of hydrogen and treatment of winery wastewater. The most important bottlenecks, in the large-scale use of single-chamber MECs, are the purity of the H_2 and the loss of hydrogen due to methane production. It will take more sophisticated techniques to prevent the production of methane and isolate the hydrogen from other gaseous products (like CO_2) in order to produce hydrogen. Additionally, the primary flaw in this technology to date is the growth of electrochemically active bacteria. To bring MFC technology closer to a practical use, it is essential to design more effective reactors that can perform during scale-up use and continue to perform for longer periods of time. To reduce building costs and energy losses, application-oriented reactor design is required. To facilitate easy integration with the existing infrastructure, the bioelectrochemical systems should be designed and configured [61, 62].

6.6 Narrow Application

MECs have not yet been applied outside of laboratories in fields like metal recovery, chemical synthesis, and pollution removal. For many applications, a performance improvement that is significant is necessary. This entails significantly greater energy efficiencies and power densities. The uses of MECs could be extended even further, such as in the recovery of nutrients. In addition to helping to reduce CO_2 emissions, MECs offer a viable and alternative method of storing electrical energy produced by solar and wind energy sources and converting it into biofuels [63, 64]. In order to overcome obstacles and mutually benefit from one another, the integration of MECs with currently used separation, convention, and treatment processes is beneficial. This increases waste conversion and hydrogen production [65].

6.7 Short-Term Stability

For the MEC system to be economically viable and maintain energy balance, it is imperative to enhance long-term stability [66]. It has been noted in numerous studies that MECs' long-term performance is declining. The reduction in the anodic biofilm's electrochemical activity, the degradation of the cathode's performance, the fouling and deformation of the separator materials, and the clogging of the system by solid pollutants and excessive biomass in wastewaters are some of the factors contributing to this performance decline [67]. Furthermore, antibiotic treatment of electrogenic microorganisms may be a different biotechnological strategy for enriching electroactive bacteria from mixed cultures especially for long-term MEC applications, thereby increasing hydrogen generation in MECs while inhibiting the generation of methane [68].

6.8 Wide Gap Between Theoritcal Hopes and Experimental Realization

Between theoretical expectations and experimental realizations, there is a significant discrepancy. In order to verify that the production of hydrogen can be achieved at a lower voltage than that required for the electrolysis of water, MECs pilots with volumes of at least several liters are used. Current densities, which are frequently less than 1 A m^{-2}, nevertheless continue to be extremely low [69]. The practice of attempting to apply MECs to wastewater treatment has been one of the causes of this poor performance. Increasing the electrochemical process's scalability with such poorly conductive electrolytes is a very difficult task because domestic wastewaters, the typical targets for MECs, usually have very low ionic conductivity (0.2 S m^{-1}) or possibly less than 0.2 S m^{-1} ionic conductivity [70]. The idea of using wastewater in an electrolyser is actually heretical to an "abiotic" electrochemical engineer. In general, two opposing goals are to maintain microbial growth in a culture medium and to use an electrolyser for large-scale hydrogen production. To prevent deposits on electrode surfaces and electrode catalyst inhibition, the most pure electrolytes with the least complex chemical composition are needed for the electrochemical processes that have been developed to this point at the industrial scale. The effectiveness of abiotic electrocatalysts in a culture medium would rapidly diminish due to their heightened sensitivity to any type of inhibition caused by chemical compound adsorption. Moreover, strongly acidic or alkaline electrolytes are used in water electrolysis procedure [71]. On the contrary, a neutral pH and rich chemical are needed to support microbial growth. Culture media need to be highly ionized, especially with Ca^{2+} and Mg^{2+}. This can lead to hydroxide deposits and a wide range of chemical compounds that can adsorb on electrode surfaces. The project's biggest obstacle is that the two goals are diametrically opposed to one another.

6.9 *Increase in Current Density-Related Concerns*

Heterogeneous processes employing a solid catalyst are generally recognized in the field of chemical engineering as more complex than homogenous processes. An electrochemical process is at least twice as complex as a homogeneous process, as evidenced by the fact that it is based on two distinct but interdependent heterogeneous catalytic reactions. Thus, it is only feasible to plan the replacement of a homogeneous process with an electrochemical process if significant benefits are anticipated. An electrochemical reactor is a complex device that requires careful planning and execution. In order to lower the ionic resistance of the electrolyte, two electrodes must be designed, usually placed face to face, with the smallest possible inter-electrode distance. It is hard and expensive to pump energy to maintain homogeneous hydrodynamics inside the small compartment(s). Should the intrinsic conductivity of the electrodes be insufficient, the reactor will need multiple electrical connections and possibly large current collectors [72]. A separator is required on occasion. According to technical requirements, this kind of setup requires a large number of seals and stiff frames secured with numerous screws or forceps. Because of electrode wear and deposit damage, dismantling them is a labor-intensive process that may need to be done frequently. In-depth technical knowledge is needed for the labor-intensive maintenance. Furthermore, it is always a delicate safety issue to pass electricity through solutions. The peripheral electrical setup and electrical connections can cause large ohmic losses. An increase in electrode surface area exacerbates each of these problems [73]. Greater electrode surface area increases the risk of clogging or leaks, makes it harder to keep the electrode flat, has more connections, requires more careful and time-consuming disassembly and reassembly procedures, and increases the possibility of stray currents. For these reasons, the primary goal in the design of electrolysers is to ensure a high current density. In order for electrolysers to be used in industrial settings for the synthesis of chemicals, their surface area must be reduced. This can be demonstrated by looking at the case of water electrolysis. It is actually feasible to design water electrolysis reactors at a low-cost operating with a low current density. Cheap electrodes made of inferior steel could function flawlessly in alkaline electrolysers at low current densities. Moreover, since there is no catalyst on the electrodes, running at low current would reduce ohmic losses and possibly offset overpotential losses. Reversed PEM fuel cells, which have a maximum operating voltage of 10,000 A m^{-2}, can produce hydrogen at a lower cost than low-cost water electrolysers when it comes to low current density. That being said, large-scale industrial development has never been considered for inexpensive water electrolysers operating at low current density. To raise current densities, however, scientists are still working to enhance electrocatalysts [74]. Not because of an insane obsession with cutting-edge technology, but rather because industrial electrolysers cannot be scaled up to produce hydrogen on a large scale—not even for a basic reaction like water electrolysis—unless high current density is guaranteed. Clearly, there should be some qualification to this fundamental guideline, which was developed for large-scale production electrolysers, especially when considering microbial electrochemical technologies. Microbial electrochemical technologies have the ability to produce innovative ideas, which is one of their fascinating features. New electrochemical processes that do not require a high current density as a prerequisite for development can therefore originate from microbial electrochemical technologies. Microbial electrochemical snorkels, e.g., can be developed on a large scale using low current because their configuration can be made incredibly simple. Scientific community could think of them as basic conductive biofilters that don't need special maintenance or a complex setup. A non-Faradaic microbial electrochemical process called electro-fermentation is another illustration. In this process, the redox state of the microbial cells is regulated by the potential. Although the current in this instance does not supply the charge required for conversion, it does indirectly orient the catalytic pathway. Can an inventive method for producing hydrogen be discovered, wherein maximizing current density might not be the primary prerequisite [75]. In conclusion, in order to develop MEC, two distinct research strategies need to be investigated. On the one hand, novel ideas could be found that do away with the requirement to maximize the current density. This path is still accessible. However, the primary goal should be to increase current density if the goal is to modify traditional electrolyser architectures for microbial catalysis.

6.10 Electrolytes-Related Issues

An "abiotic" electrochemical engineer would find it highly peculiar to contaminate the electrolyte with a medium so filthy as to foster microbial growth. Even more astonishing, it appears that wastewater and effluent treatment has accounted for the majority of MEC pilot projects. The design of efficient electrochemical cells is made exceedingly difficult by the extremely low ionic conductivity of these media, which is typically of the order of 0.2 S m^{-1}, occasionally less. Higher current densities of about 1.35 A m^{-2} can be achieved by using phosphate buffer effluent and using a synthetic medium augmented with acetate substrate in MECs pilot plants. This kind of substrate has been found to produce current densities up to 10 A m^{-2} [76].

In order to achieve a fair ionic conductivity of 9 S m^{-1}, a high concentration of salt (35 g/L NaCl) was used, which resulted in the highest current density of 42.5 A m^{-2} that has been recorded with a MEC several liters in volume [77]. A significant contributing factor to the subpar results seen thus far with large-sized MECs is the electrolytes' low conductivity. Because increasing size inevitably results in a rapid decrease in performance, it is crucial to optimize the MEC architecture at low ionic conductivity in order to allow the electrodes to operate in the least harmful possible conditions. Development of microbial anodes capable of functioning in saline and hypersaline media has been the focus of numerous attempts to get around this significant challenge [78].

A recent review states that salt marsh inocula have been used to create the most effective halotolerant microbial anodes. When polarized at 0.1 V/SCE with 40 mM acetate, these microbial anodes generated up to 85 A m^{-2}. They functioned in media containing 45 g L^{-1} of NaCl and 10.4 S m^{-1} of ionic conductivity, which is on par with the values utilized in water electrolysis [79]. For example, the ionic conductivity of a thin PEM is about 10 S m^{-1}, while the electrolytes with 20–30% KOH that are frequently used in alkaline water electrolysis can reach up to 60 S m^{-1} [80].

In order to create this halotolerant microbial anode, the electrolyte's main ingredients were phosphates (3 mM), CH_3COONa (40 mM), NH_4Cl (37 mM), and NaCl 45 g L^{-1} (770 mM). According to the transport numbers, 95% of the current in the electrolyte was carried by the movement of Na^+ to the cathode (37.9%) and Cl^- to the anode (57.5%) in the solution bulk, away from the concentration gradients that were present in the area around the electrode surfaces [81]. Within the electrolyte, NaCl performed its function as a current carrier flawlessly. As you might remember, OH^- and H^+ do not aid in the electrolyte's ability to conduct electricity. At neutral pH, their concentrations are too low to have a significant effect the transport of electricity [69]. Particularly, the addition of small concentrations of ions such as magnesium, calcium and sulfate should modify the accurate distribution of the transport numbers, as should the salt marsh inoculum added at 5–10% v/v into the medium. However, these minor disruptions shouldn't change the overall scenario that is outlined here. An expression of the ion fluxes based on the exchange of 100 electrons at the electrodes can be used to describe the ionic transport through the electrolyte. 12.5 molecules of acetate are consumed, 87.5 protons are created at the bioanode and 100 hydroxide ions are produced at the cathode when 100 electrons are transferred to the anode. In order to rectify the resulting ion imbalance, the anode receives 57.5Cl^- and 1.5 CH_3COO^- ions, while the cathode receives 37.9Na^+ and 2.5 NH^{4+} ions. This draws attention to an important issue: migration has no bearing on the pH balance at the electrodes. The phosphate species did not significantly affect the ionic transport in most of the electrolytes. The phosphate species were present in the medium at only 3 mM concentration. Both the migration of buffering species to the anode and the migration of protons generated at the anode are not sufficient to neutralize the protons produced there. The anode acidification can only be mitigated by the $H_2PO_4^-$ ion flux, which is only 0.06 ions per 100 electrons exchanged. Likewise, migration does not drive away the OH^- ions generated at the cathode. The cathode alkanization and the anode acidification can only be brought about by the diffusion of concentration gradients. The anode vicinity acidifies, while the cathode vicinity alkalinizes [82]. The rate of the hydrogen evolution reaction is negatively impacted by cathode alkalinization from a thermodynamic perspective. Additionally, because cations like Ca^{2+} and $Mg^{2+,}$ which are necessary to support the bioanode, deposit on the cathode surface as hydroxides, alkalinization of the cathode vicinity poses a unique challenge for MECs [83]. One major problem with microbial electrochemical processes is known to be the chemical fouling at the cathode surface due to the precipitation of salt. One effective way to lower the pH and overpotential associated with oxygen reduction in a microbial fuel cell is to bubble CO_2 inside the cathode compartment. The same procedure was able to lower the bulk pH in a MEC, but it had less of an impact on the cathode overpotential [84]. The possibility existed that the pH may have maintained a high local value close to the cathode surface. Because microbial anodes are severely inhibited, or even deteriorate, at slightly acid pH values, anode acidification is an even more important issue in addition to its detrimental effect on thermodynamics. The limited capacity to reduce anode acidification in the medium utilized in earlier studies to create effective halotolerant anodes may be related to the low phosphate concentration [85]. However, things would only get marginally better even if the phosphate concentration was raised to 100 mM or higher. As an example, 1.7 $H_2PO_4^-$ ions migrate toward the anode for every 100 electrons exchanged when 100 mM of phosphate species are present. Actually, the heavy addition of NaCl makes the issue worse. It is common practice to use high concentrations of salt in the supporting electrolyte to raise the electrolyte ionic conductivity, which was found to be 10.7 S m^{-1}; however, this is accompanied by a decrease in the migration flux of the buffering species to the anode. The medium's ionic conductivity would decrease to 2.6 S m^{-1} in the absence of NaCl, but the $H_2PO_4^-$ ion migration flux would rise to 8.2 [86]. High concentrations of salt have already been shown to have negative effects in experiments. Therefore, in the context of MEC, the standard procedure of adding salts to the medium might not be the best one. High salt concentrations, of course, make the electrolyte more ionic conductive, but they also reduce the migration's ability to lessen anode acidification; after all, a local pH drop is a significant rate-limiting step for microbial anodes [87]. Rather than adding large amounts of non-buffering salts, an intriguing approach to optimize the electrolyte for MEC might be to raise the buffer concentration to the highest level that microbial anodes can withstand.

Moreover, raising the phosphate concentration ought to provide two advantages: HPO_4^{2-} species migration would lessen the acidity of the bioanode. In addition to the possibility that the presence of hydrogenated phosphate species would improve the cathode's homogeneous hydrogen evolution catalysis. The ion migration fluxes in MEC will remain insignificant compared to the ideal case of PEM electrolysers due to the diversity of ions present in the electrolytes used. Protons are the only ones that migrate through the electrolyte because PEM fixes the anions in its structure. This completes the balance between proton production at the anode and proton consumption at the cathode [88]. A key problem for MECs is the sensitivity of microbial anodes to local acidification. In order to develop MECs that can operate at high current densities, it will therefore continue to be extremely difficult to master ion transport with the goal of reducing anode acidification [89].

7 Conclusion

Global warming has made it more important than ever to develop and apply sustainable and environmentally friendly new green technologies for methane production. Treatment of wastewater, hydrogen, and methane with MECs has already shown MECs to be a promising technology. Regarding the practical use of MECs to produce methane, the last 20 years of research in the field have been noteworthy. One of the best alternatives for treating wastewater that can produce methane as a significant value-added product is a single-stage methane production system that can operate at room temperature with inexpensive inputs.

Our goal when designing a MEC is to have it both ways. In essence, two antagonistic processes need to be associated for a MEC to succeed. On the one hand, clean electrolytes with high ionic conductivity, extreme pH values, and a simple chemical composition are ideal for use in large-scale production electrolysers. However, microbial processes frequently reject highly conductive solutions and need neutral pH and chemically rich culture media. The two types of processes have disjoint ranges of optimal conditions. At the low current densities needed to sustain metabolic processes, this is not an obstacle. On the other hand, the system is destroyed by the disparity in ideal conditions when the engineer attempts to push the process toward industrial production at high current densities. The MEC is hampered by some type of negative feedback loop as current density rises. Is it possible for MECs to develop into an industrial technology? This is still a challenging and fascinating question. In an attempt to provide an answer, this book chapter suggests possible research directions.

References

1. Götz M, Lefebvre J, Mörs F, McDaniel Koch A, Graf F, Bajohr S, Reimert R, Kolb T (2016) Renewable power-to-gas: a technological and economic review. Renew Energy 85:1371–1390. https://doi.org/10.1016/j.renene.2015.07.066
2. Shaner MR, Atwater HA, Lewis NS, McFarland EW (2016) A comparative technoeconomic analysis of renewable hydrogen production using solar energy. Energy Environ Sci 9(7):2354–2371. https://doi.org/10.1039/C5EE02573G
3. Zou X, Zhang Y (2015) Noble metal-free hydrogen evolution catalysts for water splitting. Chem Soc Rev 44(15):5148–5180. https://doi.org/10.1039/C4CS00448E
4. Parveen T, Umar K, Ibrahim MNM (2020) Role of nanomaterials in the treatment of wastewater: a review. Water 12(2):495
5. Rozendal RA, Hamelers HVM, Euverink GJW, Metz SJ, Buisman CJN (2006) Principle and perspectives of hydrogen production through biocatalyzed electrolysis. Int J Hydrogen Energy 31(12):1632–1640. https://doi.org/10.1016/j.ijhydene.2005.12.006
6. Khatoon A, Mohd Setapar SH, Umar K, Parveen T, Rafatullah M (2020) Outlook on the role of microbial fuel cells in remediation of environmental pollutants with electricity generation. Catalysts 10(8):819
7. Umar K, Ahmad A (2021) Modified graphene oxide anode: a bioinspired waste material for bioremediation of Pb^{2+} with energy generation through microbial fuel cells. Chem Eng J 417:128052
8. Khalid U, Tabassum P, Akil A, Lokhat D, Siti H (2021) A glimpse into the microbial fuel cells for wastewater treatment with energy generation. Desalin Water Treat 214:379–389
9. Fadzli FS, Rashid M (2021) Electricity generation and heavy metal remediation by utilizing yam (*Dioscorea alata*) waste in benthic microbial fuel cells (BMFCs). Biochem Eng J 172:108067
10. Ahmad A, Ibrahim MNM, Yaqoob AA, Setapar SHM (2022) Microbial fuel cells for environmental remediation. Springer Nature
11. Jeremiasse AW, Hamelers HV, Buisman CJ (2010) Microbial electrolysis cell with a microbial biocathode. Bioelectrochemistry (Amsterdam, Neth) 78(1):39–43. https://doi.org/10.1016/j.bioelechem.2009.05.005

12. Bandyopadhyay A, Stöckel J, Min H, Sherman LA, Pakrasi HB (2010) High rates of photobiological H_2 production by a cyanobacterium under aerobic conditions. Nat Commun 1:139. https://doi.org/10.1038/ncomms1139
13. Cheng S, Xing D, Call DF, Logan BE (2009) Direct biological conversion of electrical current into methane by electromethanogenesis. Environ Sci Technol 43(10):3953–3958. https://doi.org/10.1021/es803531g
14. Chuo SC, Mohamed SF, Mohd Setapar SH, Ahmad A, Jawaid M, Wani WA (2020) Insights into the current trends in the utilization of bacteria for microbially induced calcium carbonate precipitation. Materials 13(21):4993
15. Call D, Logan BE (2008) Hydrogen production in a single chamber microbial electrolysis cell lacking a membrane. Environ Sci Technol 42(9):3401–3406. https://doi.org/10.1021/es8001822
16. Cheng S, Logan B (2007) Sustainable and efficient biohydrogen production via electrohydrogenesis. Proc Natl Acad Sci USA 104:18871–18873. https://doi.org/10.1073/pnas.0706379104
17. Jiang Y, Liang P, Zhang C, Bian Y, Sun X, Zhang H, Yang X, Zhao F (2016) Periodic polarity reversal for stabilizing the pH in two-chamber microbial electrolysis cells. Appl Energy 165:670–675. https://doi.org/10.1016/j.apenergy.2016.01.001
18. Kadier A, Jiang Y, Lai B, Rai PK, Kuppam C, Mohamed A, Kalil M (2018) Biohydrogen production in microbial electrolysis cells from renewable resources, pp 331–356. https://doi.org/10.1201/9781351228138-12
19. Escapa A, Mateos R, Martínez EJ, Blanes J (2016) Microbial electrolysis cells: an emerging technology for wastewater treatment and energy recovery. From laboratory to pilot plant and beyond. Renew Sustain Energy Rev 55:942–956. https://doi.org/10.1016/j.rser.2015.11.029
20. Waqas M, Rehan M, Aburizaiza A, Nizami DA-S (2018) Wastewater biorefinery based on the microbial electrolysis cell: opportunities and challenges, pp 347–374. https://doi.org/10.1016/B978-0-444-64017-8.00017-8
21. Cui W, Liu G, Zeng C, Lu Y, Luo H, Zhang R (2019) Improved hydrogen production in the single-chamber microbial electrolysis cell with inhibition of methanogenesis under alkaline conditions. RSC Adv 9(52):30207–30215. https://doi.org/10.1039/C9RA05483A
22. Lu L, Hou D, Wang X, Jassby D, Ren Z (2016) Active H_2 harvesting prevents methanogenesis in microbial electrolysis cells. Environ Sci Technol Lett 3. https://doi.org/10.1021/acs.estlett.6b00209.
23. Rodríguez-Couto S, Ahmad A (2021) Preparation, characterization, and application of modified carbonized lignin as an anode for sustainable microbial fuel cell. Process Saf Environ Prot 155:49–60
24. Idris MO, Noh NAM (2023) Sustainable microbial fuel cell functionalized with a bio-waste: a feasible route to formaldehyde bioremediation along with bioelectricity generation. Chem Eng J 455:140781
25. Ibrahim MNM, Yaakop AS, Rafatullah M (2022) Utilization of biomass-derived electrodes: a journey toward the high performance of microbial fuel cells. Appl Water Sci 12(5):99
26. Fan Y, Xu S, Schaller R, Jiao J, Chaplen F, Liu H (2011) Nanoparticle decorated anodes for enhanced current generation in microbial electrochemical cells. Biosens Bioelectron 26(5):1908–1912. https://doi.org/10.1016/j.bios.2010.05.006
27. Freguia S, Rabaey K, Yuan Z, Keller J (2007) Electron and carbon balances in microbial fuel cells reveal temporary bacterial storage behavior during electricity generation. Environ Sci Technol 41(8):2915–2921. https://doi.org/10.1021/es062611i
28. Hu H, Fan Y, Liu H (2009) Hydrogen production in single-chamber tubular microbial electrolysis cells using non-precious-metal catalysts. Int J Hydrogen Energy 34:8535–8542. https://doi.org/10.1016/j.ijhydene.2009.08.011
29. Navarro-Flores E, Chong Z, Omanovic S (2005) Characterization of Ni, NiMo, NiW and NiFe electroactive coatings as electrocatalysts for hydrogen evolution in an acidic medium. J Mol Catal A Chem 226(2):179–197. https://doi.org/10.1016/j.molcata.2004.10.029

30. Olivares-Ramírez JM, Campos-Cornelio ML, Godinez J, Arco E, Castellanos RH (2007) Studies on the hydrogen evolution reaction on different stainless steels. Int J Hydrogen Energy 32:3170–3173. https://doi.org/10.1016/j.ijhydene.2006.03.017
31. Idris MO, Guerrero-Barajas C, Kim H-C (2023) Scalability of biomass-derived graphene derivative materials as viable anode electrode for a commercialized microbial fuel cell: a systematic review. Chin J Chem Eng 55:277–292
32. Ibrahim MNM, Rodríguez-Couto S (2020) Development and modification of materials to build cost-effective anodes for microbial fuel cells (MFCs): an overview. Biochem Eng J 164:107779
33. Guerrero-Barajas C (2021) Modern trend of anodes in microbial fuel cells (MFCs): an overview. Environ Technol Innov 23:101579
34. Serra A (2021) Self-assembled oil palm biomass-derived modified graphene oxide anode: an efficient medium for energy transportation and bioremediating Cd (II) via microbial fuel cells. Arab J Chem 14(5):103121
35. Kim KN, Lee SH, Kim H, Park YH, In S-I (2018) Improved microbial electrolysis cell hydrogen production by hybridization with a TiO_2 nanotube array photoanode. Energies 11(11):3184
36. Kim K-R, Kang J, Chae K-J (2017) Improvement in methanogenesis by incorporating transition metal nanoparticles and granular activated carbon composites in microbial electrolysis cells. Int J Hydrogen Energy 42. https://doi.org/10.1016/j.ijhydene.2017.06.142
37. Choi M-J, Yang E, Yu H-W, Kim I, Oh S-E, Chae K-J (2018) Transition metal/carbon nanoparticle composite catalysts as platinum substitutes for bioelectrochemical hydrogen production using microbial electrolysis cells. Int J Hydrogen Energy 44. https://doi.org/10.1016/j.ijhydene.2018.07.020
38. Pophali A, Singh S, Verma N (2020) Simultaneous hydrogen generation and COD reduction in a photoanode-based microbial electrolysis cell. Int J Hydrogen Energy 45. https://doi.org/10.1016/j.ijhydene.2020.01.053
39. Wang W, Zhang B, He Z (2019) Bioelectrochemical deposition of palladium nanoparticles as catalysts by *Shewanella oneidensis* MR-1 towards enhanced hydrogen production in microbial electrolysis cells. Electrochim Acta 318. https://doi.org/10.1016/j.electacta.2019.06.038
40. Rani G, Krishna K, Yogalakshmi KN (2021) Enhancing the electrochemical performance of Fe_3O_4 nanoparticles layered carbon electrodes in microbial electrolysis cell. J Environ Chem Eng 9(6):106326. https://doi.org/10.1016/j.jece.2021.106326
41. Kim KY, Habas SE, Schaidle JA, Logan BE (2019) Application of phase-pure nickel phosphide nanoparticles as cathode catalysts for hydrogen production in microbial electrolysis cells. Biores Technol 293:122067. https://doi.org/10.1016/j.biortech.2019.122067
42. Jayabalan T, Manickam M, Vijayarengan P, Mohamed SN (2019) Enhancing biohydrogen production from sugar industry wastewater using metal oxide/graphene nanocomposite catalysts in microbial electrolysis cell. Int J Hydrogen Energy 45. https://doi.org/10.1016/j.ijhydene.2019.09.068
43. Chaurasia AK, Goyal H, Mondal P (2020) Hydrogen gas production with Ni, Ni–Co and Ni–Co–P electrodeposits as potential cathode catalyst by microbial electrolysis cells. Int J Hydrogen Energy 45(36):18250–18265. https://doi.org/10.1016/j.ijhydene.2019.07.175
44. Jayabalan T, Matheswaran M, Radhakrishnan TK, Mohamed SN (2021) Influence of nickel molybdate nanocatalyst for enhancing biohydrogen production in microbial electrolysis cell utilizing sugar industrial effluent. Bioresour Technol 320:124284. https://doi.org/10.1016/j.biortech.2020.124284
45. Tahir M (2019) Microbial photoelectrochemical cell for improved hydrogen evolution using nickel ferrite incorporated WO_3 under visible light irradiation. Int J Hydrogen Energy 44. https://doi.org/10.1016/j.ijhydene.2019.01.067
46. Park S-G, Rajesh PR, Hwang M-H, Chu K, Cho S, Chae K-J (2020) Long-term effects of anti-biofouling proton exchange membrane using silver nanoparticles and polydopamine on the performance of microbial electrolysis cells. Int J Hydrogen Energy 46. https://doi.org/10.1016/j.ijhydene.2020.04.059
47. Jayabalan T, Matheswaran M, Preethi V, Mohamed SN (2020) Enhancing biohydrogen production from sugar industry wastewater using metal oxide/graphene nanocomposite catalysts in

microbial electrolysis cell. Int J Hydrogen Energy 45(13):7647–7655. https://doi.org/10.1016/j.ijhydene.2019.09.068

48. Rozenfeld S, Teller H, Schechter M, Farber R, Krichevski O, Schechter A, Cahan R (2018) Exfoliated molybdenum di-sulfide (MoS(2)) electrode for hydrogen production in microbial electrolysis cell. Bioelectrochemistry (Amsterdam, Neth) 123:201–210. https://doi.org/10.1016/j.bioelechem.2018.05.007
49. Kim K-Y, Logan B (2019) Nickel powder blended activated carbon cathodes for hydrogen production in microbial electrolysis cells. Int J Hydrogen Energy 44. https://doi.org/10.1016/j.ijhydene.2019.04.041
50. Tang J, Bian Y, Jin S, Sun D, Ren Z (2021) Cathode material development in the past decade for H_2 production from microbial electrolysis cells. ACS Environ Au 2(1):20–29. https://doi.org/10.1021/acsenvironau.1c00021
51. Fang X, Cheng Y, Li S, Dai H, Zhao Y, Du X, Ma X, Hao X (2021) Fabrication of durian-like CoNi/$CoFe_2O_4$ composite electrocatalysts on nickel foam for hydrogen evolution in a microbial electrolysis cell. Int J Energy Res 46. https://doi.org/10.1002/er.6791
52. Dai H, Yang H, Liu X, Jian X, Liang Z (2016) Electrochemical evaluation of nano-$Mg(OH)_2$/graphene as a catalyst for hydrogen evolution in microbial electrolysis cell. Fuel 174. https://doi.org/10.1016/j.fuel.2016.02.013
53. Jeon Y, Kim S (2016) Persistent hydrogen production by the photo-assisted microbial electrolysis cell using a p-type polyaniline nanofiber cathode. ChemSusChem 9(23):3276–3279. https://doi.org/10.1002/cssc.201600906
54. Cai W, Liu W, Han J, Wang A (2016) Enhanced hydrogen production in microbial electrolysis cell with 3D self-assembly nickel foam-graphene cathode. Biosens Bioelectron 80:118–122. https://doi.org/10.1016/j.bios.2016.01.008
55. Kundu A, Sahu J, Redzwan G, Hashim M (2013) An overview of cathode material and catalysts suitable for generating hydrogen in microbial electrolysis cell. Int J Hydrogen Energy 38. https://doi.org/10.1016/j.ijhydene.2012.11.031
56. Umar K (2021) Biomass-derived composite anode electrode: synthesis, characterizations, and application in microbial fuel cells (MFCs). J Environ Chem Eng 9(5):106111
57. Daud NNM, Hussin MH (2024) Evaluating the electrode materials to improve electricity generation with the removal of multiple pollutants through microbial fuel cells. Biomass Convers Biorefin 1–22
58. Zhang Y, Wang Y, Angelidaki I (2015) Alternate switching between microbial fuel cell and microbial electrolysis cell operation as a new method to control H_2O_2 level in bioelectro-Fenton system. J Power Sources 291:108–116. https://doi.org/10.1016/j.jpowsour.2015.05.020
59. Heidrich ES, Dolfing J, Scott K, Edwards SR, Jones C, Curtis TP (2013) Production of hydrogen from domestic wastewater in a pilot-scale microbial electrolysis cell. Appl Microbiol Biotechnol 97(15):6979–6989. https://doi.org/10.1007/s00253-012-4456-7
60. Cusick RD, Bryan B, Parker DS, Merrill MD, Mehanna M, Kiely PD, Liu G, Logan BE (2011) Performance of a pilot-scale continuous flow microbial electrolysis cell fed winery wastewater. Appl Microbiol Biotechnol 89(6):2053–2063. https://doi.org/10.1007/s00253-011-3130-9
61. Tenca A, Cusick R, Schievano A, Oberti R, Logan B (2013) Evaluation of low cost cathode materials for treatment of industrial and food processing wastewater using microbial electrolysis cells. Int J Hydrogen Energy 38:1859–1865. https://doi.org/10.1016/j.ijhydene.2012.11.103
62. Noh NAM, Hussin MH (2023) Synthesis and fabrication of palm kernel shell-derived modified electrodes: a practical step towards the industrialization of microbial fuel cells. Chem Eng J 475:146321
63. Bakar MABA, Kim H-C, Ahmad A, Alshammari MB, Yaakop AS (2022) Oxidation of food waste as an organic substrate in a single chamber microbial fuel cell to remove the pollutant with energy generation. Sustain Energy Technol Assess 52:102282
64. Al-Zaqri N, Yaakop AS, Umar K (2022) Potato waste as an effective source of electron generation and bioremediation of pollutant through benthic microbial fuel cell. Sustain Energy Technol Assess 53:102560

65. Kadier A, Kalil MS, Abdeshahian P, Chandrasekhar K, Mohamed A, Azman NF, Logroño W, Simayi Y, Hamid AA (2016) Recent advances and emerging challenges in microbial electrolysis cells (MECs) for microbial production of hydrogen and value-added chemicals. Renew Sustain Energy Rev 61:501–525. https://doi.org/10.1016/j.rser.2016.04.017
66. Escapa A, Gil-Carrera L, García V, Morán A (2012) Performance of a continuous flow microbial electrolysis cell (MEC) fed with domestic wastewater. Biores Technol 117:55–62. https://doi.org/10.1016/j.biortech.2012.04.060
67. Carmona-Martínez AA, Trably E, Milferstedt K, Lacroix R, Etcheverry L, Bernet N (2015) Long-term continuous production of H_2 in a microbial electrolysis cell (MEC) treating saline wastewater. Water Res 81:149–156. https://doi.org/10.1016/j.watres.2015.05.041
68. Escapa A, San-Martín MI, Mateos R, Morán A (2015) Scaling-up of membraneless microbial electrolysis cells (MECs) for domestic wastewater treatment: bottlenecks and limitations. Biores Technol 180:72–78. https://doi.org/10.1016/j.biortech.2014.12.096
69. Oliot M, Galier S, Roux-de Balmann H, Bergel A (2016) Ion transport in microbial fuel cells: key roles, theory and critical review. Appl Energy 183. https://doi.org/10.1016/j.apenergy.2016.09.043
70. Ruiz Y, Baeza J, Guisasola A (2016) Microbial electrolysis cell performance using non-buffered and low conductivity wastewaters. Chem Eng J 289. https://doi.org/10.1016/j.cej.2015.12.098
71. Oliot M, Poehere C, Erable, B, Bergel A (2017) Influence of the electrode size on microbial anode performance. Chem Eng J 327. https://doi.org/10.1016/j.cej.2017.06.044
72. Moscoviz R, Toledo-Alarcón J, Trably E, Bernet N (2016) Electro-fermentation: how to drive fermentation using electrochemical systems. Trends Biotechnol 34(11):856–865. https://doi.org/10.1016/j.tibtech.2016.04.009
73. Erable B, Etcheverry L, Bergel A (2011) From microbial fuel cell (MFC) to microbial electrochemical snorkel (MES): maximizing chemical oxygen demand (COD) removal from wastewater. Biofouling 27(3):319–326. https://doi.org/10.1080/08927014.2011.564615
74. Ramírez-Vargas CA, Arias CA, Carvalho P, Zhang L, Esteve-Núñez A, Brix H (2019) Electroactive biofilm-based constructed wetland (EABB-CW): a mesocosm-scale test of an innovative setup for wastewater treatment. Sci Total Environ 659:796–806. https://doi.org/10.1016/j.scitotenv.2018.12.432
75. Viggi CC, Presta E, Bellagamba M, Kaciulis S, Balijepalli DSK, Zanaroli G, Papini M, Rossetti S, Aulenta F (2015) The “oil-spill snorkel”: an innovative bioelectrochemical approach to accelerate hydrocarbons biodegradation in marine sediments. Front Microbiol 6:881. https://doi.org/10.3389/fmicb.2015.00881
76. Guo K, Prévoteau A, Rabaey K (2017) A novel tubular microbial electrolysis cell for high rate hydrogen production. J Power Sources 356:484–490
77. Nam J-Y, Logan B (2011) Enhanced hydrogen generation using a saline catholyte in a two chamber microbial electrolysis cell. Fuel Energy Abstr 36:15105–15110. https://doi.org/10.1016/j.ijhydene.2011.08.106
78. Lacroix R, Silva SD, Gaig MV, Rousseau R, Délia M-L, Bergel A (2014) Modelling potential/current distribution in microbial electrochemical systems shows how the optimal bioanode architecture depends on electrolyte conductivity. Phys Chem Chem Phys 16(41):22892–22902. https://doi.org/10.1039/C4CP02177K
79. Grattieri M, Minteer S (2018) Microbial fuel cells in saline and hypersaline environments: advancements, challenges and future perspectives. Bioelectrochemistry (Amsterdam, Neth) 120:127–137. https://doi.org/10.1016/j.bioelechem.2017.12.004
80. Carmo M, Fritz DL, Mergel J, Stolten D (2013) A comprehensive review on PEM water electrolysis. Int J Hydrogen Energy 38(12):4901–4934. https://doi.org/10.1016/j.ijhydene.2013.01.151
81. Rousseau R, Santaella C, Bonnafous A, Achouak W, Godon JJ, Delia ML, Bergel A (2016) Halotolerant bioanodes: the applied potential modulates the electrochemical characteristics, the biofilm structure and the ratio of the two dominant genera. Bioelectrochemistry (Amsterdam, Neth) 112:24–32. https://doi.org/10.1016/j.bioelechem.2016.06.006

82. Santini M, Marzorati S, Fest-Santini S, Trasatti S, Cristiani P (2017) Carbonate scale deactivating the biocathode in a microbial fuel cell. J Power Sources 356. https://doi.org/10.1016/j.jpowsour.2017.02.088
83. Sugnaux M, Happe M, Cachelin CP, Gasperini A, Blatter M, Fischer F (2017) Cathode deposits favor methane generation in microbial electrolysis cell. Chem Eng J 324:228–236. https://doi.org/10.1016/j.cej.2017.05.028
84. Ki D, Popat SC, Torres CI (2016) Reduced overpotentials in microbial electrolysis cells through improved design, operation, and electrochemical characterization. Chem Eng J 287:181–188. https://doi.org/10.1016/j.cej.2015.11.022
85. Popat SC, Torres CI (2016) Critical transport rates that limit the performance of microbial electrochemistry technologies. Biores Technol 215:265–273. https://doi.org/10.1016/j.biortech.2016.04.136
86. Nam J-Y, Logan BE (2012) Optimization of catholyte concentration and anolyte pHs in two chamber microbial electrolysis cells. Int J Hydrogen Energy 37(24):18622–18628. https://doi.org/10.1016/j.ijhydene.2012.09.140
87. Dumas C, Mollica A, Féron D, Basseguy R, Etcheverry L, Bergel A (2008) Checking graphite and stainless anodes with an experimental model of marine microbial fuel cell. Biores Technol 99(18):8887–8894. https://doi.org/10.1016/j.biortech.2008.04.054
88. Rimboud M, Pocaznoi D, Erable B, Bergel A (2014) Electroanalysis of microbial anodes for bioelectrochemical systems: basics, progress and perspectives. Phys Chem Chem Phys 16(31):16349–16366. https://doi.org/10.1039/C4CP01698J
89. Baudler A, Langner M, Rohr C, Greiner A, Schröder U (2017) Metal-polymer hybrid architectures as novel anode platform for microbial electrochemical technologies. ChemSusChem 10(1):253–257. https://doi.org/10.1002/cssc.201600814

SPRINGER NATURE

GPSR Compliance

The European Union's (EU) General Product Safety Regulation (GPSR) is a set of rules that requires consumer products to be safe and our obligations to ensure this.

If you have any concerns about our products, you can contact us on ProductSafety@springernature.com

In case Publisher is established outside the EU, the EU authorized representative is:

Springer Nature Customer Service Center GmbH
Europaplatz 3
69115 Heidelberg, Germany

The manufacturer's authorised representative in the EU is Springer Nature Customer Service Centre GmbH, Europaplatz 3, 69115 Heidelberg, Germany. If you have any concerns regarding our products, please contact ProductSafety@springernature.com

Printed and bound by CPI Group (UK) Ltd, Croydon, CR0 4YY
01/10/2026
02225857-0008